Heidegger's Confrontation with Modernity

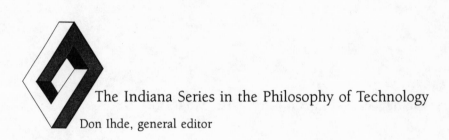

The Indiana Series in the Philosophy of Technology

Don Ihde, general editor

Heidegger's Confrontation with Modernity

Technology, Politics, and Art

Michael E. Zimmerman

INDIANA UNIVERSITY PRESS
Bloomington and Indianapolis

Manufactured in the United States of America

Library of Congress Cataloging-in-Publication Data

Zimmerman, Michael E.
  Heidegger's confrontation with modernity: technology, politics, and art / Michael E. Zimmerman.
    p. cm. — (The Indiana series in the philosophy of technology)
  Includes bibliographical references.
  ISBN 0-253-36875-8. — ISBN 0-253-20558-1 (pbk.)
  1. Heidegger, Martin, 1889–1976—Contributions in philosophy of technology. 2. Technology—Philosophy—History—20th century.
I. Title. II. Series.
B3279.H49Z552   1990
193—dc20   89-45415
                                                                     CIP

    1 2 3 4 5 94 93 92 91 90

For Teresa,
who has always known

Wo aber Gefahr ist, wächst
Das Rettende auch.

(But where danger is, also
 grows the saving power.)

Friedrich Hölderlin
Patmos

Contents

Editor's Foreword ix
Acknowledgments xii
Introduction xiii
List of Abbreviations xxiii

Division One *Heidegger and the Politics of Productionist Metaphysics: The Longing for a New World of Work*

Chapter 1: Germany's Confrontation with Modernity 3
Chapter 2: Political Aspects of Heidegger's Early Critique of Modern Technology 17
Chapter 3: Heidegger, National Socialism, and Modern Technology 34
Chapter 4: Jünger and the *Gestalt* of the Worker 46
Chapter 5: Heidegger's Appropriation of Jünger's Thought, 1933–34 66
Chapter 6: Jünger's Thought in Heidegger's Mature Concept of Technology 77
Chapter 7: National Socialism, Nietzsche, and the Work of Art 94
Chapter 8: Hölderlin and the Saving Power of Art 113

Division Two *Heidegger's Critique of Productionist Metaphysics*

Chapter   9: Equipment, Work, World, and Being   137
Chapter 10: *Being and Time:* Penultimate Stage of Productionist
            Metaphysics?   150
Chapter 11: The History of Productionist Metaphysics   166
Chapter 12: Production Cycles of the "Laboring Animal":
            A Manifestation of the Will to Will   191
Chapter 13: How Modern Technology Transforms the Everyday World—
            and Points to a New One   205
Chapter 14: Authentic Production: *Techne* as the Art of Ontological
            Disclosure   222
Conclusion: Critical Reflections on Heidegger's Concept of Modern
            Technology   248

Notes   275
Index   299

Editor's Foreword

Indiana University Press is proud to launch the Indiana Series in the Philosophy of Technology with the following trio: *John Dewey's Pragmatic Technology*, by Larry A. Hickman; *Technology and the Lifeworld: From Garden to Earth*, by Don Ihde; and *Heidegger's Confrontation with Modernity*, by Michael E. Zimmerman.

The Indiana Series is the first North American series explicitly dedicated to the philosophy of technology. (There are other series relating philosophy and technology, particularly those which collect interdisciplinary articles, but none of these have been devoted to the development of a new subdiscipline within philosophy.) Broadly conceived, it nevertheless will address a wide variety of issues relating to technology from distinctly philosophical perspectives. Philosophically, our approach will be pluralistic; and this is evidenced in our first round of books. The traditions both of American pragmatism and of Euro-American trends are represented.

Our trio is a timely one. We begin with radical reappraisals and interpretations of the two early twentieth-century philosophers who made questions of technology central to their thought, John Dewey and Martin Heidegger. And we are also beginning with a systematic reformulation of a framework and set of questions regarding technology in its cultural setting in *Technology and the Lifeworld*. Later we will be adding volumes of a more topical and thematic nature, including books on "Engineering Birth," "Big Instrument Science," "Media and Rationality," "Technological Transformations of Perception," and many others.

Our goals have been to include philosophically critical, historical, and interpretive studies as well as original and topical ones within a perspective which is balanced, reasoned, and rigorous regarding the emergent field of the philosophy of technology.

In *Heidegger's Confrontation with Modernity* Michael Zimmerman undertakes a critical and timely reappraisal of Martin Heidegger on technology.

There can be little doubt that Heidegger is one of the most prominent philosophical giants of the twentieth century. In continental Europe he is the self-acknowledged influence, often the most primary such thinker, behind the main figures in many of today's debates. The two primary hermeneutic philosophers, Hans Georg Gadamer and Paul Ricoeur, have noted their debts, as have both the leaders of contemporary critical theory in Germany through Jürgen Habermas and the late post-structuralist Michel Foucault, and the deconstructionist Jacques Derrida of France.

In North America, philosophers as diverse as William Barrett of existentialist fame and Richard Rorty as a "post-analytic" philosopher have confirmed Heidegger as one of the "top three." The first names Heidegger along with Wittgenstein and James in this category, the latter substitutes Dewey for James.

Even more to the point for this series, Heidegger stands as the single most important philosopher to have directed and made central to his thought the age of *technology*. No historical or critical look at the philosophy of technology could afford to ignore Heidegger. The broad claim that the entire history of metaphysics from Plato to the present is a trajectory into the age of technology combined with the paradigm inversion of the relation between science and technology wherein science becomes the instrument of a technological revealing of world, raises the issues of technology to highest metaphysical and ontological status.

Yet Heidegger's position as a philosophical giant has been repeatedly marred and distorted by his relationship with National Socialism prior to and through World War II. The reopening of that debate has once again brought forth both extreme and spectacular denunciations and close-to-blind defenses of Heidegger.

That is why Michael Zimmerman's daring and balanced reappraisal in Division One is so important. He faces the issues directly and, through tracing a contextual history of the times, locates Heidegger with respect both to the German intellectual climate in which Heidegger sides with other "reactionary modernists" of the period, and to the then common theme of Germanic destiny as a third way between capitalism and communism. But more, Zimmerman also finds that Heidegger adapts aspects of the often chilling pro-war technology stance of Ernst Jünger, a younger contemporary. This account is often as dramatic as a novel, at least from the retrospective stance of our present knowledge of what happened in Hitler's Germany.

Zimmerman's account makes the situation understandable—but in no way forgivable. The theme of a third way between what Nicolas Berdyaev called the "metaphysical equivalence" of the United States and the USSR in the early twenties, speaking at the same time as Heidegger, later came to a head in Germany's National Socialism. Yet this same alternative between two "devils" has not disappeared. It is echoed today in the speeches of Ayatollah Khomeini, among others. Zimmerman, better than anyone known to me, undertakes an in-depth analysis of this most troubling question related to Heidegger scholarship, and no one will be able, after reading Zimmerman, to simply return to things as they were.

In Division Two, the role of technology and the coming of the age of technology is examined in the light of Heidegger's overall philosophical project. It is the history of being which ends up focused upon the Western technological way of viewing the world. Here the spectrum ranging from the nascent romanticism in much of Heidegger to the implications for a deep ecology is examined. For beyond reappraisal, Heidegger's role for subsequent philosophy of technology also needs examination and critique.

Heidegger's Confrontation with Modernity brilliantly fulfills the goals of the Indiana Series in the Philosophy of Technology. It critically and unflinchingly faces the hardest issues; it balances itself between the too often facile utopianism of many authors and the gloomy dystopianism of too many philosophers. Instead, Zimmerman carefully and deeply deals with the ambiguities of technology and its interpreters—in this case, Heidegger.

DON IHDE

Acknowledgments

I owe my thanks to several people who encouraged and assisted me with this book. First, I want to thank my wife, Teresa Toulouse, a scholar who gave generously of her time and energy to provide exceptional editorial assistance which greatly improved the present work. She has also supported me in so many other ways.

To Don Ihde, editor of Indiana University Press's series on philosophy and technology, I owe thanks for having convinced me to write this book and for having recommended that it be published. Janet Rabinowitch, senior sponsoring editor at Indiana University Press, was very supportive throughout the years during which this work was being prepared.

My thanks also go to Charles P. Bigger, John D. Caputo, Hubert L. Dreyfus, William J. Richardson, Charles M. Sherover, Thomas J. Sheehan, Gene D'Amour, Parvis Emad, Carla Fishman, Manfred Frings, John D. Glenn, Michel Haar, Jeffrey Herf, Eric Mack, Frank Schalow, and Kathleen Wright for their contributions.

I would not have been able to complete this book in a timely manner without a six-month sabbatical leave from Tulane University, and without a subsequent six-month fellowship from the National Endowment for the Humanities. I acknowledge Tulane and NEH for their financial support during the academic year 1988–89. Francis L. Lawrence, Academic Vice-President and Provost of Tulane, was instrumental in making possible the arrangements which allowed me to accept the NEH grant. My thanks to him for his support over the years.

Shirley Pratt, for many years an administrative assistant in the Newcomb College philosophy department, provided much-appreciated support during the time I composed this work. I thank her for her loyalty and concern. My thanks also to Carol Jobtanski and Carol Russell for their help.

Finally, there are many other colleagues and friends to whom I am grateful for their insight and love over the years.

Introduction

Our age is not a technological age because it is the age of
the machine; it is an age of the machine because it is the
technological age. [WHD: 54/24]

This book discusses how a major twentieth-century philosopher, Martin Heidegger, interpreted and evaluated modern technology. For Heidegger, "modern technology" had three interrelated meanings: first, the techniques,
devices, systems, and production processes usually associated with *industrialism*; second, the rationalist, scientific, commercialist, utilitarian, anthropocentric, secular worldview usually associated with *modernity*; third, the
contemporary *mode of understanding or disclosing things* which makes possible
both industrial production processes and the modernist worldview. Heidegger
maintained that the third meaning of "modern technology" is most important. Both industrialism and modernity are symptoms of the contemporary
disclosure of things as raw material to be used for expanding the scope of
technological power for its own sake. This one-dimensional disclosure of
things as raw material, in Heidegger's view, has resulted not from human
decision, but instead from developments within "the history of being" itself.
The technological stage in that history has so transformed how things are
understood that people have been more or less compelled to take part in the
industrial order and to adopt the modernist worldview related to it.

　　In this work, the major sense of "modern technology" will always be the
contemporary disclosure of all things as raw material. The reader should keep
in mind, however, that "modern technology" will also mean the industrialism
and the modernist worldview made possible by that disclosure. At times, for
the sake of clarity, I shall use the term "industrial technology" to refer specifically to those industrial processes and technical devices called forth by the
technological disclosure of things.

　　Heidegger's account of modern technology as a way of disclosing things can be clarified when placed within the history of German philosophy. Like Kant, Heidegger believed that the task of the philosopher was to discover the transcendental conditions which made possible human knowledge and action. These conditions are not themselves *things*, but instead make possible our objective *experience* of things. In his analysis of modern technology, Heidegger attempted to discover the conditions necessary for the possibility of our one-dimensional experience of entities as raw material. Like Hegel, Heidegger maintained that human activities are by and large not self-referring and self-originating, but instead are guided and shaped by a historical "play" of language and concepts which is not under human control. This conceptual-linguistic play determines the categories which shape the possibilities for human action, knowledge, and belief in determinate historical epochs. While Heidegger rejected Hegel's view that history was a progressive movement toward divine self-consciousness, both thinkers attempted to discover the nature of the conceptual and ontological "movements" which stamped the various historical epochs into which Western humanity has been "thrown." Heidegger believed that these epoch-stamping movements were not themselves worldviews, but instead were the ontological conditions necessary for the emergence of a particular worldview. For Heidegger, then, the worldview called "modernity" was never a final term in explaining the contemporary situation, but instead was a symptom of a deeper movement that was hidden from view. He maintained that the movement of these historical epochs began with Plato's metaphysics and culminated in the technological era. The major periods in Western history—Greek, Roman, medieval, Enlightenment, technological—mark, in Heidegger's view, the stages of a long *decline* in Western humanity's understanding of what it means for something "to be." In the technological age, in particular, for something "to be" means for it to be raw material for the self-enhancing technological system.

Heidegger's interpretation of modern technology differs radically from the much more familiar interpretation offered by naturalistic anthropology. According to such anthropology, consciousness is an evolutionary development which has made the human animal particularly adaptive to a wide variety of climates and material conditions. Humans have survived because they learned how to make and use tools and symbols. For such an anthropology, modern industrial technology is simply a sophisticated version of the tools used by primitive humanity. The major difference between earlier and later technology is simply that newer tools are designed and built in accordance with scientific principles unknown to earlier periods of human life.

Heidegger maintained that this naturalistic and instrumentalist view of technology has a very limited validity. He contested assumptions about the "clever animal" by refusing to conceive of humans as merely animals with high intelligence. Countering the argument that industrial technology resulted from the historical trials and errors of material practices, moreover, he maintained instead that industrial technology has arisen because of a one-dimensional mode of understanding what things *are*. According to Heidegger, for something "to be" means for it "to be disclosed" or "to be manifest." In a particular epoch, human behavior is shaped by the way in which things *manifest* themselves. If things manifest themselves as creatures of God, people treat things in one way; if things reveal themselves as nothing but raw material, people treat them in another way.

Heidegger's concern with the technological mode of understanding things was intimately involved with his lifelong investigation of what is meant by "the being of entities." By "being," he did not mean an eternal "foundation" for entities, such as Plato's forms or the Christian God. Instead, he defined "being" as the history-shaping ways in which entities reveal themselves. One of his major goals became to understand the nature and character of the process through which "the being of entities" changed over the centuries. He came to view the one-dimensional, technological understanding of the being of entities as the final stage in a history initiated not tens of thousands of years ago in prehistoric humanity, but instead far more recently—in ancient Greece. The history of the West, according to Heidegger, is the story of how the "productionist metaphysics" of the ancient Greeks gradually degenerated into modern technology.

Because the technological era comes at the *end* of this history of metaphysics, Heidegger's commentators have not always granted to the more general question of "technology" itself the central role which it plays in that history. But Heidegger in fact maintained that the technological era was prefigured from the very beginning of the history of metaphysics; indeed, modern technology was, so he believed, the inevitable outcome of that history. The Greek founders of metaphysics defined the being of entities in a proto-technological way. For them, "to be" meant "to be produced." Hence, according to Heidegger, the history of metaphysics became the history of the unfolding of *productionist* metaphysics. The technological understanding of being, the view that all things are nothing but raw material for the ceaseless process of production and consumption, is merely the final stage in the history of productionist metaphysics. Heidegger read Plato as the initiator of this metaphysics. Fascinated like other Greeks by human making and producing, Plato conceived of the being of entities in terms drawn from human manufacturing. Indeed, Heidegger argued, his concept of the ideal "form"—that which is eternally present and ultimately real—was drawn from the role played by the blueprint or model in the work of a craftsman. Just as the craftsman's blueprint provides the structure for the thing he makes, so too the eternal form provides the structure for things which come to be in the temporal-empirical world. Plato considered the eternal, nonempirical form to be the unchanging, permanently present foundation or basis for things. In a way that was to prove decisive for the entire history of productionist metaphysics, he presupposed that for something "to be," it must rest upon some "ultimate" foundation. Productionist metaphysics thus also became a *foundationalist* metaphysics.

One major task of this work will be to explain Heidegger's analysis of how the productionist and foundationalist metaphysics inaugurated by Plato devolved into modern technology: the view that for something "to be" means for it to be raw material for the subject who is compelled to produce more and more solely for the sake of producing. From the beginning of his career, Heidegger was centrally concerned with the nature of working and producing—and with its relation to the question of the being of entities. It was no accident that he began *Being and Time* with an account of the world of the workshop and equipment.

A second task of this work will be to show how the history of productionist metaphysics encouraged ways of producing which Heidegger regarded as increasingly blind to the nature of working, to the materials used in working, and to the nature of the "works" produced by working. A third, related task will be to show how such "inauthentic" working and producing has transformed the everyday world. For example, the homogenizing production processes of industrial technology and modernist political ideologies have destroyed the uniqueness of individual peoples and places. A fourth task will be to analyze Heidegger's attempt to conceive of an *authentic* mode of working and producing that would provide an alternative to the modes which have constituted the history of productionist metaphysics.

According to Heidegger, the discovery of an authentic mode of production would be tantamount to the inauguration of an entirely new, post-metaphysical era for the West. Authentic producing in this new era would be akin to what the Greeks originally meant by *techne*: a knowing and careful pro-ducing, a drawing-forth, a letting-be of things. In Heidegger's view, the attentive activity of "letting things be" was not "work" as it is known under productionist metaphysics, but instead the essence of *art*. The nagging question, however, was whether this new era's artistic understanding of things could avoid the pitfalls which led the Greeks into the productionist metaphysics which culminated in modern technology. Heidegger believed that the new era would be possible only if humanity were enabled to "produce" a work of art that would restore meaning to the things which had been made meaningless in the technological era.

Having completed this description of Heidegger's views about modern technology, we shall then evaluate his account of productionist metaphysics, his conception of the nature and the impact of "modern technology," and his assertion that art may provide the way beyond the technological era.

At an early stage in my research, I believed that this work would be complete once I had explained Heidegger's analysis of the origin and effects of and the alternatives to productionist metaphysics. It seemed enough that I explain how modern technology arose within the history of productionist metaphysics, and how a new history could be initiated by a new, post-metaphysical understanding of things. Yet as I pursued my research, I began to see that Heidegger's account of the history of the Western world, as shaped by productionist metaphysics, was in many ways consistent with the reactionary vision of many Germans during the first part of this century. Instead of assuming that Heidegger had been able to attain a privileged position, sheltered both from political presuppositions and from the vicissitudes of the Western history he analyzed, I came to realize that Heidegger viewed that history from *within* the limits imposed by his own historical circumstances. I concluded that his conceptual analysis of "the history of productionist metaphysics," however insightful it seemed, was inevitably colored and shaped by the cultural conversation into which he himself was "thrown." Three related events reinforced my guiding belief that an internal relation existed between Heidegger's interpretation of the history of being and his relation to his own historical circumstances.

First, the recent disclosures about the nature and extent of Heidegger's involvement with National Socialism led me to examine critically the interpretation of that episode offered by Heidegger and his defenders, who claimed that it was a brief and unfortunate episode unrelated to his thinking.[1] In fact, however, Heidegger's decision to support National Socialism was deeply related to his political and philosophical understanding of Germany's situation in the early twentieth century. He did not regard himself as an "ordinary" reactionary, however, nor later as an "ordinary" Nazi, contending that such people were incapable of understanding the *metaphysical* issues involved in the question of modern technology. Rather, he believed that only a gifted and high-minded thinker could comprehend the metaphysical origins of the deadly "symptoms" afflicting Germany.

These symptoms included movements which may seem incompatible to many, but which to Heidegger and other conservative Germans were all aspects of "modernity": commercialism, industrialism, liberalism, socialism, Bolshevism, American democracy, individualism, materialism, scientism, positivism, and rationalism. Heidegger became convinced that National Socialism represented a "third way" between the related evils of industrial capitalism and industrial communism which were threatening to crush Germany between them. He believed that his own philosophical version of National Socialism would make it possible for Germany to initiate a "new beginning" comparable in scope to the beginning initiated by the ancient Greeks. This new beginning, so he hoped, would bring an end to the alienating and destructive modes of working and producing associated with industrial technology. It is no accident that he sought to become the spiritual *Führer* of the National Socialist Democratic *Workers'* Party, for Heidegger longed for the emergence of a new social order in which work and workers would both regain their integrity and importance. His provocative interpretation of the history and character of productionist metaphysics can be fully understood only in the light both of his despair about the drastic changes wrought in the workplace and throughout the German nation by modern technology, and of his hope that a saving alternative to it could be found for Germany.

A second factor in my decision to examine the historical context of Heidegger's concept of modern technology was Jeffrey Herf's study *Reactionary Modernism: Technology, Culture, and Politics in Weimar and the Third Reich*.[2] Herf demonstrates that Heidegger's account of modern technology was by no means as original as I had once thought, but was in fact greatly indebted to and formed an important part of a long-standing German debate about how industrialism and the related phenomenon of modernity were affecting the German *Geist*. Many participants in this debate were militant nationalists and political reactionaries, who regarded industrial technology, Enlightenment rationalism, and the political ideals of the French Revolution as interrelated aspects of the same "dark force" destroying Germany's traditional values. Other contemporary participants in the technology debate, equally concerned about the fate of the German *Geist*, wanted to restore an authoritarian political order, but also wanted to retain the power-giving fruits of industrial technology. Heidegger shared many of these political views. His reading of the works of important "reactionary modernists" greatly influenced both his philosophical conception of modern technology and his decision to become involved in a *political* struggle against it.

The most important of these reactionary modernists was Ernst Jünger. Discovering his writings was the third factor which shaped my changing historical and contextual approach to Heidegger's concept of modern technology. While Herf briefly discusses Heidegger's debt to Jünger and other reactionary modernists, I analyze this debt in detail, because I regard it as crucial not only for understanding the development of Heidegger's understanding of the technological epoch, but also for offering some explanation of his support for National Socialism. Jünger's writings, I contend, form a link that mediates between Heidegger's thought, on the one hand, and his engagement with National Socialism, on the other. In *Der Arbeiter* [DA] and other publications in the early 1930s, Jünger forecast that the *Gestalt* of the worker was "stamping" humanity in the mode of the worker-soldier compelled to mobilize the earth into a planetary industrial foundry. Believing that National Socialism provided the only available alternative for preventing the realization of Jünger's grim predictions for the future of Germany and the West, Heidegger chose to support Hitler's "workers' revolution." Heidegger maintained that National Socialism would restore Germany to its authentic origins, thereby making possible a new relationship to work on the part of the German *Volk*. To the end of his career, Heidegger not only continued to believe that Jünger's vision was the best description of the final, technological stage in the history of productionist metaphysics, but also maintained that his own version of National Socialism had held out genuine promise for movement beyond the technological stage of Western history.

Reading about Heidegger's engagement with National Socialism, his links with reactionary modernists, and especially his debt to Ernst Jünger, I discovered that the political and historical dimensions of Heidegger's concept of modern technology could not be ignored. As a result of this discovery, both the emphasis and the structure of my original project changed. In order to understand Heidegger's theories about the history and character of productionist metaphysics, I realized that I had to tell the concurrent narrative of his personal and philosophical involvement with the political issues of his own day. This narrative thus became Division One of my study, and the description and critique of his conception of modern technology, as seen in the light of the history of productionist metaphysics, became Division Two. Let me stress, however, that the contents of these two divisions are, in fact, intertwined. Heidegger attempted to apply his thinking to Germany's desperate situation because he believed that only his interpretation of the *work-dimension* of Western metaphysics could provide the guidance needed to establish a new world of working and producing for Germany. For him, very simply, the misery of the German *Volk* during the 1920s and 1930s had arisen because the history of being had led to the degradation of work. The task was to found a new mode of working and producing, a mode which emphasized the relationship between art (especially poetry) and producing. For his model of authentic producing, or "letting things be," Heidegger took the disclosive event involved in the work of art. One reason Heidegger was attracted to National Socialism was that it, too, proposed to overcome the alienation of the modern worker by transforming labor into a form of art.

I have divided this complex material into two divisions for expository purposes. The already involved political narrative of the first division would have become overcomplicated had it been interrupted by frequent references to details of Heidegger's conception of productionist metaphysics. Other commentators have already brought to light important features of that conception, but they have not explained satisfactorily the key to the relation between his philosophy and his politics, viz., his concern with the nature of working and producing. Hence, I chose first to examine Heidegger's reactionary political application of his understanding of the history of productionist metaphysics. Second, in view of what that examination revealed, I turned to an exposition of that history. The reader is thus asked to read Division Two as internally related to Division One. Both divisions end with a chapter on authentic producing. Division One arrives at that topic by way of examining Heidegger's attempt to "apply" his thought politically; Division Two arrives at the topic of producing by examining aspects of his conception of the history of being. The two divisions, in the end, constitute moments in a hermeneutic circle. Understanding Heidegger's political vision of work informs our understanding of his critique of the metaphysics of production; in turn, insight into that critique will help us to comprehend his motivation for supporting Hitler's revolution in the realm of work and worker.

Having described this new structure of my book and the importance that discoveries about Heidegger's politics had in effecting it, I wish to raise a crucial question to be addressed later in this work: should Heidegger's regrettable political application of his thought influence our assessment of the *validity* of that thought? Some critics maintain that, in effect, Heidegger's philosophy was essentially reactionary and even fascist, the ideological "reflex" of a man belonging to a social class threatened by the advent of social and economic changes associated with the industrial age. While I cannot agree with such a reductionist reading of Heidegger's thought, a reading which recommends that we cease taking his writings seriously, I also cannot agree with those who pretend that Heidegger's thought was not colored in a significant way by his own strongly held political opinions. Read in its own historical context, Heidegger's concept of modern technology emerges as a critique of the *legitimacy* not only of industrial production processes, but of the whole of modernity as well. For him, nihilistic modern culture arose from the same one-dimensional disclosure of entities that simultaneously gave rise to the industrial forms of working and producing. According to Heidegger, technological humanity, far from being the autonomous agent in control of the technological conquest of nature, had itself become the "subject" of the self-directing work processes of modern technology. In the totally administered technological world, talk of individual "autonomy" or "freedom" made little sense, for in that world people had become indistinguishable ciphers shaped by the demands of industrial modes of production. Moreover, no mere change in the "ownership" of the means of production could alter the alienating and destructive character of industrial work; capitalism and socialism alike were, from Heidegger's viewpoint, manifestations of the limitless Will to Power associated with the technological disclosure of things.

Jünger's writings provided Heidegger with the vocabulary to describe humanity's paradoxically servile yet heroic status in the technological age. Jünger viewed world history as an aesthetic phenomenon, a spectacle produced by the eternal Will to Power at work behind all things, including human actions. Only great visionaries such as himself, Jünger believed, could see that the extraordinary achievements of technological humanity were both inspired by and in the service of this hidden Will to Power. Jünger's view that this Will to Power mobilized humanity in terms of the *Gestalt* of the technological worker influenced Heidegger in two crucial ways. First, it helped convince him that only a radical new beginning, like that proposed by National Socialism, could help Germany to escape Jünger's technological forecast. Second, it led Heidegger to look for the essence of that new beginning not so much in philosophy but in art, especially as art was understood by Nietzsche and Hölderlin.[3] He sought to provide an alternative interpretation to Jünger's interpretation of Nietzsche's aesthetic view of history.

Whereas for Jünger the Will to Power was the eternal metaphysical force at work shaping the protean face of world history, for Heidegger the Will to Power was itself a historical phenomenon associated with the history of productionist metaphysics. While acknowledging that technological humanity was shaped by the *Gestalt* of the Will to Power, Heidegger also believed that an artistic revolution inspired by the poetry of Hölderlin could initiate a new, non-technological future for humanity. For Heidegger, art meant authentic pro-ducing. Not the act of a subject who "works" on something in the sense of "causing" it to be built, art was instead the process of enabling things to disclose themselves in accord with their own possibilities. Throughout his career, Heidegger believed that not only "high" art, but also "art" as authentic handiwork or craftwork provided insight into the mode of producing that could replace the alienating and mechanical mode of producing present in modern industrial technology. As an ontologically disclosive mode of working or producing, art could give rise to new forms of humanity and community, no longer stamped by the metaphysical Will to Power, and to a new mode of working, no longer guided by the history of productionist metaphysics.[4]

While both divisions of this study raise critical questions about Heidegger's concept of modern technology, in the conclusion I offer a more extended critique of issues that the entire work has addressed. Since his most important contribution to the "philosophy of technology" was his claim that modern technology resulted from the history of productionist metaphysics, I begin by examining the extent to which his deterministic conception of history was indebted to Hegel and Marx. Next, in the light of this analysis, I analyze the contention that Heidegger's own interpretation of the history of metaphysics employs some of the very metaphysical categories that he sought to overcome. Moreover, I inquire into the political dangers involved in his deconstruction of metaphysics, especially insofar as such deconstruction undermined the Enlightenment doctrine of individual liberties and helped to justify his support for a disastrous "post-Enlightenment" political movement. Third, I address the fact that there are a number of alternatives to Heidegger's analysis of the origins of modern technology. One such alternative is offered by those feminists who argue that modern technology is a particularly powerful manifestation of the patriarchal domination of woman, the body, and nature. While Heidegger focused on the history of being, in other words, he ignored the history of patriarchy. I conclude that the phenomenon of modern technology is too complex to be explained by any one conceptual approach.

My academic interest in Heidegger's concept of modern technology is directly tied to my personal concern about the fate of humanity and the earth in the technological age. The issues facing us today are similar to the ones which faced Heidegger half a century ago: Is a stable human community possible in a politically pluralistic world that is constantly transformed by technological innovation? Have traditional political discourse and practices been eclipsed or marginalized by technological developments in production, transportation, and communication? Are genuine individuality and freedom still possible in the technological world of nonstop producing and consuming? Do we retain any valid "measure" that would provide limits for what may be done with technological breakthroughs in areas such as genetic engineering? Are we, in short, *permitted* to do anything we become *capable* of doing? Does talk of "*human progress*" in connection with industrial technology obscure the fact that only certain classes, nations, races, and often only one sex of "humanity" benefit from such "progress," while many other people suffer from it? Even if we manage to avoid ecological catastrophe, will the human "spirit" be eroded by the loss of everything wild and free as the result of planetary technology? What are the psychological and social consequences of an age in which everything is evaluated and measured in terms of the human quest for control and power?

This work does not pretend to offer any final answers to such crucial questions. More important than answers at this point in time is our willingness to take the questions themselves seriously. Only by questioning the presuppositions, perils, and promises of the technological age will humanity have any hope of discovering authentic ways of living within the dangerous and the wondrous possibilities opened up by that age.

TEXTUAL AND TERMINOLOGICAL ISSUES

This book takes into account all of Heidegger's published writings, but focuses particular attention on remarks he made about modern technology in the untranslated volumes of his *Collected Works* (*Gesamtausgabe*). These volumes, especially those from the 1930s, contain many important references to Heidegger's conception of the political situation of his day and of the political dimension of his own thoughts about modern technology.

Translating Heidegger's writing is a notoriously difficult matter. I have chosen to translate his key terms *Sein* and *Seiende* as "being" and "entity." This translation fails to convey the fact that there is a close relation between the two German words. *Seiende* is a participial form of the German verb *Sein*, "to be." I use this translation because it has the virtue of allowing me to distinguish between *Sein* and *Seiende* without adopting the usual practice of translating *Sein* by "Being" and *Seiende* by "being." Spelling "being" with a capital *B* tends to reify the phenomenon that it names. For Heidegger, *Sein* meant the *event* in which an entity reveals or shows itself. Hence, "to be" meant not an eternal metaphysical foundation or first principle, but instead the "presencing" or "appearing" of a thing. Such presencing or appearing, so Heidegger argued, could not take place without the "clearing" which constitutes the essence of human existence. Without humans, entities would be unable to reveal or to display themselves; in this sense, they would not "be."

As is customary, I have not translated the word *Dasein*, an ordinary German term to which Heidegger gave special meaning. For Heidegger, *Dasein* names humanity's unique capacity for understanding the being of entities. This capacity is not a possession or faculty of humanity. Instead, according to Heidegger, *Dasein* possesses humanity as the "there" (*da-*) in which being (*-sein*) can manifest itself.

I have made two other terminological decisions that are not related to translation. First, throughout the text, I have tried wherever possible to use gender-neutral terminology. My decisions regarding use of such terminology have been complicated by the fact that Heidegger, Jünger, and almost every other writer considered in this book frequently used masculine-biased terminology. In my commentaries on quotations which utilize such terminology, I have, for example, sometimes used terms such as "humankind" or "humanity" instead of "man." At other times, however, I have used the masculinist vocabulary of the quotations when the authors themselves had only male humans in mind.

Second, I frequently refer to "early" and "later" Heidegger. He himself acknowledged that his thought underwent some sort of "change" (*Kehre*), but attempts to explain or to date it have proved to be very controversial. In some respects, Heidegger's thought changed in 1929; in other respects in 1934, and in 1938, and in 1943. Indeed, his thought may be regarded as a continuous unfolding of a basic intuition which never essentially changed. Many commentators agree, however, that an important development took place in Heidegger's thinking between 1929 and 1935. Hence, by "early" Heidegger I shall generally mean his pre-1929 writings. There will, however, be some occasions in which I shall use the term "early" Heidegger to refer to a view which he held about a topic prior to developing a subsequent, modified position about that same topic. Context will help to clarify what I mean in each case by "early" Heidegger.

List of Abbreviations

HEIDEGGER'S WORKS

References to Heidegger's works will use the following abbreviations. All references will include a cross-reference to the English translation, where such exists. Thus [HW: 248/91] indicates that I am quoting p. 248 of *Holzwege*, a translation of which is found on p. 91 of the English edition listed in the *Holzwege* entry found below. For the most part, the translations are my own, though I have sometimes used (and frequently modified) existing translations.

I begin the listings with volumes from Heidegger's *Gesamtausgabe* (*Collected Works*), all of which are published by Vittorio Klostermann, Frankfurt am Main. In the case of some books which are published in editions outside the *Gesamtausgabe* and in the *Gesamtausgabe* as well—especially *Sein und Zeit*, *Holzwege*, and *Wegmarken*—I have generally cited the non-*Gesamtausgabe* editions, since they are more generally available.

Gesamtausgabe

GA, 1   *Frühe Schriften*, ed. Friedrich-Wilhelm von Herrmann (1978).
GA, 2   *Sein und Zeit*, ed. Friedrich-Wilhelm von Herrmann (1977).
GA, 4   *Erläuterungen zu Hölderlins Dichtung*, ed. Friedrich-Wilhelm von Herrmann (1982).
        Pp. 9–31: "Remembrance of the Poet," trans. Douglas Scott, in *Existence and Being*, ed. Werner Brock. Chicago: Henry Regnery Co., 1949.
        Pp. 33–48: "Hölderlin and the Essence of Poetry," trans. Douglas Scott, in *Existence and Being*.
GA, 13 *Aus der Erfahrung des Denkens, 1910–1976*, ed. Hermann Heidegger (1983).
        "Why Do We Remain in the Province?" trans. Thomas J. Sheehan, *Listening*, 12 (Fall, 1977), pp. 122–125.
GA, 20 *Prolegomena zur Geschichte des Zeitbegriffs* (Summer Semester, 1925), ed. Petra Jaeger (1979).
        History of the Concept of Time: Prolegomena, trans. Theodore Kisiel. Bloomington: Indiana University Press, 1985.

GA, 21   *Logik: Die Frage nach der Wahrheit* (Winter Semester, 1925/26), ed. Walter Biemel (1976).

GA, 24   *Die Grundprobleme der Phänomenologie* (Summer Semester, 1927), ed. Friedrich-Wilhelm von Herrmann (1975).
The Basic Problems of Phenomenology, trans. Albert Hofstadter. Bloomington: Indiana University Press, 1982.

GA, 25   *Phänomenologische Interpretation von Kants Kritik der reinen Vernunft* (Winter Semester, 1927/28), ed. Ingtraud Görland (1977).

GA, 26   *Metaphysische Anfangsgrunde der Logik im Ausgang von Leibniz* (Summer Semester, 1928), ed. Klaus Held (1978).
The Metaphysical Foundations of Logic, trans. Michael Heim. Bloomington: Indiana University Press, 1984.

GA, 29/30   *Die Grundbegriffe der Metaphysik: Welt-Endlichkeit-Einsamkeit* (Winter Semester, 1929/30), ed. Friedrich-Wilhelm von Herrmann (1983).

GA, 31   *Vom Wesen der menschlichen Freiheit, Einleitung in die Philosophie* (Summer Semester, 1930), ed. Harmut Tietjen (1982).

GA, 32   *Hegels Phänomenologie des Geistes* (Winter Semester, 1930/31), ed. Ingtraud Görland (1980).

GA, 33   *Aristoteles: Metaphysik V 1–3* (Summer Semester, 1931), ed. Heinrich Hüni (1981).

GA, 39   *Hölderlins Hymnen "Germanien" und "Der Rhein"* (Winter Semester, 1934/35), ed. Susanne Ziegler (1980).

GA, 40   *Einführung in die Metaphysik* (Summer Semester, 1935), ed. Petra Jaeger (1983).

GA, 41   *Die Frage nach dem Ding: Zu Kants Lehre von den transzendentalen Grundsätzen* (Winter Semester, 1935/36), ed. Petra Jaeger (1984).

GA, 43   *Der Wille zur Macht als Kunst* (Winter Semester, 1936/37), ed. Bernd Heimbüchel (1985).

GA, 45   *Grundfragen der Philosophie: Ausgewählte "Probleme" der "Logik"* (Winter Semester, 1937/38), ed. Friedrich-Wilhelm von Herrmann (1984).

GA, 48   *Nietzsche: Der Europäische Nihilismus* (Second Freiburg Trimester, 1940), ed. Petra Jaeger (1986).

GA, 51   *Grundbegriffe* (Summer Semester, 1941), ed. Petra Jaeger (1981).

GA, 52   *Hölderlins Hymne "Andenken"* (Winter Semester, 1941/42), ed. Curd Ochwadt (1982).

GA, 53   *Hölderlins Hymne "Der Ister"* (Summer Semester, 1942), ed. Walter Biemel (1984).

GA, 54   *Parmenides* (Winter Semester, 1942), ed. Manfred S. Frings (1982).

GA, 55   *Heraklit.* (1) *Der Anfang des Abendlandischen Denkens (Heraklit)* (Summer Semester, 1943); (2) *Logik: Heraklits Lehre vom Logos* (Summer Semester, 1944), ed. Manfred S. Frings (1979).

GA, 56/57   *Zur Bestimmung der Philosophie.* (1) *Die Idee der Philosophie und das Weltanschauungsproblem* (War Emergency Semester, 1919); (2) *Phänomenologie und Transzendentale Wertphilosophie* (Summer Semester, 1919), ed. Bernd Heimbüchel (1987).

GA, 61   *Phänomenologische Interpretationen zu Aristoteles. Einführung in die phänomenologische Forschung* (Winter Semester, 1921/22), ed. Walter Bröcker and Kate Bröcker-Oltmanns (1985).

Other Works by Heidegger

AED   *Aus der Erfahrung des Denkens*, 1910–1976, ed. Hermann Heidegger. Frankfurt am Main: Vittorio Klostermann, 1983.

D   *Denkerfahrungen*, ed. Hermann Heidegger. Frankfurt am Main: Vittorio Klostermann, 1983.

DR   *Das Rektorat 1933/34. Tatsachen und Gedanken*, ed. Hermann Heidegger.

Frankfurt am Main: Vittorio Klostermann, 1983. "The Rectorate 1933/
34: Facts and Thoughts," trans. Karsten Harries, *The Review of Metaphysics*, 38, No. 3 (March, 1985), pp. 481–502.

G   *Gelassenheit*. 2d ed. Pfullingen: Günther Neske, 1960. *Discourse on Thinking*,
trans. John M. Anderson and E. Hans Freund. New York: Harper &
Row, 1966.

HW   *Holzwege*. 5th ed. Frankfurt am Main: Vittorio Klostermann, 1972.
Pp. 7–68: "The Origin of the Work of Art," in *Poetry, Language, Thought*,
trans. Albert Hofstadter. New York: Harper & Row, 1971.
Pp. 69–104: "The Age of the World Picture," in *The Question concerning
Technology*, trans. William Lovitt. New York: Harper & Row, 1977.
Pp. 193–247: "The Word of Nietzsche: 'God Is Dead,' " in *The Question
concerning Technology*.
Pp. 248–295: "What Are Poets For?" in *Poetry, Language, Thought*.
Pp. 296–343: "The Anaximander Fragment," in *Early Greek Thinking*, trans.
David Farrell Krell and Frank Capuzzi. New York: Harper & Row, 1975.

ID   *Identity and Difference*, trans. Joan Stambaugh. New York: Harper & Row,
1969. The German text, *Identität und Differenz*, appears in the Appendix
of this book.

KPM   *Kant und das Problem der Metaphysik*. 2d ed. Frankfurt am Main: Vittorio
Klostermann, 1951.
Kant and the Problem of Metaphysics, trans. James S. Churchill. Bloomington:
Indiana University Press, 1968.

MHZ   *Martin Heidegger, zum 80. Geburtstag von seiner Heimstadt Messkirch*. Frank-
furt am Main: Vittorio Klostermann, 1969.
Pp. 36–45: "Homeland," trans. Thomas Franklin O'Meara, *Listening*, Vol. 6,
No. 3 (Autumn, 1971), pp. 231–238.

N I   *Nietzsche*, Volume One. 2d ed. Pfullingen: Günther Neske, 1961.
Pp. 11–254: *Nietzsche*, Vol. I, *The Will to Power as Art*, trans. David Farrell
Krell. New York: Harper & Row, 1979.
Pp. 254–472: *Nietzsche*, Vol. II, *The Eternal Recurrence of the Same*, trans.
David Farrell Krell. New York: Harper & Row, 1984.
Pp. 473–658: *Nietzsche*, Vol. III, *The Will to Power as Knowledge and as Metaphys-
ics*, "The Will to Power as Knowledge," trans. Joan Stambaugh, David
Farrell Krell, and Frank A. Capuzzi. New York: Harper & Row, 1987.

N II   *Nietzsche*, Volume Two. 2d ed. Pfullingen: Günther Neske, 1961.
Pp. 7–29: *Nietzsche*, Vol. III, *The Will to Power as Knowledge and as Metaphys-
ics*, "The Eternal Recurrence of the Same and the Will to Power."
Pp. 31–256: *Nietzsche*, Vol. IV, *Nihilism*, "European Nihilism," trans. Frank
A. Capuzzi, ed. David Farrell Krell. New York: Harper & Row, 1982.
Pp. 257–333: *Nietzsche*, Vol. III, *The Will to Power as Knowledge and as Meta-
physics*, "Nietzsche's Metaphysics."
Pp. 335–398: *Nietzsche*, Vol. IV, *Nihilism*, "Nihilism as Determined by the
History of Being."
Pp. 399–457: *The End of Philosophy*, "Metaphysics as the History of Being,"
trans. Joan Stambaugh. New York: Harper & Row, 1973.
Pp. 458–480: *The End of Philosophy*, "Sketches for a History of Being as
Metaphysics."
Pp. 481–490: *The End of Philosophy*, "Recollection in Metaphysics."

SA   *Schellings Abhandlung über das Wesen der Menschlichen Freiheit (1809)*. Tübin-
gen: Max Niemeyer, 1971.
Schelling's Treatise on Human Freedom (1809), trans. Joan Stambaugh. Athens:
Ohio University Press, 1985.

SDU   *Die Selbstbehauptung der deutschen Universität*, ed. Hermann Heidegger.
Frankfurt am Main: Vittorio Klostermann, 1983.

"The Self-Assertion of the German University," trans. Karsten Harries, *The Review of Metaphysics*, 38, No. 3 (March, 1985), pp. 467–480.

SG   *Der Satz vom Grund*, 4th ed. Pfullingen: Günther Neske, 1971.
       Pp. 191–211: "The Principle of Ground," trans. Keith Hoeller. *Man and World*, VII (August, 1974), pp. 207–222.

Sp   "Nur noch ein Gott kann uns retten." *Spiegel*-Gespräch mit Martin Heidegger am 23 September, 1966. *Der Spiegel*, No. 26, May 31, 1976, pp. 193–219.
       "Only a God can save us: *Der Spiegel*'s interview with Martin Heidegger," trans. Maria P. Alter and John D. Caputo, *Philosophy Today*, XX (Winter, 1976), pp. 267–284.

SZ   *Sein und Zeit*. 11th ed. Tübingen: Max Niemeyer, 1963.
       Being and Time, trans. John Macquarrie and Edward Robinson. New York: Harper & Row, 1962.

TK   *Die Technik und die Kehre*. Pfullingen: Günther Neske, 1962.
       Pp. 1–36: "The Question concerning Technology," in *The Question concerning Technology*.
       Pp. 37–47: "The Turning," in *The Question concerning Technology*.

US   *Unterwegs zur Sprache*. 3d ed. Pfullingen: Günther Neske, 1965.
       Pp. 9–33: "Language," in *Poetry, Language, Thought*.
       Pp. 35–end: *On the Way to Language*, trans. Peter D. Hertz. New York: Harper & Row, 1971.

VA I, II, III   *Vorträge und Aufsätze*. 3d ed. Pfullingen: Günther Neske, 1967.
VA I   Pp. 5–36: "The Question concerning Technology," in *The Question concerning Technology*.
       Pp. 37–62: "Science and Reflection," in *The Question concerning Technology*.
       Pp. 63–91: "Overcoming Metaphysics," in *The End of Philosophy*.
       Pp. 93–118: "Who Is Nietzsche's Zarathustra?" trans. Bernd Magnus, *The Review of Metaphysics*, XX (March, 1967), pp. 411–431.
VA II Pp. 19–36: "Building Dwelling Thinking," in *Poetry, Language, Thought*.
       Pp. 37–59: "The Thing," in *Poetry, Language, Thought*.
       Pp. 61–78: " . . . Poetically Man Dwells . . . ," in *Poetry, Language, Thought*.
VA III Pp. 3–26: "Logos (Heraclitus, Fragment B 50)," in *Early Greek Thinking*.
       Pp. 27–52: "Moira (Parmenides VIII, 34–41)," in *Early Greek Thinking*.
       Pp. 53–78: "Aletheia (Heraclitus, Fragment B 16)," in *Early Greek Thinking*.

WG   *Wom Wesen des Grundes*, appearing in a bilingual edition as *The Essence of Reasons*, trans. Terrence Malick. Evanston: Northwestern University Press, 1969.

WGM   *Wegmarken*. Frankfurt am Main: Vittorio Klostermann, 1967.
       Pp. 1–20: "What Is Metaphysics?" trans. David Farrell Krell, in *Basic Writings*. New York: Harper & Row, 1977.
       Pp. 73–98: "On the Essence of Truth," trans. John Sallis, in *Basic Writings*.
       Pp. 99–108: "Postscript to 'What Is Metaphysics?' " trans. R. F. C. Hull and Alan Crick, in *Existence and Being*.
       Pp. 109–144: "Plato's Doctrine of Truth," trans. John Barlow, in *Philosophy in the Twentieth Century*, Vol. III, *Contemporary European Thought*, ed. William Barrett and Henry D. Aiken. New York: Harper & Row, 1971.
       Pp. 15–191: "Letter on Humanism," trans. Frank A. Capuzzi, with J. Glenn Gray and David Farrell Krell, in *Basic Writings*.
       Pp. 195–212: "The Way Back into the Ground of Metaphysics," trans. Walter Kaufmann, in *Existentialism from Dostoevsky to Sartre*. Cleveland: The World Publishing Company, 1956.
       Pp. 273–308: "Kant's Thesis about the Thing," trans. Ted E. Klein and William E. Pohl, *The Southwestern Journal of Philosophy*, IV, 3 (Fall, 1973), pp. 7–33.

"On the Being and Conception of *Physis* in Aristotle's Physics B, 1," trans. Thomas J. Sheehan, *Man and World*, IX, 3 (April, 1976), pp. 219–270.

WHD   *Was Heisst Denken?* Tübingen: Max Niemeyer, 1954.
What Is Called Thinking? trans. Fred D. Wieck and J. Glenn Gray. New York: Harper & Row, 1972.

ZFB   *Zur Frage nach der Bestimmung der Sache des Denkens*, ed. Hermann Heidegger. St. Gallen: Erker-Verlag, 1984.

ZSB   *Zollikoner Seminare: Protokolle-Gespräche-Briefe*, ed. Medard Boss. Frankfurt am Main: Vittorio Klostermann, 1987.

ZSD   *Zur Sache des Denkens*. Tübingen: Max Niemeyer, 1969.
On Time and Being, trans. Joan Stambaugh. New York: Harper & Row, 1972.

ZSF   *Zur Seinsfrage*, appearing in a bilingual edition as *The Question of Being*, trans. Jean T. Wilde and William Kluback. New Haven: College and University Press, 1958.

JÜNGER'S WORKS

AB   *Das Abenteurliche Herz. Werke*. Volume 7. *Essays III*. Stuttgart: Ernst Klett.

DA   *Der Arbeiter: Herrschaft und Gestalt. Werke*. Volume 6. *Essays II*.

FB   *Feuer und Bewegung. Werke*. Volume 5. *Essays I*.

KIE   *Der Kampf als inneres Erlebnis, Werke*. Volume 5. *Essays I*. Klett Verlag.

SS   *The Storm of Steel (In Stahlgewittern)*, trans. Basil Creighton. London: Chatto & Windus, 1929.

TM   "Die totale Mobilmachung." *Werke*. Volume 5. *Essays I*.

US   Über den Schmerz." *Werke*. Volume 5. *Essays I*.

DIVISION ONE

Heidegger and the Politics of Productionist Metaphysics:
The Longing for a New World of Work

Chapter 1. Germany's Confrontation with Modernity

> The danger in which stands the "holy heart of the peoples" of the West is not that of a decline, but instead that we, ourselves bewildered, *yield ourselves to the will of modernity* and drive it on. [GA, 55: 181]

For Martin Heidegger, "modernity" constituted the final stage in the history of the decline of the West from the great age of the Greeks to the technological nihilism of the twentieth century. He believed that the Greeks initiated "productionist metaphysics" when they concluded that for an entity "to be" meant for it to be produced. While what they meant by "production" and "making," for Heidegger, differed from the production processes involved in industrial technology, still it was the Greek understanding of the being of entities which eventually led to modern technology. In this chapter, we shall discover that Heidegger's critique of modern technology was shaped in important ways by the explosive situation of turn-of-the-century Germany. Deeply threatened by the advances of modernity and industrial technology, many Germans wanted to defend the traditional ways of life which they believed were essential to German identity. Heidegger shared these concerns about the fate of Germany in the face of modern technology.

A. HEIDEGGER'S AMBIGUOUS RELATIONSHIP TO REACTIONARY POLITICS

It is not unfair to say that Heidegger was a political reactionary, although he frequently regarded others of the same conviction with scorn, because of their superficiality. Before going any further, we must define what is meant by "reactionary" consciousness. One critic has defined it as

> a state of awareness that is in rebellion against the loss of community, the loss of identity, and the loss of a sense of transcendence. Reactionary consciousness rebels against loneliness and anomie, against rationalism and materialism, and against the artifices of human progress and technology. In seeking to ameliorate the condition of spiritual and emotional impoverishment brought on by modern life in industrial societies, reactionary consciousness marks an atavistic flight from nearly any aspect of experience that may be called modern. In this flight, the pursuit of a return to nature, or the embrace of instinct over reason, or the quest to recognize links between people which are racial rather than historical, are common.[1]

Heidegger shared many of these reactionary attitudes toward modernity and industrialism. He hated materialism, scientific reductionism, the decline of community, the evils of urban life, spiritual decay, atomistic individualism, and alienation from the transcendent dimension. Like other reactionaries, he rejected the economic and political values of the Enlightenment and called for a new social order that could arise only by returning to Germany's primal roots. Unlike other reactionaries, however, he wanted to renew Germany's roots not in order to restore a pastoral pre-technological world, but instead in order to inaugurate a world order in which Germans could establish a new mode of working and producing. While yearning for a renewal of *Volksgeist*, Heidegger dismissed those reactionaries who claimed that such a renewal required a return to the instinctual, atavistic roots of the *Volk*. Such appeals to instinctual renewal and racism pretended to fight against the evils of the scientific worldview, Heidegger argued, but in fact they were (perhaps unwittingly) in league with the scientific naturalism that interpreted humans as merely clever animals. Heidegger regarded most reactionary ideologues rather contemptuously, despite his sympathy for many of their ideals, because they were philosophically naive and shallow.

Some of those ideologues maintained that Germany's problems began with the Enlightenment and could presumably be cured by a return to a pre-Enlightenment social order. Others claimed that the problem was that Germany had been corrupted by the insinuation of Roman ideals and the intrusion of Latin languages. Only a return to Germany's alleged roots in the great age of the Greeks could renew the German *Geist*. While sharing such scorn for everything Roman and admiration for all things Greek, Heidegger nevertheless maintained that the decadent history of the West began with *Greek* metaphysics. The primal Greek encounter with the being of entities had quickly degenerated into the productionist metaphysics which culminated in modern technology. What Germany required in its time of crisis, in Heidegger's view, was not a superficial cultivation of the Greek classics, but instead a new encounter with the being of entities, an encounter that would issue forth in a radically new, non-metaphysical era for Germany and the West. In the 1930s, he concluded that his thinking could provide the spiritual leadership necessary for the success of the National Socialist "revolution," which promised to initiate such a new beginning.

In his quest to renew Germany, Heidegger used almost all of the important terms—including degeneration, nihilism, decline, the need for rootedness, *Volksgeist*, decision, spiritual transformation, martyrdom, revelation, renewal, achievement of Germany's salvific mission—and debated with the works of many of the important authors—including Spengler, Jünger, and Klages—of the reactionary political movement which helped bring Hitler to power. Yet, he almost always spoke of these ideas and thinkers in ways which were different from how they were usually understood, because he interpreted them not as a politician, nor even as an ideologue, but instead primarily as a philosopher. The crisis facing Germany, he maintained, was metaphysical; it resulted from the decline in Western humanity's understanding of being. Germany's plight could not be alleviated by the social and political, or even by the "spiritual" changes proposed by ordinary reactionary ideologues. Only he himself, Heidegger insisted, apprehended the nature of the new beginning required for Germany's salvation.

Despite Heidegger's antipathy toward many reactionary writers, to understand his own political stance and its relation to his concept of modern technology, we must have some grasp of the reactionary political scene with which his thought was so ambiguously intertwined. To that end, we shall examine in this chapter the history and character of the political and cultural milieu which helped to shape his view of things. At the outset, however, I want to make clear that there were many Germans—including followers of cultural heroes such as Kant and Goethe—who shared in the Enlightenment's vision of human freedom made possible by elevating reason to a primary role in human affairs.[2] In the early twentieth century, liberal German politicians and economists gave one interpretation of Enlightenment "freedom" and "reason," while German socialists gave yet another. The short-lived Weimar Republic was a parliamentary democracy, a fact which shows that many Germans favored at least the appearance of liberal political ideals. Moreover, the fact that Communists and Social Democrats fought on equal terms with National Socialists at the polls and in the streets of Germany during the 1920s indicates that another significant segment of the population was committed to some version of Marx's vision of the Enlightenment. Evidently, however, political and economic developments in the early 1930s, including the fear of a communist takeover and the misery caused by the Great Depression, attracted enough Germans to the reactionary attitude, that Hitler was able to attain power. In promising to protect Germans from the insidious forces of modernity—commercialism, communism, capitalism, industrialism, materialism, and social atomism—the politically astute Hitler knew he was saying what millions wanted to hear.

The following account does not pretend that either National Socialism or Heidegger's affiliation with it can be "deduced" from earlier conservative and reactionary tendencies within German history. Human affairs are simply too complex to be interpreted in such a manner. Nevertheless, a brief outline of the historical trends of reactionary German politics will prove helpful in understanding why Heidegger and so many others believed in National Socialism's promise to renew German life.

B. THE PAINFUL EMERGENCE OF MODERN, INDUSTRIAL GERMANY

One historian has argued that around the time of the Renaissance, suspicion of modern trends had already begun to be something like a "national style."[3] Luther's Reformation played an important role both in reinforcing Germany's anti-modernist attitude and in strengthening authoritarian views as well. His belief that the "justification" needed for personal salvation could come only by God's grace and not by human works went against modernist optimism about humanity's capacity to become master of its own fate and may have helped to promote among Germans what some have called the "authoritarian character." As Edmond Vermeil has noted,

> A Germany that had remained Catholic or become converted to Calvinism would have had very different destinies; she would have become more distinctly Western, for Lutheran religious ideas, which embraced forms of territorialism, generally caused Germans in the various regions to abandon political thought and action for what they called *Obrigkeit*, a sort of monarchism with an authoritarian bent founded upon a mixture of civil and religious power.[4]

Many Lutherans concluded that the ways of modernity—including commercialism, social leveling, and rationalism—were dangerous. Germans who remained Catholic were also suspicious of modernity, including Luther's own destruction of ancient sacred customs and institutions. Outside Germany, in Holland, France, and England, the Protestant Reformation was successful in fostering a new respect for individual self-determination in the political realm of the emerging nation-state. Germany, however, was much slower in adopting such politically liberal views, in part because Germany was so long in achieving nationhood. It remained divided into hundreds of separate principalities until the mid-nineteenth century.

Lack of political unity, along with authoritarian governments, led many Germans to search for freedom, individuality, and national "identity" in the realm of language—philosophy and poetry—instead of in the realm of politics. While the small burgher class was attracted to the liberal political ideals of eighteenth-century France and England, broader public suspicion of "Western" modernity prevented the Enlightenment from obtaining a firm grip. Many Germans were nevertheless infatuated with the French Revolution and internalized its doctrines of political freedom. Germans who hoped the Enlightenment would lead to political liberty in their own country were shocked that the French Revolution led first to the Terror and then to the conquest of their homeland by Napoleon. The fruits of the Enlightenment's scientific rationalism, abstract individualism, and universal humanism turned out to be war and domination, not peace and freedom. In his patriotic addresses to the German people (1807–1808), Fichte rallied his countrymen against the occupying French. In these stirring addresses, he utilized not only the revolutionary language of universal freedom, but also the language of nationalism which was so central to the French Revolution, especially in its Jacobin phase. Fichte helped to shape Germany's subsequent view of itself as a superior people gifted both with a superior language and with a fateful mission to preserve the highest ideals of humankind.

For many Germans, these ideals had much in common with the ideals of the French Revolution, but these ideals existed in an uneasy relationship. On the one hand, there was the original revolutionary ideal of universal personal liberty; on the other hand, the Jacobin ideal of the self-worshipping nation-state, a particular expression of Rousseau's doctrine of the "general will" of the people. Some critics have argued that the national self-worship inspired by the Jacobin phase of the French Revolution was ultimately to lead to National Socialism more than a century later.[5]

The failure and perversion of the French Revolution proved to be a rallying point for conservative forces across Europe, especially in Germany. After the post-Napoleonic reaction of 1815, however, the Revolutions of 1848 signaled the emergence of another period of European optimism about the human capacity to change the world in light of modernist political, economic, scientific, and industrial practices. Even agriculturally dominated and socially conservative Germany could not long resist the influx of the Industrial Revolution, aspects of which were already noted by Hegel in the early nineteenth century. Anticipating Marx's critique of the alienation of labor, Hegel remarked that in the machine age "work becomes thus absolutely more and more dead, it becomes machine-labour, the individual's own skill becomes infinitely limited, and the consciousness of the factory worker is degraded to the utmost level of dullness."[6]

Today, it may be difficult to imagine the disorienting and disheartening effect that the rapid industrialization of the late nineteenth century had on Germans, many of whom remained closer to the Middle Ages than to modern times. Old social and economic structures, institutions, personal relationships, forms of life, the natural environment itself—all were rapidly changed by the relentless processes of the Industrial Revolution. V. R. Berghahn argues that

> the development of modern Germany is *best understood* against the background of the Industrial Revolution which affected Central Europe with full force in the final decades of the nineteenth century. . . . [N]owhere else in Europe did the transition from an economy based on agriculture to one dominated by industry occur with the same rapidity as in Germany. Inevitably, the Industrial Revolution also had a profound effect on social structures, on life-styles and political behavior of people as well as on their perceptions of the world around them. These, too, changed more rapidly in Germany than in other European countries. . . . While it is not easy fully to appreciate the highly dynamic situation which had developed in Germany by the turn of the century, it is nonetheless fundamental to an understanding of the subsequent course of the country's history. . . . [7]

From 1880 to 1913, German coal output *quadrupled*; during the same period, steel production increased *tenfold* and outstripped British production by 1913.[8] Such statistics are especially noteworthy when we recall that Britain, the first industrialized country, had once enjoyed a wide industrial lead over Germany. Germany's late industrial revolution was accompanied by a demographic revolution. In addition to rapid population increase, millions of Germans left farms and small towns to find work in the new industrial centers. By 1907, only half of the sixty million Germans lived at their places of birth. Industrial modes of production threatened the livelihoods of small artisans and the lower-middle class. For industrial workers crowded into cities unprepared for their arrival, moreover, living conditions were dreadful. Alienated from their previous social relationships, overburdened by the work they were required to do, and shocked by their material circumstances, many workers became increasingly attracted to various socialist organizations. By way of contrast, middle- and lower-middle-class Germans, fearing that the workers' movement would bring anarchism and "Godless communism," became increasingly authoritarian and urged the government to control the unruly industrial masses.

Support for the largely authoritarian, Prussian-dominated government by some members of the growing middle and capitalist classes was ambiguous, however. As their own socio-economic status rose, many burghers retained their traditional unpolitical stance and were willing to appreciate their economic gains; but others complained that the political system should become more open and democratic to accommodate the burgher's new importance in society. Simultaneously, the rise of the political power of the working class by the early twentieth century made the Social Democrats the largest party in the Reichstag, a fact that greatly concerned conservative elements, and which indicated that at least in some respects German society was "developing" in line with what was happening in other industrialized countries. Nevertheless, Germany remained authoritarian and paternalistic in most social and economic matters. Having little experience of and insufficient appreciation for the principles of liberal democracy, Germany lacked the political structures and traditions that sustained France, England, and the United States during the difficult days which began in 1914 and ended in 1945.

C. THE *VÖLKISCH* MOVEMENT'S ATTACK ON MODERNITY

Perhaps the most significant ideological trend which arose in reaction to modernity and industrialization was the *völkisch* movement, which came on the scene in the late nineteenth century. Repelled by the egoistical, commercial, and spiritless mentality of modern economic society, *völkisch* writers called for renewed contact with the natural and cosmic forces which, while inaccessible to the rational mind, were capable of rejuvenating and transforming the increasingly mechanized German spirit. According to *völkisch* ideologues, these cosmic forces were at work in the common language, traditions, art, music, social customs, religion, blood, and soil which united a particular *Volk*. An individual was nothing on his own and could find fulfillment only by becoming attuned to and preserving the cosmic life force animating the "soul" of the *Volk*. George Mosse has commented about how the depersonalizing forces of industrialism favored the rise of *völkisch* thinking:

> Bewildered and challenged, men attempted to re-emphasize their own personality. But since the rate of industrial transformation, as well as its effects, seemed to evade the grasp of reason, and men could not easily make themselves a part of the new social order, many turned from rational solutions to their problems and instead delved into their own emotional depths. The longing for self-identification, the individual's desire to fulfill his capacities, ironically heightened by the process of alienation, was accompanied by the contradictory urge to belong to something greater than oneself. . . . The human condition was conceived as straddling two

spheres—that of the individual on earth as well as a larger unit outside society in which man could find a universal identity. The latter came to represent the "real" reality . . . [which] was defined in terms of the cosmos. . . . [9]

Völkisch claims that a *Volk* was rooted in a particular soil, ground, or place lent themselves to racist, usually anti-Semitic, doctrines. Developing their own version of social Darwinism, many *völkisch* writers concluded that some races were more vital, intelligent, passionate, spiritual, noble, aristocratic, and open than other races, which were sterile, calculating, mechanistic, cold-hearted, abstract, debased, and devious. The Jews, who were allegedly responsible for the crass materialism and commercialism undermining the German *Volk*, were readily labeled as one of these inferior but clever races whose influence had to be eradicated. The influence of the rootless Jews was especially strong in the big cities, which *völkisch* leaders often condemned as cancers upon the *Volk* soul. Believing that spiritual strength came from rootedness in the natural soil of their homelands, *völkisch* thinkers called for reconciliation with nature, not for the technological domination of it:

> Man was seen not as a vanquisher of nature, nor was he credited with the ability to penetrate the meaning of nature by applying the tools of reason; instead, he was glorified as living in accordance with nature, at one with its mystical forces. In this way, instead of being encouraged to confront the problems cast up by urbanization and industrialization, man was enticed to retreat into a rural nostalgia. Not within the city, but in the landscape, the countryside native to him, was man fated to merge with and become rooted in nature and the Volk. And only in this process, taking place in the native environment, would every man be able to find his self-expression and his individuality.[10]

Völkisch ideologues maintained that scientific rationalism, economic and political individualism, and industrial technology were behind such rootlessness. These evils were the products of the French "ideas of 1789" and the English "spirit of Manchester." Houston Stewart Chamberlain, author of the enormously influential book *The Spirit of the Nineteenth Century* (1899), summarized the attitude adopted by many Germans toward recent Western history: "Toward the end of the eighteenth century, the great transformation took place which will probably one day be recognized as the most terrible catastrophe to have befallen mankind, so that one may ask oneself whether the dignity of man can still be saved; I am speaking of the mechanization and the resulting industrialization of life."[11]

Many of the most important *völkisch* authors, including Julius Langbehn and Moeller van den Bruck, were lonely men who regarded themselves as outsiders, cut off from participation in what had once been an allegedly profound German community.[12] They interpreted their loneliness not as a personal or psychological problem, but instead as a widespread cultural phenomenon resulting from secularism, materialism, and industrialism. While many people were successful in making the transition to a secular and modernist worldview, many others yearned for something to replace the moribund

Christian God. The latter experienced the loss of the old God as a threat to their own sanity; as the world lost the basis for its meaning, they became personally disoriented and terrified. To compensate for the loss of God, they began to deify the *Volk* and to make it the source of meaning and purpose for their lives. Only the radical renewal of the spirit of the *Volk*, so they believed, could halt Germany's slide into nihilism and restore greatness to the German nation. Such renewal would require that Germans commit themselves to high-minded values of personal sacrifice, commitment to the communal whole, and martial courage.

Völkisch writers particularly scorned the *Besitzburgertum*, "men of property," who organized life not according to idealist martial virtues, but instead solely according to commercial, instrumental, and utilitarian values. Sharing in the condemnation of such men was the conservative poet Stefan George, whose elitist prewar circle attracted many creative, romantic, and politically conservative followers. He concluded that modern people, supposedly having been "freed" from the hierarchical relations required for an authentic community (*Gemeinschaft*), had in fact become trapped by the impersonal economic relations of abstract commercial society (*Gesellschaft*). The gold lust basic to the business mentality, so George argued, transformed everything into calculable quantities, thereby sucking the living substance out of all people and things. In 1911, George predicted:

> After fifty years more of continuous progress even the final remains of all substance will have disappeared, if nothing else comes to the world other than the stain [*Makel*] of progress, if through commerce, newspapers, schools, factories, and barracks the urban-progressive contamination has driven into the furthest corner of the world, and the world [that has been] satanically overturned, the world of America, the ant-world, has finally established itself.[13]

Not everyone agreed that money lust was the animating force behind what Oswald Spengler had described in his best-selling book as "the decline of the West" (1918, 1920). Max Scheler, for example, argued that the commercialist tendencies of modern humanity were a function of a more basic will to acquire and to control: "Capitalist *economy* is based in the *will to endless acquiring* (as an act), *not* in the will to *acquisition* (as a growing *possession* of things)."[14] This "will to acquire" was another expression of the "unrestricted tendency of the city-bourgeoisie toward a *systematic*, and not only occasional, *control over nature* and an endless accumulation and capitalization of knowledge for controlling nature and the soul."[15] Like Spengler and Scheler, the economist Werner Sombart sought to discover an alternative to Marx's economic interpretation of Western history.[16] While Marx proclaimed that economic activity was the foundation for the cultural "superstructure," conservative, anti-communist writers hoped to show that there was another foundation beneath the economic one—a foundation which resisted analysis by calculating rationality. German conservatives found communist collectivism equally unsatisfactory as liberal individualism: both were expressions of the overzealous rationalism of the Enlightenment. In reaction against the clamor for a communist revolution, on the one hand, and a liberal revolution, on the other, prewar conservatives and reactionaries began mounting a campaign for what was to be called the "conservative revolution."[17] In *The Politics of Cultural Despair*, Fritz Stern argues,

> The ideologies of the conservative revolution superimposed a vision of national redemption upon their dissatisfaction with liberal culture and with the loss of authoritative faith. They posed as the true champions of nationalism, and berated the socialists for their internationalism, and the liberals for their pacifism and their indifference to national greatness. At the very least they demanded greater national authority and cohesion, and usually they were partisans of imperialism or national aggrandizement as well. Often their longing for national heroism led them to worship violence, which in turn they justified by arguments drawn from social Darwinism or racism.[18]

The rise of German nationalism was spurred by the perceived need to fashion a German identity in opposition to the "progressivism" which produced material wealth at the price of spiritual bankruptcy. The critique of progressivism arose in the late nineteenth century, about the time Wilhelm Dilthey was developing the view that there is no underlying plan or purpose to the epochs of world history. Rather, according to Dilthey's "historicism," each historical epoch and culture must be understood on its own terms. The true historian attempts to discern "forms" (*Gestalten*) or "morphological structures" which may be shared by more than one culture. Historicism may be understood, in part, as an attempt to justify the study of human history and culture in the face of "disenchantment of the world" (Max Weber), i.e., the move by science to discount all grand narratives (including the Biblical one) about the "meaning" of world history and to explain human behavior in materialistic terms which had no recourse to concepts such as "purpose" or "meaning."

Conceding the outmoded character of traditional doctrines about transcendent meaning in history, Dilthey said, "We carry no meaning [*Sinn*] of the world into life. We are open to the possibility that meaning and significance first arise in man and his history."[19] For Dilthey, then, because humans were unique among entities in being self-interpreting and meaning-bestowing, human life and culture should be studied not by the natural sciences, but by the *human sciences* (*Geisteswissenschaften*). In his "life philosophy" (*Lebensphilosophie*), he maintained that human history could be understood only by gaining insight into the unique life experience (*Erlebnis*) of individuals within a particular historical epoch. While the rational intellect was useful for comprehending regularities of material reality, only intuitive, sympathetic understanding could disclose the mysterious dimension of human creativity and culture.[20]

Dilthey's rigorous formulations were gradually popularized. First, *völkisch* writers conceived of his attempts to limit the scope of scientific reasoning and to explore the nature of intuition as outright affirmations of the irrational and the mystical. Second, they claimed that the historicist critique of the idea of a transcendental foundation for or direction to Western history was consistent with the view that each historical *Volk* was a unique expression of the eternal life force. Third, they used the contrast between theoretical rationality and lived experience to support the view that rationality had sapped the strength of German blood, thus making it degenerate and effeminate—incapable of competing with the rising tide of the "colored peoples" of the world. The racial strength of German blood had been enervated not only by the influence of rationalism, which cut people off from the mystical and transcendental dimensions of life, and which undermined the importance of sexual and martial passion. Also to blame for racial "tired blood" were both Marxism and liberalism, whose prattle about "equality" turned society into an undifferentiated herd by ignoring the fact that some people are more noble, more capable, more powerful than others. Industrialism, too, was to blame, for it destroyed the unity of the *Volk* by promoting class warfare and abstract social relations.

Such anti-modernist, anti-industrial ideas were by no means restricted to Germany, but influenced other European countries and the United States as well. Artists and intellectuals alike complained about the degeneracy, meaninglessness, and emptiness of bourgeois life in *fin de siècle* Europe. Traditional folkways were undermined and trivialized by the dynamism of commercialism; life lost its adventure, novelty, and challenge when confronted with the explanations of mechanistic science; and both the skills of the artisan and the courage of the soldier became insignificant when compared with the achievements of the assembly line and the firepower of modern weapons. Such trends discounted those aspects of life that many people prized: intuition, emotion, spontaneity, spiritual and bodily ecstasy. Around the turn of the century, Nietzsche in Germany, Freud in Austria, and D. H. Lawrence in England all emphasized the importance in human affairs of instinctual, biological, and sexual motives that were concealed and denied by the overly rational, repressive bourgeoisie.

Paradoxically, despite the antipathy of many *völkisch* ideologues toward industrialization and technology, an increasing number of them—especially after Germany's defeat in World War I—believed that only by appropriating modern industry and armaments could Germany both defend itself against hostile neighbors and achieve its national mission of world dominion. Such reactionaries, while welcoming the military and economic fruits of modernity, rejected the liberal values, institutions, and rationalism that had made industrial technology possible in the first place. Writers such as Ernst Jünger and Oswald Spengler called for Germany to combine an elitist, authoritarian social order with a complete commitment to industrial technology. Jeffrey Herf has aptly described this view as "reactionary modernism."[21]

While many *völkisch* thinkers retained a more idealistic vision of Germany's world mission, conservatives of all stripes agreed that the state was the embodiment of the spirit of the *Volk*; hence, the aims of the state had to take precedence over those of the individual. Indeed, the life of the individual was said to gain personal meaning only if sacrificed for the achievement of the state's, and hence of the *Volk*'s, high purposes. Talk about self-sacrifice appealed to many idealistic Germans, who sincerely believed that their world-historical task was to save the West from the evils of selfishness, materialism, rationalism, and industrialism.[22] Given the importance of individual economic freedom for the burgher class, it is remarkable that this class increasingly supported calls for eliminating "liberal" ideals of individual independence and personal autonomy. Clearly, however, the promise of exaltation by alliance with something greater than the atomistic ego overcame objections about the need to defend the liberty of that ego. It may be argued that turn-of-the-century Germans increasingly projected the power drives of their own egos onto the state. Some nationalists, dropping all pretense to idealistic self-justification, claimed that the very "meaning" of the state was nothing but the Will to Power. Shortly before World War I, Erich Kaufmann maintained that the essence of the modern industrial state "is the unfolding of power, is the will to assert itself and to establish itself in history."[23] And further, "the victorious war [is] the final means to that highest goal," the unfolding of power.[24]

During the prewar era, however, Germany's romantic idealism lived on in the Youth Movement (*Jugendbewegung*), a striking embodiment of the longing for personal fulfillment in the context of communal solidarity. Wandering around the German countryside, many young people experimented with alternative forms of life, praised communion with nature, and speculated about the possibility of a vital, non-bourgeois community. Young people quite self-consciously proclaimed themselves to be a "new generation" which would overthrow the old generation, seduced by crass materialism, commercialism, and self-interest. The Youth Movement was only one of many signs that Germans were searching desperately for something that would exalt their narrow lives, that would free them from social isolation, that would restore meaning to their lives, and that would unite them in a primal bond that must have been available to Germans in a simpler, more genuine era. That longed-for event, so they concluded, was the outbreak of the Great War in 1914.

D. THE EXTRAORDINARY IMPACT OF THE GREAT WAR

Although war enthusiasm raged in all the countries involved, it was particularly strong in Germany. The war provided an opportunity to vent prewar political tension, to break out of rigid social mores, to give expression to long-suppressed aggressive instincts, and to achieve a longed-for social solidarity. War enthusiasm swept up everyone, including intellectuals such as Hermann Cohen and Paul Natorp, with whom Heidegger would become a colleague at the University of Marburg. Before the war, speaking out against the capitalist system's abuse of working people, Cohen and Natorp agreed that "the discrepancy between the reality of life in the industrial age and the ideals of a humanistic *Kultur* must be, if not totally eliminated, at least greatly ameliorated."[25] When the Great War came, however, professors and intellectuals suspended criticism of Germany's internal conditions and adopted Fichte's role of spiritual defenders of the German homeland. One commentator points out that Natorp and Cohen utilized [their neo-Kantian] philosophy in an effort to impose some meaning on the conflagration before them. Both were involved in the formulation of what became known as the "ideas of 1914," the conscious effort to construct philosophically an alternative to what great numbers of German intellectuals saw as the ideology of the French Revolution and the English "doctrine of Manchesterism." Particularly Paul Natorp was deeply involved in this movement. His *Kriegsschriften*, an attempt to bring Kant and philosophy into the service of the German cause, were widely read in Germany during the war years.[26]

Natorp argued that to the east, Germany faced the morally inferior, backward masses of Russia, and to the west, it faced the materialistic, rationalistic, and individualistic powers of Britain and France. It was Germany's "sacred cultural mission" to preserve the German *Seele* (soul) from such alien forces. In his best-selling book *The Genius of the German War*, Max Scheler praised this mission. As the war progressed, his views became less nationalistic and more temperate, but like so many others caught up in the heat of the moment he had already lent his "pen and reputation" to the service of his nation in crisis.

Germany's defeat in the Great War exacerbated prewar social tensions, including the conflict between the left-wing industrial masses, on the one hand, and the middle classes and reactionary forces, on the other. Hitler saw, however, that the war had accomplished one important thing: it had discredited the corrupt bourgeois order blocking the way to the primal encounter needed for Germany's regeneration. The Weimar Republic instituted shortly after the end of the war was a last-ditch attempt to institute a Western-style democracy in Germany. But it was doomed from the start. Many Germans resented parliamentary politics to begin with, since parliamentary stalemate and infighting had allegedly helped to start the war and to lose it. To conservatives, reactionaries, socialists, and communists alike, Weimar was merely an expedient to be discarded once the revolution (from left or right) took place. The political chaos provoked by left- and right-wing coups and bloody street fighting and the economic misery caused by the hyperinflation of the early 1920s helped to undermine the standing of the Weimar Republic. Right-wing fanatics regarded as a traitor anyone who cooperated with it. In 1922, they assassinated Walther Rathenau, an industrialist and politician who hoped that Germany could "succeed in bringing about the great synthesis between a most dynamic industrialism and a most refined culture, between the rule of mechanization and 'the realm of spirit.' "[27] Rathenau's belief that only by fulfilling the war reparations agreement could Germany be reinstated in the European community was anathema to his right-wing enemies, who were doubly opposed to him because he was a Jew. Gordon A. Craig has emphasized the anti-Semitic dimension of the attack on Rathenau:

> With the murder of Walther Rathenau, a border had been crossed, and Germany had entered a new and forbidding territory in which to be Jewish was more than a handicap or a social embarrassment; it was a danger and, not impossibly, a sentence to death. For Rathenau's only crime had been that he was a Jew who had dared not only to pretend that he was a German but to represent Germany to the

outside world. His death caused a wave of horror in the land, but there is no doubt that many people privately considered it a not unjustified punishment. The obscene ravings of prewar anti-Semites had been given their first concrete expression. It was deplored but rationalized. The unthinkable had become thinkable. *Kristallnacht* was only sixteen years away. . . . [28]

During most of the Weimar era (1919–1933), Germany was a perplexing mixture of disillusionment and hope, moral depravity and spiritual yearning, social destructiveness and artistic creation, political discord and economic development.[29] With the infusion of American capital into the country in the mid-1920s, a period of economic stability and relative political calm replaced the anarchistic postwar years. German industrialists began applying the principles of American "scientific management" not only to achieve greater efficiency in production processes but also to manipulate industrial relationships in ways that avoided strikes and slowdowns. Despite the protests of organized labor, management intensified the process of rationalizing the workplace by introducing assembly-line work, or "Fordism." Fordism indicated a widening of the scope of scientific management:

> While Taylorism concerned only the management of labor, Ford's doctrine stressed reorganization of the entire productive process. In part this was a rationalization, for Taylorism, too, had earlier been interpreted in the widest sense. But now for practical reasons European apostles of scientific management and rationalization chose to confine Taylorism to its original concern with labour efficiency. Conversely, the contributions of Ford—the moving assembly line, standardization, and the enlargement of a mass market by low prices and high wages—were seized upon to prove the social potential open to capitalism and large-scale industry, as they existed.[30]

While Taylorism and Fordism were used to strengthen the grip of private capital on labor, many German engineers and technicians believed that industrial technology could be used in a way which inspired and uplifted, rather than alienated and degraded, the German *Volk*. As Herf maintains, "The cultural dilemma of Germany's engineers was the following: How could technology be integrated into a national culture that lacked strong liberal traditions and that fostered intense romantic and anti-industrial sentiments?"[31] Many engineers believed that technology was an authentic manifestation of, and not a threat to, the German *Geist*; that private capitalist interests were to blame for the socially damaging effects of industrial technology; and that only a strong central state could guarantee the welfare of the community by guiding the development and employment of industrial technology.[32]

Friedrich Dessauer, an engineer and a leading figure in Germany's *Streit um die Technik* in the first third of the twentieth century, shared the view that profit-oriented capitalism was responsible for perverting the service-oriented and creative capacities of industrial technology.[33] Many engineers sought to portray industrial technology as akin to the productions of those great German artists who were so important for fostering authentic *Gemeinschaft*.[34] As op-

posed to the spirit-corroding forces of Americanism and commercialism, so
argued the son of the inventor of the Diesel engine, technology revealed hints
of a "nobler race . . . of stronger life instincts."[35] The task was to unite Germany's idealistic heritage with its innate capacity for technical development,
in a way that avoided the destructive courses followed by America and Russia.
By rejecting the revolutionary ideals which found expression in both capitalism and socialism, while simultaneously exalting technological achievements
(which many have regarded as the product of liberal capitalism) and calling
for a corporatistic, authoritarian German state, many German engineers adhered to the paradoxical ideal of reactionary modernism. As we shall see,
Heidegger shared some of these attitudes, since he, too, called for a new, non-
liberal social order that would establish new modes of work that would presumably transform the nature of industrial technology.

    While industrialism continued to make enormous changes in German
society during the Weimar years, that era also witnessed an extraordinary
flood of creativity on the part of artists, poets, writers, scientists, technicians,
musicians, architects, engineers, academicians, and philosophers. This willingness to experiment, at least in the social-political realm, took on a desperate character during the Great Depression, triggered by the collapse of the
stock market in New York in October, 1929. As Germany headed into a
terrible economic and social crisis, which the stalemated Weimar parliament
was incapable of resolving, increasing numbers of people listened to the
promise of Adolf Hitler to save Germany from the evils of modernity, including the Jewish capitalists and Jewish communists who were allegedly orches-
trating Germany's economic and social miseries. Hitler portrayed himself as
the longed-for savior of the German *Volk*, the man who would renew German
Geist and enable it to fulfill its world-historical mission as redeemer of a fallen
world. Although his National Socialist Democratic Workers' Party never received a majority vote in national elections, Hitler was unexpectedly made
chancellor of Germany early in 1933. Shortly thereafter, he consolidated his
power and became Germany's dictator.

    In this chapter, I have outlined some of the cultural, political, and social
circumstances which formed the background for Heidegger's life and thought
during the first part of this century. The topics and attitudes described here
were constitutive of the "cultural conversation" in which Heidegger himself
inevitably took part. My aim here has been to show that his concern with
modernity and industrialism was by no means idiosyncratic, but instead was
widely shared. Moreover, as we shall see in chapters to come, his descriptions
and accounts of modern technology were heavily indebted to his interlocutors
in the *Streit um die Technik*. In the next chapter, we will examine how early
Heidegger's critique of modern technology was influenced by the reactionary
politics of his day.

Chapter 2. Political Aspects of Heidegger's Early Critique of
Modern Technology

Viewing industrial technology through the optic of anti-modernist political
views, early Heidegger regarded it neither as a neutral instrument for human
ends, nor as a sign of humanity's evolution to a higher stage, but instead as a
symptom of the final epoch in the long decline of humanity's understanding of
the being of entities. During the mid-1930s, however, he began moving toward a somewhat less negative evaluation of industrial technology. Germany's
mission and opportunity, so he argued, was to undergo a transformation that
would make possible a new relationship to technology, a relationship in which
the worker was no longer a slave to its demands, but instead became an
authentic producer of things.

In this chapter we shall begin to show how Heidegger's early thinking
and his political behavior were influenced by his critical appropriation of
many of the reactionary attitudes toward modernity and industrialism which
we discussed in chapter one. First, we shall see that already by 1919 Heidegger was arguing that his generation was destined to make a radical new beginning in the face of the insidious influences of modern technology. Next, we
shall see that *Being and Time*'s phenomenological "description" of everyday
life was in part a negative political evaluation of industrial society. This political slant is partly responsible for the ambiguity present in *Being and Time*'s
account of everydayness. On the one hand, it purports to reveal the essential,
timeless, or transcendental features of everyday life; on the other hand, those
"descriptions" are in some ways politically charged interpretations of everyday life in the specific historical circumstances of urban-industrial society.
Early Heidegger's analysis did not make sufficiently clear the distinction between what were essential or universal features of everyday life and what were
historical and specific aspects of it.

Finally, we shall examine the extent to which Heidegger's view of technology was influenced by the views of Oswald Spengler, who argued that
contemporary Europeans had become degenerate and were thus increasingly
enslaved to a technological system which had escaped their control. Despite
his debt to Spengler, however, Heidegger argued that the decline of the West
was not biological but metaphysical in character. If there occurred a metaphysical transformation of humanity, so Heidegger increasingly believed, then
humanity could adopt a new, non-servile relationship to industrial technology.
In lectures of 1929–30, given during the collapse of the German economy

after the nightmare on Wall Street, Heidegger began calling for a "manager" who would accomplish such a transformation, thereby restoring mystery and meaning to the lives of the *Volk*, which had been devastated by the unconstrained forces of modernity and industrialism. From here, it was but a step for him to embrace Hitler's "solution" to the threat posed by modernity and industrialization. While Heidegger at first embraced this solution, he finally regarded it as a particularly virulent manifestation of the nihilism which Hitler claimed to be fighting.

A. HEIDEGGER'S EARLY CRITIQUE OF THE ENLIGHTENMENT TRADITION

Born on September 26, 1889, in the little Swabian town of Messkirch, Heidegger was the son of lower-middle-class parents.[1] His grandfather was a shoemaker, and Heidegger's own father was a part-time cooper and sexton of a local Roman Catholic church. His mother was descended from a long line of peasants from a nearby locale. Raised a strict Roman Catholic, he attended a Catholic *Gymnasium* in Constance before going to Freiburg to complete his studies. He spent two weeks in a Jesuit seminary before being dismissed for reasons of health (a heart condition). Subsequently, he studied theology for two years at the highly Catholic University of Freiburg. During this era, the Roman Catholic church was extremely critical of the godlessness and sinfulness of urban life, of the social destructiveness of industrial civilization, and of the agnosticism and materialism of modern science. Deeply attached to his childhood countryside and its peasant inhabitants, Heidegger was shocked by the transformations wreaked by economic and industrial "progress." Moreover, he developed an attitude of contempt for all those who were blind enough to promote, or even to tolerate, such a soul-destroying progress. As I have argued elsewhere, his early theological interest in personal salvation was intertwined with the urgent question of how a person could be "justified" in such a fallen world.[2]

After he abandoned his theological studies for graduate study in philosophy, Heidegger began to translate his concern for theological "justification" into a concern for philosophical and personal "authenticity." His readings in Kierkegaard and Nietzsche heightened his sense of despair regarding the emptiness of the "modern age" and "herd society." Encounters with expressionist poetry and Dostoevsky's novels reminded him of the ultimate importance of the transcendent domain which lies beyond the material world. With the outbreak of the Great War, Heidegger served off and on both as a mail censor and as a staffer at a military meteorology station. Although he spent some time near the front, his health did not permit him to be an infantryman. Like other members of the "war generation," he believed that the Great War's baptism of fire radically separated him from the world and mores of those members of the previous generation who had been seduced by commercialism and materialism. Around 1919, having married a Lutheran woman, Elfride Petri, Heidegger ceased being a confessing Catholic and oriented himself first toward evangelical Protestantism, and then away from Christianity altogether, although he never really abandoned his ties to the Roman Catholic church.

Heidegger's doctoral dissertation on psychologism (1914) and his *Habilitationsschrift* on a problem attributed at the time to Duns Scotus (1916) suggest that his philosophical concerns were of a strictly "academic" nature. [GA, 1] Both of these early works, however, were part of Heidegger's lifelong attempt to show that the transcendental structure of human existence could never be reduced to or understood in terms of the principles of natural science. Human existence, he argued, was incapable of being objectified. From our point of view, we might say that within the confines of philosophy he was arguing for a spiritual regeneration of a humanity alienated from itself by modernity and industrialization. For Heidegger, however, philosophy was not an ivory-tower discipline, but instead a leading factor guiding the spiritual development of the German *Volk*.

Already in his early Freiburg lectures (1919–23), he was explicitly conceiving of his own work as a challenge to modernity. In his lectures on value-philosophy (1921–22), he argued that people typically used the term *Kultur* to describe those historically self-conscious peoples who had committed themselves to the goals of shaping (*Gestaltung*) the world, of accomplishing great things through science and technology: "At the end of the nineteenth century, specific achievements were held to be technology [*Technik*] and the theoretical basis making it first possible: natural science. One speaks of the age of natural science, of the century of technology." [GA, 56/57: 130] Unlike some conservatives, who spoke the language of *Realpolitik* and imperialism, Heidegger regarded achievement-oriented *Kultur* not as an authentic alternative to, but instead as the culmination of, the despised Enlightenment *Zivilization*.

Also unacceptable to Heidegger was the ideal of the "cultured man" (*gebildeter Mensch*), typically an academic trained for years in the classics and devoted to upholding the standards of "high culture" in opposition to the commercialism of "mass culture." While Heidegger himself was in some ways such a "cultured man," he also resented what he regarded as the artificiality of such men, as well as their supposedly superficial understanding of the classics.[3] All too often their idea was to shore up the existing society by encouraging study of Latin and Greek. While himself an excellent student of classical Greek, Heidegger was not interested in "conserving" the existing society; from his point of view, the whole thing had become hollowed out and rotten, apart from a few pockets of peasant life. Moreover, as he was to see ever more clearly, one could not hope to challenge modernity by revitalizing Greek art and culture, for Greek metaphysics itself had made modern technology possible. To save the West, the Germans would have to have an original encounter with the same ontological "primal" which was the source for Greek culture.

The Enlightenment portrayed itself as something new: the self-assertion of human reason and freedom. According to Heidegger, however, the Enlightenment was in fact a major event on the way to the nihilism of modernity. Far from being an agency of freedom, the Enlightenment was itself completely dominated by "mathematical natural science and . . . rational thinking in general. . . . " [GA, 56/57: 132] Rationalism's universalist view of history, the stages of which were described by Rugot and Comte, reduced the individual to an atom in a monochromatic society. In a remark which anticipated what he would later say about the historical role of the poet, Heidegger remarked that "the poet was not evaluated [by rationalists] as a shaper [*Gestalter*] within a genuine experiential world, but instead as an improver of language, which in its refinement and polish brought public and social life to an elevated stage. . . . " [GA, 56/57: 133] Even Kant fell prey to Enlightenment thinking when he spoke of the rational form and goals of human history! German romanticism attempted to reverse this trend, however. Herder, for example,

> saw historical reality in its manifold and irrational fullness and above all the independent self-worth of every nation, every age, every historical manifestation in general. . . . Historical reality was no longer seen exclusively in a schematic-rulebound, rationalistic-linear direction of progress. . . . The goal of progress is also no longer an abstract, rational happiness and goodness. . . . There awakens the view for individual, qualitatively original centers of reality and contexts of reality; the categories of "ownness" ["*Eigenheit*"] become meaningful and related to all shapes of life, i.e., *this* [life] first becomes visible as such. [GA, 56/57: 133–134]

Heidegger pointed out that Schlegel turned toward original and independent forms of literature; research into sagas and myths began; one learned to appreciate folksongs, instead of regarding them as signs of barbarism; history of individual states became important; Schleiermacher saw for the first time the proper value of *Gemeinschaft*; and Fichte, Hegel, and Schelling deepened the intuitions of Herder in the triumph of German idealism. Subsequent social and economic problems undermined profound speculation and dragged humanity toward merely practical activities, expressed in "the modern life-direction toward . . . the formation of technology in its widest sense." [GA, 56/57: 136] The objectifying experience (*Erfahrung*) of science de-vivifies (*ent-lebt*) the lived experience (*Erlebnis*) of people. [GA, 56/57: 84ff.] Consider, Heidegger argued, how the rational-geometrical view of space and time, and the objectifying scientific view of objects, transforms tools of the environing world (*Umwelt*) into objects disconnected from lived experience. As an alternative to objectifying tendencies of rationality, Heidegger recommended the intuitive "seeing" of a phenomenology that was sympathetically attuned to life. Philosophy, more primordial than the busy activity (*Betrieb*) of normal science, was the passionate inquiry into the authentic origin (*Ursprung*) and beginning (*Anfang*) of things. [GA, 61: 37–41] Philosophy was an existential affair, "a *way of behaving.* (Plato would never define philosophy as *techne* [in the sense of a technical discipline within the structure of the modern university])." [GA, 61: 50][4]

In "What Is Metaphysics?" (1929), Heidegger attacked the modern university for its overspecialization and positivism, both of which were destructive to the very being of the university and to the community. [WGM: 1–20/

95–112] Commentators have sometimes portrayed this essay as signaling his turn from purely philosophical toward more practical concerns. In fact, however, radical universal reform was his concern a decade earlier. Speaking in the same language he would later use to support the Nazi "revolution" in the German university, he proclaimed in 1919: "Today, we are not ripe for *genuine* reform in the sphere of the university. Becoming ripe for it is the matter of a *whole generation*. Renewing the university signifies rebirth of genuine scientific consciousness and life-context. Life relations, however, *are renewed only in regress into the genuine origins of spirit.* . . . " [GA, 56/57: 4; my emphasis in last sentence]

Two years later, he argued that his generation was like no other in German history because its members knew that they stood at the end of one tradition and at the beginning of another. [GA, 61: 73–74] The decline of the West was an inevitable result, he believed, of the intrinsic human movement toward "ruination" (*Ruinanz*) and away from primal insights. The task of philosophy was to fight against this decline and to summon the *Volk* back to an originary encounter with the being of entities. [GA, 61: 131ff.] The Great War had in effect deconstructed the rigid traditions concealing the truth. The issue was whether Heidegger's generation could initiate a new beginning that would restore meaning to a world from which God seemed increasingly absent. Heidegger shared with *völkisch* ideologues the conviction that the German *Volk* was being corrupted and weakened by the influences of rationalism, liberalism, commercialism, and other "foreign" trends. He offered a metaphysical interpretation of the meaning of the *Volk*, however, which was clearly at odds with the view of most *völkisch* writers. Yet most parties agreed that the new generation would have to act in a decisive way if Germany was to be saved.

In *Being and Time*, written during a period of relative calm between the two wars, he expressed in somewhat less dramatic terms the importance of decisions made by entire generations: "Only in communicating and in struggling does the power of destiny become free. *Dasein*'s fateful destiny in and with its 'generation' goes to make up the full authentic historizing of *Dasein*." [SZ: 385/436] Six years later, Heidegger concluded that his generation's fateful destiny was to support Hitler. While *Being and Time* itself contains no *overt* political message or direction, a careful reading of its account of everyday life reveals its anti-modernist orientation.

B. THE CRITIQUE OF MODERNITY AND MASS CULTURE IN *BEING AND TIME*

Being and Time, seeking to explain the relationship between temporality and the human understanding of the being of entities, argues that human existence forms the temporal "horizon" or "clearing" in which entities can manifest themselves or "be." Heidegger wanted to develop a conception of human existence which attempted to avoid arbitrary preconceptions about what it means to be human. Hence, he derived his conception of human existence from his phenomenological description of everyday life. In everyday life, a person exists as a socially defined "anyone self" (*das Man*). Individuals are constituted by social and cultural relationships that can never, in principle, be fully comprehended. As Charles Guignon has observed, "Far from being an autonomous and isolated subject, the self is pictured as the 'Anyone' (*das Man*), a 'crossing point' of cultural systems unfolding through history. To be human, in Heidegger's view, is to be a place-holder in a network of internal relations, constituted by a public language, of the communal world into which Dasein is thrown."[5]

Heidegger was greatly influenced by Wilhelm Dilthey's revival of the notion of "objective mind," the view that so-called individual consciousness is a nodal point in a set of internal relations constituting the whole of culture.[6] Knowledge, language, customs, and everyday practices are not private matters contained inside a separate subject, but instead are *public* affairs. Dilthey, like Hegel and Marx before him, was influenced by the ancient Greek idea that the individual is essentially a social entity. Drawing on such predecessors, Heidegger argued that human existence is a relational, contextual, holistic, participatory event which cannot be constituted by isolated egos existing independently of each other. For him, questions about whether there are other "selves" apart from his own were misguided.

According to Heidegger, people willingly conform to social norms, partly because of anxiety, partly because they gain self-definition only in relationship to a social whole into which they have been "thrown" at birth. Self-conscious deliberation about "choices" is relatively unimportant when compared with the choices that have already been made for us. We find ourselves "pressing ahead" into the possibilities that are laid open for us at the beginning. Our possibilities are largely circumscribed by family, gender, community, socioeconomic class, race, and culture. We are so deeply constituted by our social relations that "even *authentic* existence is not something which floats above falling everydayness; existentially, it is only a modified way in which such everydayness is seized upon." [SZ: 179/224]

While Heidegger claimed to reveal something essential about human existence in his account of everydayness, in fact many of his "descriptions" were critical evaluations of everyday life in industrial-urban society. He conceded this fact up to a point when he remarked that everydayness is "a mode of *Dasein*'s Being, even when that *Dasein* is active in a highly developed and differentiated culture—and precisely then." [SZ: 50/76] Further, "the extent to which [the] dominion [of the anyone self] becomes compelling and explicit may change in the course of history." [SZ: 129/167] Heidegger was convinced that in industrialized society—with its capacity for shaping opinion, rationalizing behavior, creating needs, and standardizing values—the anyone self had gained complete control:

> In utilizing public means of transport and in making use of information services such as the newspaper, every other is like the next. This being-with-one-another dissolves one's own *Dasein* completely into the kind of being of "the other," in such

a way, indeed, that the others, as distinguishable and explicit, vanish more and more. In this inconspicuousness and unascertainability, the real [*eigentliche*] dictatorship of the anyone self is unfolded. We take pleasure and enjoy ourselves as *the anyone self* [*man*] takes pleasure; we read, see, and judge about literature and art as *the anyone self* sees and judges; likewise, we shrink back from the "great mass" as *the anyone self* shrinks back; we find "shocking" what *the anyone self* finds shocking. The anyone self, which is nothing definite, and which all are, though not as the sum, prescribes the kind of being of everydayness. [SZ: 126–127/164]

Heidegger argued that one of the basic features of everyday life was its "publicness" (*Öffentlichkeit*), a term often used in connection with "civil society" (*bürgerlichen Gesellschaft*), i.e., the business world. Heidegger did not emphasize the bourgeois connotation of this term, possibly to avoid revealing the extent to which his analysis of everyday life was in fact an analysis of life in degenerate commercial society.[7] In such a society, nothing truly original can be revealed because language has degenerated to the state of idle talk (*Gerede*), the passing around of ungrounded opinions. Even in academic circles, idle talk "spreads in wider circles and takes on an authoritative character. Things are so because one says so. Idle talk . . . becomes aggravated to complete groundlessness." [SZ: 168/212] Idle talk frees the anyone self from having to understand anything in a genuine, original way. "Idle talk . . . is the kind of being which belongs to *Dasein*'s understanding when that understanding has been uprooted. . . ." [SZ: 170/214] Heidegger offered no *evidence* for these claims. He assumed that honest and competent readers would agree with him that Germans had been uprooted and made homeless.

Heidegger experienced this homelessness profoundly, so much so that his sanity seems to have been threatened by the loss of familiarity and meaning in a world devoid of God. In 1929–30, he commented approvingly on Novalis's statement that

philosophy is authentic homesickness [*Heimweh*], a drive at all times to be at home. . . . A remarkable definition, naturally romantic. Homesickness—is there still something like this in general today? Has it not become an incomprehensible word, even in everyday life? For has not the contemporary urban man and ape of civilization long since abolished homesickness? And [to think of] homesickness as the absolute determination of philosophy! [GA, 29/30: 7]

Heidegger's thought was motivated in part by his profound homesickness, by his yearning to overcome isolation, alienation, and meaninglessness. Here, his proximity to major themes of *völkisch* authors was very close indeed. He believed that his task as a philosopher was to help the German *Volk* find a home for itself, although this was an uphill task. Big-city Germans were distracted by curiosity about new places, faces, and products. [SZ: 172/216] Having lost touch with its highest possibilities, curiosity-stricken German *Dasein* is "everywhere and nowhere. This mode of being-in-the-world reveals a new kind of being of everyday *Dasein*—a kind in which *Dasein* is constantly uprooting itself." [SZ: 173/217] Much of Heidegger's account of everydayness,

then, was of the *inauthentic* everydayness of urban life, which *völkisch* authors
also decried. Heidegger sometimes spoke as if everydayness and inauthenticity
were identical, but elsewhere he spoke as if inauthenticity were an aggravated
version of the everyday tendency toward falling. "Undifferentiated" everydayness involves a certain "drifting" into prevailing customs, traditions, aspira-
tions, and norms. Presumably even peasants are caught in this drift. When it
becomes a plunge, however, then one has moved into inauthenticity—the
craving for distractions to conceal one's mortality. Anxiety (*Angst*), which
according to Heidegger is the most basic of all moods, reveals the groundlessness of the anyone self and its culture. In the face of anxiety, one can
either accept one's mortal existence (thereby becoming authentic, *eigentlich*,
"owned") or else flee into distractions (thereby becoming inauthentic,
uneigentlich, "disowned"). Heidegger's account of anxiety, which is much indebted to Kierkegaard, has proved to be valuable in the field of psychotherapy. Some critics have remarked, however, that *Being and Time*'s discussion of
anxiety resembles the psychotic experience of complete depersonalization, an
experience with which Heidegger seems to have been threatened much like
his heroes, Hölderlin and Nietzsche.[8]

Marxist critics have argued that such depersonalization resulted not from
psychological but from economic factors associated with monopoly capital
and its "anonymous" corporations. According to Coletti, for example, even
Heidegger's account of everydayness was shaped by his experience with life in
capitalist society:

> *Sein und Zeit* is a work upon which are indelibly stamped the signs of the crisis of
> the German society of the period. [. . .] The "enterprise" takes on an independent
> life, as if it belonged to no one—the object becomes the subject, and the subject
> becomes the object of its object. The uncontrolled forces of society exacerbate to
> the extreme the nature of those forces extensively analyzed by Marx, which oper-
> ate "behind men's backs" with the peremptory necessity of natural events.[9]

Similarly, Theodor Adorno maintains that "socially, the feeling of meaninglessness [or anxiety] is a reaction to the wide-reaching freeing from work
which takes place under conditions of continuing social unfreedom [under
capitalism]."[10] Marxists argue that Heidegger's *description* of depersonalized
life in the modern world was accurate; his mistake lay in his failing to see the
economic *explanation* for that depersonalization. Like many reactionary thinkers, he regarded as reductionistic the Marxist argument that depersonalization
and alienation resulted from economic factors such as class struggle. According to Heidegger, economic analysis, whether liberal or socialist, could shed
no light on the modern rootlessness because such analysis was itself a product
of modernity!

Heidegger maintained that depersonalization was an ontological condition arising from *Dasein*'s tendency toward falling, a tendency which had
become aggravated in modern times. Far from being the *result* of economic-
material causes, then, depersonalization resulted from the human fallenness

which itself made possible a world dominated by (inauthentic) economic and material concerns. Later Heidegger would claim that human rootlessness was not a result of the individual's or humanity's flight from the truth, but instead the result of the self-concealment of that truth from humanity.

Marxists were not the only ones to criticize early Heidegger's account of anxiety. Max Scheler, for example, also argued that anxiety was a historical, not an essential, feature of human existence. Commenting on *Being and Time*, he remarked,

> I am convinced that ever since Judaism and Christianity defined Western man, he has lived under a disproportionately greater burden of anxiety than any other type of man in the world. . . . [T]his weight of anxiety in great measure conditions [modern man's] enormous world-activity, his hunger for power and his never-ending thirst for "progress" and technological transformation in the world; and furthermore that this anxiety has emerged in a very peculiar and strong way in Protestantism.[11]

Later, in his account, of the history of productionist metaphysics, Heidegger was to agree with Scheler and Weber that the Protestant yearning for certainty in regard to salvation helped to form the control-oriented and calculating personality structure necessary for the emergence of capitalism and industrial technology. The drive by modern *homo economicus* to control all things in order to maximize profit was a hopeless attempt to reduce the anxiety and meaninglessness of life in a world from which the gods had fled. Similarly, early Heidegger regarded as pointless all political attempts to improve the social order within the context of the Weimar Republic. That republic itself, in his view, was a manifestation of the spiritual emptiness of life in modern industrial "democracies."

According to Winfried Franzen, *Being and Time* appealed to conservative intellectuals because it addressed them theoretically, personally, and existentially without calling upon them to *do* anything specific. It let them think of themselves as "insiders" with regard to the sorry state of German culture, and it also permitted them and even encouraged them to remain "outsiders" with no obligation to help improve the existing situation. Indeed, in *Being and Time*, "total secession was commended as a possibility of self-assertion. Publicness, politics, society were held here not only as beside the point, but also as corrupting. With this the insecure subject, feeling powerless, could best identify itself: evil could be placed abstractly on society, without thereby deriving the necessity of improving the *current* social order."[12]

Being and Time, then, condemned the existing society and in effect called for a radical break from it, although that book did not draw the outlines of the alternative. Heidegger concluded only that that upon which one was to resolve could be decided only by the factical circumstances in which one found oneself. As Franzen remarks, "A mere decade after Thomas Mann's *Reflections of an Unpolitical Man* (1918), existential ontology could be read as the emphatic theory of anti-political existence, and indeed without having to characterize itself as such."[13] While *Being and Time* praised the resolute individual and condemned the impersonality of mass culture in a way that would at first seem consistent with liberal-democratic ideals, the authentic individual would seem to be unpolitical, at least in the unacceptable political context of Weimar. In *Being and Time*, Heidegger spoke as if he awaited the moment when, in solidarity with other members of the existential elite, he could form an alternative to the decadent state of affairs. Indeed, when he wrote of the necessity of achieving an authentic "repetition" (*Wiederholung*) of Germany's historical possibilities, many of his readers understood that he had in mind something like the radical social transformation envisaged by members of the "conservative revolution."

C. HEIDEGGER'S CRITICAL APPROPRIATION OF SPENGLER IN THE FIGHT AGAINST MODERN TECHNOLOGY

Heidegger's Freiburg lectures of 1929–30, *The Basic Concepts of Metaphysics*, show that he believed that moment was coming. Here, he claimed that Germany's prevailing mood was one of deep, metaphysical *boredom*. He reminded his audience that "moods are the 'presupposition' and the 'medium' of thinking and acting. This amounts to saying: they reach back more primordially into our essence, in which we first come upon ourself—as a *Da-sein*." [GA, 29/30: 102] To disclose Germany's mood at the end of the 1920s, he examined the works of four representative authors—Oswald Spengler, Ludwig Klages, Max Scheler, and Leopold Ziegler. Heidegger's comments on Spengler and Scheler are particularly important.

In *The Decline of the West*,[14] the "essential" feature of Spengler's "prophesying" was this: "The decline of life in and through spirit [*Geist*]. What spirit, primarily as reason (*ratio*), has formed and created in technology, economics, and world-commerce, in the whole formation of *Dasein*, symbolized by the great city—that turns against the soul, against life and smothers it and compels culture to decline and collapse." [GA, 29/30: 105] Heidegger largely agreed with Spengler's assessment of the current state of affairs. Even after World War II, Heidegger said: "That people today tend once again to be more in agreement with Spengler's propositions about the decline of the West, lies in the fact that (along with various superficial reasons) Spengler's proposition is only the negative, though correct, consequence of Nietzsche's word, 'The wasteland grows.' We emphasize that this word is thoughtful. It is a true word." [WHD: 14/38] The "wasteland" refers to the devastation of the earth, the flight of the gods, and the darkening of the world that Heidegger believed were aspects of the culmination of productionist metaphysics in modern technology. For Heidegger, however, the decline of the West from the great age of the Greeks occurred not for biological or racial reasons, but for metaphysical and spiritual ones. By way of contrast, Spengler—influenced by Schopenhauer, Nietzsche, and Darwin—argued that the West's original drive for technological control and the subsequent decline of the West were related to cycles involved in the "struggle for life." Heidegger countered that such "naturalistic" interpretations of human life were products of the misguided metaphysical conception of man as the "rational animal."

Despite disdaining Spengler's propositions and methods, early Heidegger began to conceive of his own work as an attempt to provide a philosophically sound account for the *symptoms* of decline popularized by Spengler. In 1921–22, he said that there had never been a more unphilosophical age than the present one: "The talk of decay, technologization (Bergson, Spengler) is so far confused that the phenomena in which and for which and by virtue of which the decay is carried out are not positively made the problem." [GA, 61: 26] While Spengler was attuned to the current *Zeitgeist* and thus had to be taken seriously, Heidegger observed: "Spengler's lack: philosophy of history without the historical, *lucus a non lucendo*." [*Ibid.*, 74] In his famous account of the "history of being," Heidegger sought to provide the authentic philosophy of history missing in Spengler.

Both writers claimed that modern technology could be explained not in terms of material causation, but instead only in terms of factors which transcended the ken of natural science. For Spengler, even instinctual behavior was finally shaped by a non-biological metaphysical force: Will. Heidegger, however, argued that Spengler's conception of history was essentially biologistic, the result of a superficial interpretation of Nietzsche's doctrines. Spengler's attempt to discover the "morphological types" of history indicated that he was gripped by a quasi-scientific, and thus misguided, view of history. [GA, 39: 227] As a result, Spengler "understands 'decline' [*Untergang*, 'going-under'] in the sense of mere going-to-the-end, i.e., as biologically represented perishing [*Verendung*]. Animals 'go under,' since they perish. History goes under, insofar as it goes back into the hiddenness of the beginning—i.e., though in the sense of perishing, [history] does not go under, because it *can* never *thus* 'decline.' " [GA, 54: 168]

For Heidegger, then, the West could not "decline" because history was not constituted by human *existence*, not by animal *life*. The decline of the West was not an aspect of the biological repetition of cultural formations, but instead a dimension of the creative cycle of repetition (*Wiederholung*) which involves a movement back toward the primal "source" (*Ursprung*) of history and a going-forth into a new historical beginning.

Despite his critique of Spengler, Heidegger was much indebted to his interpretation of modern technology. Spengler conceived of that phenomenon as Faustian *Geist*, the modern manifestation of the Will to Power. The drive for power originated when Western man conceived the "monstrous thought" of yoking the very forces of nature.[15] Foreshadowings of Heidegger's own account of modern technology are present in Spengler's dramatic statements from *Man and Technology* (*Mensche und Technik*, 1932):

> To construct a world for himself, himself to be God—that was the Faustian inventor's dream, from which henceforth arose all projects of the machines, which approached as closely as possible to the unachievable goal of perpetual motion.

The concept of the booty of the beast of prey gets thought to the end. Not this or that, like fire, which Prometheus stole, but instead the world itself with the mystery of its forces gets dragged into the structure of this culture. Whoever was not himself seized by this will to almighty power over nature must experience it as devilry, and one constantly perceived and feared the machine as the discovery of the devil.[16]

In his audacious quest, Faustian man turned nature into a stockpile of raw materials whose only value lay in their usefulness for his titanic purposes. Speaking in a remarkably prescient way, Spengler predicted that changes of climate (such as the current trend toward global warming) would result from the transformation of great forests into newsprint. Modern society, he went on, has become a grasping, calculating machine which poisons the environment and compels us to see everything mechanistically: *"One no longer sees a waterfall without transforming it into the thought of electric power.* One doesn't see land full of pasturing herds without thinking of the evaluation of their meat-stock, no beautiful handiwork of a native inhabitant without the wish to replace it by a modern technical procedure."[17]

Compare all this to what Heidegger was to say years later about the technological disclosure of nature:

> The hydroelectric plant is set into the current of the Rhine. . . . In the context of the interlocking processes pertaining to the orderly disposition of electrical energy, even the Rhine itself appears as something at our command. . . . What the river is now, namely, a water power supplier, derives from out of the essence of the power station. . . . But, it will be replied, the Rhine is still a river in the landscape, is it not? Perhaps. But how? In no other way than as an object on call for inspection by a tour group ordered there by the vacation industry. [VA I: 15/16]

Spengler maintained that the age of reason led man to worship technology as a materialistic savior, which is as "inaccessible as God the Father. . . . And its worshipper is the progress-philistine of modern times, from Lamettrie to Lenin."[18] Heidegger, too, regarded the Enlightenment as a crucial historical period in which man began the final stage of trading his essence for material comfort and security. Spengler contended that authentic Faustian man created things not for the sake of security, craved by the masses, but instead for the sheer life-affirming experience of victory. But at the very pinnacle of technological culture, the decline sets in. Just as Faustian man rebelled against nature, so his machine technology now rebels against him: "The master of the world becomes the slave of the machine. It compels him, us, and indeed all without exception, in the direction of its course, whether we know and want it, or not."[19] Like Spengler, Heidegger warned that the technological system had in many ways achieved mastery over humanity, even though humanity had set out to use technological means to become lord of the universe:

> As soon as what is unconcealed no longer concerns man even as object, but does so, rather, exclusively as standing-reserve, and man in the midst of objectlessness

is nothing but the orderer of the standing-reserve, then he comes to the very brink of a precipitous fall; that is, he comes to the point where he himself will have to be taken as standing-reserve. Meanwhile man, precisely as the one so threatened, exalts himself to the posture of lord of the earth. In this way the impression comes to prevail that everything man encounters exists only insofar as it is his construct. [VA I: 26–27/26–27]

For Heidegger, humanity had declined to the status of a power-craving animal only because it had lost touch with its authentic ontological capacity for "letting entities be." For Spengler, the decline of Western man's vitality was the tragic end that befell all great cultures. European humanity was now losing its grip on industrial technology, which was passing to the Japanese and other "colored" peoples. But such peoples used technology only as a weapon of defense against the white race. Only the Europeans understood the Faustian impulse of technology and could shape and extend it creatively. Just so, Heidegger argued that modern technology was essentially Western in origin, and could be transformed only by events originating in the West.

For Heidegger, the "decline" of the West had begun not recently, as Spengler and Chamberlain and other reactionary ideologues claimed, but instead twenty-five hundred years ago with the dawn of metaphysics at the hands of Plato. While Spengler believed that the high point of the West occurred four hundred years ago with the emergence of industrial-Faustian man, Heidegger regarded the past four centuries as increasingly blind to the being of entities. Germany's "decline" could be attributed not to a loss of instinctual vitality, but instead to the elevation of such animal vitality to the status of the "highest value"! Spengler argued that technological man, tired of his fight against nature, sought freedom from "the compulsion of soul-less activities, from the slavery of the machines, from the clear and cold atmosphere of technological organization. . . . It is the mood of Rome at the time of Augustine. . . . "[20] Spengler did hold out hope that new industrial Caesars might arise to stave off the present decline for a time, but ultimately time was running out on the West. According to Heidegger, the fact that Spengler drew back from the pessimistic conclusion of *The Decline of the West*, and spoke instead of the possibility of a "Prussian socialism" that might temporarily forestall decline, was a sign that he didn't know what he wanted.[21] "All was not meant so darkly," Heidegger remarked sardonically, "the restless activity [*Betrieb*] can go on undisturbed. . . . " [GA, 61: 74] Unlike Spengler, Heidegger was not yet willing to sign the death sentence of Germany and the West. He believed that a radical new beginning could transform Germany's relation to work and producing.

Heidegger also commented on the views of Ludwig Klages, author of *Spirit as Adversary of the Soul* (1929), who argued that calculating rationality had undermined the organic vitality at work in the soul, the seat of life itself.[22] Heidegger commented: "Liberation from spirit here means: back to life! Life, however, here gets taken in the sense of the dark boiling [*Brodelns*] of the instincts [*Triebe*], which simultaneously get taken as the fertile soil of the myth-ical." [GA, 29–30: 105] For Heidegger, Klages's book was a good example of what ought *not* to be meant by "returning to the origins" in order to avoid being destroyed by rationality. The true origin of the *Volk*, Heidegger main-tained, was not organic or biological at all; indeed, the conception of human-ity as an instinctual animal was no better than the view of humanity as a rational animal, for man was not an animal at all, in Heidegger's view. Such naturalistic views, including *völkisch* talk of "instincts," were the products of Enlightenment naturalism. Later, we shall consider in more detail Heidegger's critique of naturalistic interpretations of humanity.

Despite acknowledging his debt to Scheler elsewhere, in 1929–30 Hei-degger portrayed Scheler's ideas only slightly more sympathetically than he portrayed Klages's. Scheler "attempts . . . a balancing [*Ausgleich*, 'compensa-tion'] between life and spirit." [GA, 29/30: 106] For Heidegger, such a "bal-ancing act" missed the point, namely, that *humanity could not be understood in terms of its organic nature.* Considering his debt to Scheler's views on modern technology, Heidegger's remarks about Scheler's works are disappointing. In *Problems of a Sociology of Knowledge,* Scheler offered a history of Western civili-zation that is in many respects far more textured and nuanced than Heideg-ger's "history of being." Moreover, Scheler anticipated many of Heidegger's own conclusions about the character of modern technology. For example, Scheler pointed out the voluntarism peculiar to Western man, as exemplified by Scotus, Occum, Luther, Calvin, Descartes, Kant, Fichte, and Hegel, not to mention Spengler's Faustian man. Further, he argued that the aim of technol-ogy is not to invent "useful" machines for already existing purposes, but instead to construct all possible machines in order to express the human drive for power over nature. This drive cannot be conceived in utilitarian terms, but instead is ultimately purposeless: man dominates because he wills to domi-nate. In his concept of modern *Herrschaftswissen* (knowledge for the sake of domination), Scheler also anticipated Heidegger's conception of the inner relation between the *methodology* of modern science, on the one hand, and the *power drive* of modern technology, on the other. Science, Scheler said,

> is *primarily* a *will to "methods,"* from which, it was once thought, ever new ma-terial knowledge comes forth—almost on its own—in an unlimited fashion, in endless processes, and as specialized. . . . Everywhere a new *spirit of competition* emerges, surpassing every given phase (unlimited "progress"). And every person partaking in production seeks to surpass every other person by a wholly new *ambitiousness* for inquiry and research, unknown to a medieval "scholar," who—at least in his intention—preserved knowledge only as a good in itself.[23]

Heidegger described Spengler and Scheler as "higher journalists," whose work was based on a "truly vulgar" interpretation of Nietzsche's psychology. [GA, 29/30: 111] In other words, in reading Nietzsche they failed to make *Heidegger's* version of the distinction between "life" in general and "human existence." Hence, they allegedly interpreted Nietzsche's thinking in biological terms, while Heidegger believed that Nietzsche often used such terms as metaphors for non-biological factors. Moreover, their "diagnoses" of culture did not sufficiently illuminate inauthentic German culture, but instead were sensational literary events that offered comforting prognoses as well as diagnoses: "Who among men would not gladly know what is coming, so that one can prepare oneself for it, in order to be still less burdened and taken into claim and attack! These world-historical diagnoses and prognoses of culture do not strike us, they are *no attack upon us*." [GA, 29/30: 112]

The martial language of the passage just cited would increasingly find its way into Heidegger's writings during the 1930s, during which time he saw himself as mounting an authentic attack against German civilization, mired as it was in deep boredom (*tiefe Langeweile*).[24] Note that the literal translation of *Langeweile* is "long while." As Heidegger maintained,

> This *deep boredom* is the *basic mood*. . . . If [time] becomes long for us, we drive it and this boring-character [*Langewerden*] away! We want to have no long time and yet also have it. Boredom, long time [*Langeweile, lange Zeit*]—in Alemannian linguistic usage especially—"to have a long time" means, not accidentally, much the same as "to have homesickness." . . . Deep boredom—a homesickness. . . . Homesickness—a basic mood of philosophizing. [GA, 29/30: 120]

Once again, we encounter the theme of "homesickness." Here, we learn that it arises from deep, metaphysical boredom which cuts one off from one's ontological roots. In boredom, things manifest themselves indifferently, primarily as commodities. And the deeper the boredom, the less one is aware that one *is* bored. [GA, 29/30: 201] Heidegger characterized Germany's *deep* boredom by the phrase "It is boring for one" [*Es ist einem langweilig*]. The impersonal "it" of deep boredom had transformed the very being of *Dasein*. [GA, 29/30: 204] Yet deep boredom could also *compel* people to understand their actual situation, which was one of *complete indifference* to everything. This indifference was manifest in the widespread prattle about Germany's "role" in the world:

> Have we become for ourselves so devoid of meaning [*unbedeutend*] that we need a role? Why do we find for ourselves no meaning, i.e., no more essential possibility of being? . . . But who wants to speak thus, where world-commerce, technology, the economy seizes man and keeps him moving? . . . *Is it in the end with us thus, that a deep boredom draws in and out in the abysses of Dasein like a silent fog?* [GA, 29/30: 115]

In deep boredom, the temporal horizon of presence is so wide (i.e., the "while" became so "long") that entities can present themselves only in an undifferentiated and indifferent way. In effect, they "refuse" or "withhold" themselves. With this idea, Heidegger anticipated his later claim that in the technological era, the being of entities withdraws or conceals itself. Paradoxically, he maintained in 1929–30, the infinite widening of time in deep boredom secretly harbored the moment of eternity, the *Augenblick*, which

constitutes authentic existence. The more *Dasein* became oppressed by boredom, the more did *Dasein* become compelled to experience the moment of
truth that would lead to authenticity. Already in 1929, then, Heidegger was
thinking according to Hölderlin's phrase "Where danger is, also grows the
saving power."[25] Yet this saving moment of truth could occur only if German
Dasein were willing to experience the affliction (*Bedrängnis*) involved in deep
boredom. The difficulty lay in understanding the true character of this affliction and the profound need (*Not*) arising from it.

Everywhere, Heidegger saw signs of crisis, catastrophe, and need: "the
contemporary social misery, political confusion, the impotence of science, the
undermining of art, the groundlessness of philosophy, the weakness of religion." [GA, 29/30: 243] And everywhere there were parties, programs, and
leagues organizing themselves to redress those needs. But by trying to alleviate
particular needs, Germans refused to become aware of their most profound
need: the need for binding themselves to an *existential task*. Only in such
binding, so Heidegger stated on many occasions, was genuine freedom possible. Such a view was wholly in accord with the attitude expressed in 1927 by a
leading figure in the "conservative revolution," Hugo von Hofmannsthal: "It is
not freedom [the people] are out to find but communal bonds [*Bindung*]. . . .
Never was a German struggle for freedom more passionate yet more tenacious
than this struggle in thousands of souls in the nation for true coercion [*Zwang*],
this refusal to surrender to a coercion that was not coercive enough."[26]

Only by binding themselves to the dangerous task of regenerating their
own existence, so Heidegger proclaimed, could Germany move beyond the
security and lack of danger [*Gefahrlosigkeit*] which accompanied deep boredom. By stealing away from the "danger zone," Germans showed that "today
presumably no one presumes too much of *Dasein*, but instead . . . we everywhere complain about the misery of life." [GA, 29/30: 247] If man is to
become what he *is*, however, he must "throw the *Dasein* itself on his shoulder"; he must, in other words, *live dangerously!* [GA, 29/30: 247] For those
living in the danger zone, things revealed themselves as once more charged
with meaning; for those capable of thinking only in economic terms, by way of
contrast, everything inevitably appeared within the indifferent light of monetary categories. Years later, Heidegger was to say that the man who lives within
the medium of "business" and "exchanges" is a "merchant" who "weighs
and measures constantly, yet does not know the real weight of things. Nor
does he ever know what is in himself truly weighty and preponderant." [HW:
299/135] The constant quest for an ever higher material living standard
brought spiritual devastation:

> Devastation is more uncanny than mere destruction [*Verwüstung ist unheimlicher
> als Vernichtung.*] Destruction only sweeps aside all that has grown up or been built
> up so far; but devastation blocks all future growth and prevents all building. . . .
> The devastation of the earth can easily go along with a guaranteed supreme living
> standard for man, and just as easily with the organized establishment of a uniform
> state of happiness for all men. [WHD: 11/29–30]

Sounding like a religious prophet, Heidegger proclaimed that people were damned if they clung to economic goals and saved only if they surrendered themselves to the higher power at work through them. The German people, he argued, must take on the burden of opening a historical world in which entities could manifest themselves anew. Only in self-abandonment to the disclosive power at work through them could Germans become truly free: *freedom is affirmation of necessity.* By comparison with the infinite expanse of entities, the historical moment for Heidegger's generation was but a *point*, but only in such a temporal epiphany could a people achieve its destiny. An authentic *Gemeinschaft*

> stands in the rooted unity of an essential deed. All and everyone, we are those assigned by a slogan, followers of a program, but *no one is the manager* [*Verwalter*] of the inner greatness of *Dasein* and its necessities. This leaving-empty in the end vibrates in our *Dasein*, its emptiness is the staying away of an essential affliction. The *mystery* is lacking in our *Dasein*, and thereby stays away the inner terror which every mystery bears within itself and which gives to *Dasein* its greatness. [GA, 29/ 30: 244]

Three years before Hitler came to power, Heidegger was calling for a powerful leader, a "manager," who would restore the inner greatness of *Dasein* by renewing the mystery and terror of existence. The situation had clearly changed from that of 1921–22, when he maintained that people had to learn to see the situation philosophically, "without prophets and *Führer*-illusions." [GA, 61: 69] Now, however, he believed that only a great leader could prevent the slide into two forms of madness: total social chaos, on the one hand, or total technological order (especially Soviet Communism), on the other. Jünger's writings depicted the horrifying prospect of the second alternative: planetary technological totalitarianism. Many Germans, so Heidegger believed, were already gripped by metaphysical boredom and thus could experience entities only one-dimensionally as raw material. It was only a matter of time before Germany succumbed to the same fate which gripped Russia and America: urban-industrial frenzy, meaninglessness, metaphysical damnation. Only one chance remained: to make a radical new beginning that would free Germany from the weight of the exhausted tradition in order to attain a new relationship to the industrial technology made possible by that tradition. Here was the occasion for Heidegger's allegiance to Hitler and National Socialism, a topic which we shall examine in the next chapter.

Chapter 3. Heidegger, National Socialism, and Modern Technology

From the very beginning of his career, Heidegger was concerned about the relation between burning issues of "factical existence," on the one hand, and problems in the history of metaphysics, on the other. He saw an internal relationship between the decline of the West into nihilism and the decline of Western humanity's understanding of being. During the 1920s, however, his critique of modern technology tended to be overshadowed by the ontological analyses that were clearly in the purview of his role as an academic philosopher. His decision to begin focusing on the phenomenon of modern technology cannot be understood apart from two interrelated events: first, his practical engagement with National Socialism, and second, his confrontation with Ernst Jünger's striking predictions about the technological future.

Jünger's thesis, that the *Gestalt* of the worker was mobilizing the entire planet into a technological frenzy, was in many ways similar to what Spengler, Scheler, and others had already said. Nevertheless, Heidegger concluded that Jünger's writings gave the clearest expression to the metaphysical condition of the West at the end of the history of metaphysics. Jünger, like other reactionary authors, argued that modernity and industrial technology could not be explained either in terms of Marxist economic theories or in terms of the liberal free-market ideology. The industrial transformation of the earth was merely the empirical manifestation of a hidden, world-transforming power. Jünger maintained that this power took the form of the *Gestalt* of the worker, the latest historical manifestation of Nietzsche's cosmic Will to Power. Heidegger transformed what Jünger regarded as the history of the Will to Power into what he was to call the "history of being." Moreover, influenced once more by Jünger, he also formulated his own highly controversial claim that Nietzsche's metaphysics calls for humanity to dominate the earth through technological means.

Resolved to forestall Jünger's fearsome predictions, but equally attracted to his masculinist rhetoric of courage and hardness, Heidegger used his own philosophical vocabulary and personal magnetism to support the National Socialist "revolution," which promised to provide an authentic "third way" between the twin evils of capitalist and communist industrialism. In this chapter, I argue that Heidegger believed National Socialism would renew and discipline the German spirit, thereby saving Germany from technological nihilism. I shall also address the following question: If Heidegger believed that his own philosophy could provide spiritual direction for National Socialism, does this mean that his philosophy is essentially fascist? In the subsequent chapter, I explain Jünger's conception of modern technology. Then, I shall describe in detail Jünger's influence both on Heidegger's engagement with National Socialism and on his mature concept of technology.

A. HEIDEGGER'S ASSESSMENT OF THE IMPORTANCE OF JÜNGER'S WRITINGS

Heidegger's lectures and writings from 1934 to 1944 contain virtually all the basic elements of his mature concept of technology. He developed this concept in constant dialogue with Jünger's writings, especially *Der Arbeiter: Herrschaft und Gestalt (The Worker: Dominion and Gestalt)*. [DA][1] After the end of World War II, when this book had gone out of print, Heidegger personally encouraged Jünger to reprint it. Heidegger's most extensive published commentary on Jünger is found in an essay originally entitled "About 'The Line' " ("*Über 'Die Linie'* "), which was published in 1955 as part of a *Festschrift* for Jünger. [ZSF][2] This essay was published separately the following year under the title "The Question of Being." The original title was a play on words of the title of Jünger's essay "Over the Line" ("*Über die Linie*"), which he had written for a *Festschrift* honoring Heidegger's sixtieth birthday. The German word *über* means several things, including "over," "about" or "concerning," and "above." By his phrase "over the line," Jünger had meant that humanity was faced with the need to cross beyond the "line" separating the current age of technological nihilism from a new age yet to come. Heidegger, however, suggested that it was important to think "about" this so-called dividing line if we are to understand the essence of nihilism. Writing almost two decades after the appearance of *Der Arbeiter*, Jünger believed that nihilism was being brought to completion and that we were now at the "line" separating us from a non-nihilistic era. But Heidegger replied, "The fulfillment of nihilism is, however, not already its end. With the fulfillment of nihilism only *begins* the final phase of nihilism. Its zone, because it is dominated throughout by a normal state and its consolidation, is presumably unusually broad. This is why the zero-line where fulfillment approaches the end, is not yet at all visible at the end." [ZSF: 48/49]

Heidegger concluded that the whole metaphor of "passing over a line" (*trans lineam*) between two "zones" was inadequate. Limited by metaphysical language, Jünger could not see that humanity exists as the clearing in which historical epochs or "zones," including the nihilistic technological era, can emerge in the first place. Heidegger, too, was looking for a "reversal" that would initiate the beginning of a post-metaphysical, post-nihilistic era, but he argued that only an adequate "topology" of being would let us retain the insight "enkindled" by Jünger's talk of passing "over the line" without forcing us to retain the metaphysical presuppositions present in his concept of the "line." By the topology of being, Heidegger meant the various configurations of beingness which constituted the history of being. The technological era was one such configuration, in which entities could reveal themselves only as standing-reserve to be used up in the endless quest for power for its own sake.

Despite his criticism of Jünger, Heidegger conceded that his own definitive essay, "The Question concerning Technology," "owes enduring advancement to the description in *Der Arbeiter*." [ZSF: 44/45] He added that Jünger's writings "will *last* because by them, insofar as they speak *the language of our century*, the discussion of the essence *[Wesen]* of nihilism, which has by no means yet been accomplished, can be newly enkindled." [Ibid.] In the early 1930s, Heidegger believed that National Socialism offered an alternative to the technological nihilism forecast by Jünger, though gradually Heidegger concluded that the nihilistic era would endure for much longer than he had at first anticipated. In a sense his admonition, supposedly addressed to Jünger, that the "culmination" of the technological era meant not that it was about to end but instead that it was beginning its long endurance, was also addressed to Heidegger himself, who in 1933 believed that the technological era had practically reached its culmination and that National Socialism offered Germany a radical alternative to it.

In his 1955 essay, Heidegger failed to acknowledge the extent to which his encounter with Jünger helped to shape his interpretation of and engagement with National Socialism. Only in a posthumously published essay, *The Rectorate 1933/34: Facts and Thoughts*, originally written during Heidegger's de-Nazification hearings in 1945, was such an acknowledgment forthcoming:

> In the year 1930 Ernst Jünger's article on "total mobilization" ["*Die totale Mobilmachung*"] had appeared; in this article the basic features of his book *Der Arbeiter*, which appeared in 1932, announced themselves. Together with my assistant [Werner] Brock, I discussed these writings in a small circle and tried to show how they express a fundamental understanding of Nietzsche's metaphysics, in so far as the history and present of the Western world are seen and foreseen in the horizon of this metaphysics. Thinking from these writings and, still more essentially, from their foundations, *we thought what was coming, that is to say, we attempted to counter it, as we confronted it.* [DR: 24/484; my emphasis]

B. THE FATEFUL DECISION: HEIDEGGER'S SUPPORT
FOR NATIONAL SOCIALISM

In 1933, when Heidegger was elected Rektor of the University of Freiburg and when he affirmed his own support for Hitler's "revolution," he understood his own action as continuous with his own philosophy. With all of his intelligence, will, and energy, he attempted to counter what Jünger had predicted was coming: the epoch of total mobilization, in which humanity would be reduced to the status of the worker laboring to transform the earth into a gigantic technological organization. Defending himself against detractors in 1945, Heidegger asked:

> Was there not enough reason and essential need to think in primordial reflection toward a surpassing [*Überwindung*] of the metaphysics of the Will to Power

and that is to say a confrontation with Western thinking by returning to its begin-ning [*Anfang*]? Was there not enough reason and essential need for the sake of such reflection on the spirit of the Western world by us Germans, to awaken and to lead into battle that place which was considered the seat of knowledge and in-sight—the German university? [DR: 25/485]

Just as Natorp, Cohen, and Scheler (among many others) joined the politi-cal fray to articulate "the ideas of 1914," so Heidegger became politically in-volved in 1933 in order to articulate what he regarded as the authentic spirit of National Socialism. He sought not only to revive Fichte's influential role, but to go beyond it as well. Instead of merely rallying Germans against an occupying army, Heidegger wanted to change the whole course of German and Western history. There is no reason to doubt the sincerity of his belief in 1933–34 that Hitler was the statesman required to carry out Heidegger's spiritual vision. In his speeches supporting Hitler, Heidegger mixed his own philosophical vocabu-lary with the street language of National Socialism. Reading these documents leads one to ask whether Heidegger's thought is intrinsically fascist.[3]

For Marxist critics, the answer is obviously yes. They argue that Heideg-ger's "explanation" of modern technology in terms of the history of being mystifies the fact that *humanity*, not being, shapes human history. Because humans developed the capitalist social class structure, humans can also abolish that structure and inaugurate a world without class structure and the exploita-tion connected with it. For Marxists, Heidegger's links with fascism are clear.

Non-Marxist critics, too, including people greatly indebted to Heidegger's thought, such as Jacques Derrida, conclude that it is impossible to separate clearly the political from the philosophical in Heidegger's writings.[4] Jürgen Habermas has suggested that even in early Heidegger there can be discerned important foreshadowings of the authoritarianism that was to bloom only in the 1920s.[5] Recently, Otto Pöggeler has asked: "Was it not through a definite orientation of his thinking that Heidegger fell—and not merely accidentally—into the proximity of National Socialism without ever truly emerging from this proximity?"[6] And Philippe Lacoue-Labarthe has concluded that Heidegger's engagement of 1933 was "absolutely coherent with his thought."[7]

Despite the fact that Heidegger *the man* committed himself and his writ-ings to National Socialism, there are reasons for questioning the allegation that his writings are intrinsically linked to that movement. First, one can study his writings without suspecting that they could be or were used to support the violently anti-Semitic movement which we now associate with National So-cialism. Certainly Heidegger's Jewish graduate students from the 1920s, in-cluding Herbert Marcuse and Hannah Arendt, were shocked by his political decision in 1933. In 1928, Marcuse wrote that *Being and Time* provides the ontological foundation for Marx's theory of history.[8] That *Being and Time* could be used as theoretical support for both left- and right-wing movements under-mines the claim that Heidegger's philosophy is essentially fascist. This defen-sive ploy, however, exposes Heidegger's writings to a different accusation: that they were amenable to both of this century's major totalitarian movements.

Heidegger's political orientation, especially his contempt for Enlightenment values, profoundly shaped his interpretation of Western history. Pierre Bourdieu rightly argues, however, that Heidegger's texts are "polysemic."[9] They can be read profitably without regard to their political implications, but they can and should *also* be read in terms of those implications. His thought cannot be reduced to the level of an ideological "reflex" of socio-political conditions, but on the other hand it cannot be regarded as wholly detached from such conditions. Heidegger argued that because creative works—including philosophical ones—have a measure of autonomy, the author's views about those works are not privileged. Hence, the fact that *he* chose to interpret his own texts as consistent with National Socialism does not mean that others must interpret them in the same way. Still, while Lacoue-Labarthe's assertion that Heidegger's political engagement was "absolutely coherent with his thought" must be tempered, it must not be entirely dismissed. When reading Heidegger's writings, we must keep in mind that they were sufficiently colored by conservative, even reactionary political views that Heidegger *himself* could interpret them as compatible with a "higher" version of National Socialism. Because Heidegger often defined modern technology as the final outcome of the decline in Western humanity's understanding of being, and because he named as symptoms of this decline the democratic political ideals and institutions which were pushed aside by Hitler, we must learn to read Heidegger with a deeper concern about how his thought *may* be appropriated and applied politically.

In evaluating Heidegger's enthusiastic endorsement of Hitler, we should recall that Roosevelt and many other Western politicians spoke well of Hitler's accomplishments for years after 1933. Moreover, the world came to Berlin in 1936 to take part in the Olympic Games. Heidegger was not alone in believing at first that Hitler could be persuaded to adopt a "nobler" version of Germany's future than the one represented by racist ideologues and political hacks such as Rosenberg, Krieck, and Kolbenheyer. But many of these men also believed that they were taking part in a truly "spiritual" revolution. As one commentator explains, "Moeller van den Bruck's *The Third Reich* demonstrates quite clearly how the revolution to come was viewed essentially in spiritual, rather than political terms. Those who shared Moeller's position 'recognize a spiritual process at work behind the revolution, a process that accompanies it, into which it is transformed, or from which it emanates.' "[10]

Heidegger's decision to use his thought for political ends and to make certain "compromises" at a time of crisis was not unusual for German philosophers. Moreover, as Hans Sluga has shown, many philosophy professors became fellow travelers of National Socialism.[11] The members of one noted philosophical society, *Deutsche Philosophische Gesellschaft* (DPG), ended up supporting Hitler. The goal of the society was "the cultivation, deepening, and preservation of the German character [*Eigenart*] in the field of philosophy following the spirit of the German idealism founded by Kant and carried further by Fichte."[12] The political goals of DPG were stressed a few years later: "Internal forces failed [during World War I] when the troubles from the outside came upon us and when many of the old forms of life fell apart. 'A transformation, rebirth, and renewal of the spirit in its deepest roots . . . ' (Fichte) is now the task of those who put themselves into service to the German mission. . . . "[13] Most revealing for our purposes is the following statement from DPG's journal in 1934:

> Much of that for which we have aimed and labored has come closer to realization through the national revolution. That does not, however, absolve us from duty but in this immediate present imposes on us even more *the duty to use the power of German philosophy for the construction of the German ideology*. We must collaborate and we want to collaborate on the whole life of the German spirit, in which the timeless content of our people's mission has gained concrete shape and historical reality.[14]

At its meeting in Magdeburg in 1933, Hitler cabled the society to say, "May the strength of real German philosophy contribute to the foundation and strengthening of the German ideology. . . . "[15] In 1945, a German academician named Werner Rings (who had been dismissed from his post by Rektor Heidegger for being "un-German") spoke bitterly of the willingness of philosophers from various camps to support Hitler: "Yes, the Kantians now struck hard and with assurance, the Fichteans gave free rein to their fanaticism of the will, the philosophers of nature conjured up the demonic forces of old German pantheism, the Nietzscheans outdid Zarathustra, and the philosophers of existence torpedoed the existence of un-German Germans."[16]

Lacking a coherent body of doctrines, National Socialism invited people of many different stripes to define its spiritual direction. There were even "left-wing" Nazis, including the young Joseph Goebbels, who sometimes portrayed themselves as "National Bolsheviks." They called on Communists to join them in the struggle against the evils of big capital. Consider the remarks from a Nazi poster authored by Gregor Strasser in 1926:

> Our call goes out to you who earn your bread through honest work. If you don't want your children, and your children's children, to be damned for all eternity as *slaves of world capitalism*, if you don't want to be made into the protectors of Stock Exchange bandits and other bloodsuckers by your treacherous leaders, if you are on the contrary filled with a fanatical will of freedom, then join the ranks of [the] *National Socialist German Workers Party*.[17]

The fact that Nazis sometimes became Communists, and vice versa, should not conceal the fact that most mainline National Socialists regarded Bolshevist internationalism as the real threat to the German *Volk*. Rootless Jewish Bolsheviks were allegedly manipulating the stock exchanges to bring about the destruction of Germany. Only a proper kind of socialism, one with the interests of the working people of the German nation in mind, could save Germany from communist materialism. This socialism was "National Socialism." The socialist dimension of the Nazi critique of big capital, a critique downplayed by Hitler when he made peace with the great industrialists, should not be surprising if we recall that resentment against monopoly capitalism and industrialism was a feature of many versions of reactionary thinking. In fact, so Kenneth D. Barkin argues, "The abstractness of human relationships under capitalism which is constantly emphasized by Marx and his followers was originally a discovery of observers from the conservative camp."[18] Hence, Heidegger's affiliation with National Socialism may be understood, in part, in terms of his belief that only a corporatist, fascist community could protect German working people from the evils of wage slavery and atomistic individualism in capitalism, on the one hand, and from the ills of materialism and massification in communism, on the other.

Critics point out that even if Heidegger's motives were more high-minded than those of Hitler and his anti-Semitic cronies, he damned himself because he was willing to associate himself with such people. In his own defense, Heidegger and his followers composed the following "official story," drawing on his postwar explanations for his actions.[19] Disturbed by Germany's political crisis, the politically naive Heidegger briefly (ten months) supported National Socialism, but after seeing his error he went into a kind of "inner emigration," and became an increasingly bitter opponent of the evils of National Socialism. The title of his inaugural address as Rektor of Freiburg University, "The Self-Assertion of the German University," indicates his desire to protect the academic world from becoming totally co-opted by politics. He resisted party demands to fire anti-Nazi professors, forbade the posting of anti-Jewish posters on university grounds, and finally resigned when he could no longer tolerate political interference in university life. He was subjected to scathing criticism by party ideologues, who claimed that his was a kind of "personal" National Socialism. His classes were watched and even shut down by the Gestapo. Finally, in 1944 he was declared one of the "most expendable" professors and was sent to work repairing the Rhine dikes.

C. RECENT CRITICISM OF THE "OFFICIAL STORY"

Recently, however, the official story has come under severe criticism because of the findings of Hugo Ott[20] and Victor Farias.[21] Ott and Farias have proved that Heidegger misrepresented his Nazi past. For example, far from dissociating himself from the party in 1934, Heidegger continued being a dues-paying member of the Nazi Party (card number 312589, Gau Baden) until 1945. Certainly, it is difficult under a totalitarian regime to withdraw support for it, but Heidegger's involvement seems to have been more than *pro forma* for years after 1934. According to Dr. Carl Ulmer, for example, Heidegger made the following statements in his lectures on Schelling during the summer of 1936: "The two men who have led a counter movement against nihilism, each in different ways, Mussolini and Hitler, have both learned from Nietzsche, both essentially differently. But the authentic metaphysical domain of Nietzsche has not come to validity thereby."[22] That same year, Karl Löwith reports that Heidegger, while in Rome to give a lecture, sported a swastika pin on his lapel, announced "National Socialism was the right way for Germany," and concluded that his support for the Nazi movement was consistent with his conception of history.[23]

If by the mid-1930s Heidegger had in fact concluded that National Socialism was not the vehicle for moving beyond modern technology but instead its most virulent manifestation, surely by 1942 he would have welcomed Germany's defeat as a deliverance. Instead, he remarked: "Today, we know that the Anglo-Saxon world of Americanism has decided to destroy Europe, and that means the homeland, and that means the incipient event [*Anfang*] of the West. What is incipient is indestructible. The entry of America into this planetary war is not the entrance into history, but is already always the final American act of American history-lessness and self-destruction." [GA, 53: 68]

In the same lecture course, Heidegger announced, "One does not at all serve the knowledge and appraisal of the historical uniqueness of National Socialism if one now interprets Greek humanity [*Griechentum*] such that one could suppose that the Greeks have already all been 'National Socialists.'" [GA, 53: 106, 98] Even in 1951, Heidegger could still conclude that World War II had "decided nothing," because it brought about no spiritual transformation but was instead the victory of one technological nation over another. [WHD: 65/67] In his interview in *Der Spiegel* (given in 1966, published posthumously in 1976), Heidegger expressed doubt that democracy could "accommodate itself to the technological age." [Sp: 206/276] Moreover, he still portrayed his postwar meditation on modern technology as an extension of the task he undertook in 1933:

> I see the task of thought to consist in helping man in general, within the limits allotted to thought, to achieve an adequate relationship to the essence of technology. *National Socialism, to be sure, moved in this direction.* But those people were far too limited in their thinking to acquire an explicit relationship to what is happening today and what has been underway for three centuries. [Sp. 214/280; my emphasis]

In 1935, Heidegger echoed the party line by saying that Germany was being crushed between Russia and America, which were metaphysically the same despite their political differences: "the same dreary technological frenzy, the same unrestricted organization of the average man." What was the meaning of human existence when "the farthermost corner of the globe has been conquered by technology and opened up to economic exploitation" and when "mass meetings attended by millions are looked on as a triumph . . . ?" [GA, 40: 40–41/31] Situated in the center of Europe, Germany was under the greatest pressure. Nevertheless,

> it is the most metaphysical of nations. We are certain of this vocation, but our people will be able to wrest a destiny from it only if *within itself* it creates a resonance [*Widerhall*], a possibility of response for this vocation, and grasps its tradition creatively. . . . Precisely if the great decision regarding Europe is not to fall upon the way of annihilation, then it can occur only through new historical

spiritual energies unfolding from out of the center [namely, Germany]. [GA, 40: 41–42/31–32]

Heidegger's reference to the meaninglessness of gathering "millions" at mass meetings in Nuremberg should not conceal the fact that his statements were in many respects consistent with Nazi ideology. For example, we read the following on a poster from a Nazi election campaign:

> The German farmer stands in between two great dangers today: The one danger is the American economic system—Big capitalism! It means "world economic crisis"; it means "eternal interest slavery"; . . . it enslaves man under the slogans of progress, technology, rationalization, standardization, etc.; . . . it wants to make the world into a giant trust; it puts the machine over man; it annihilates the independent earth-rooted farmer. . . .
> The other danger is the Marxist system of BOLSHEVISM. It knows only the State economy; . . . it brings in the controlled economy; it doesn't just annihilate the self-sufficient farmer economically—it roots him out . . . ; it brings the rule of the tractor; it nationalizes the land and creates mammoth factory-farms. . . . [24]

In the 1930s and beyond, Heidegger interpreted Americanism and Bolshevism as different versions of the nihilism of modern technology—a nihilism which could be overcome only by a complete transformation of German *Dasein*. At first believing that Hitler could carry out this task, Heidegger supposedly withdrew support for National Socialism after 1934. Nevertheless, in his 1935 lecture course *An Introduction to Metaphysics*, Heidegger made the following remark which he reprinted (allegedly) verbatim in 1953: "The works that are being peddled about nowadays as the philosophy of National Socialism but have nothing whatever to do with *the inner truth and greatness of this movement* (namely, the encounter between global technology and modern man)—have all been written by men fishing in the troubled waters of 'values' and 'totalities.' " [GA, 40: 208/166]

In 1955, the young Jürgen Habermas, who at the time was greatly influenced by Heidegger's thought but who had been unaware of Heidegger's involvement with National Socialism, wrote an anguished critique of this sentence.[25] In reply to such criticisms, Christian Lewalter said that the sentence was *not* a sign of Heidegger's "political-moral agreement with the program and praxis of the NSDAP. . . . "[26] Instead, in 1935 Heidegger viewed the National Socialist movement as "a symptom for the tragic collision [*Zusammenprall*] of technology and man, and as such a symptom it has 'greatness,' because its effect reaches out to the whole West and threatens to drag it into decline [*Untergang*]."[27] Heidegger publicly approved of Lewalter's interpretation.[28] Later, in his *Spiegel* interview, he claimed that the statement was printed in 1953 just as it was read in 1935 (although the portion in parentheses was not read aloud, because even without it "I was sure that my listeners understood correctly"). Recently, however, critics charge that Heidegger did *not* reproduce exactly what he said in 1935. Otto Pöggeler, one of Heidegger's ablest commentators, concludes that *only in 1953* did Heidegger add the parenthetical remark "namely, the encounter between global technology and modern man," after his assertion about "the inner truth and greatness of this movement."[29]

For years after 1934, Heidegger continued to play an active role as part of a group of academicians who wanted to "complete" the revolution by transforming the German university system.[30] Moreover, he blacklisted people he considered to be "un-German" for somehow consorting with Jews. Heidegger's reasons for refusing to fire two Jewish professors during his rectorate were tactical, not ethical: their dismissal would not look good on the international scene. The question of whether he was deeply, or only opportunistically, anti-Semitic is complex. None of his public statements can be read as anti-Semitic, but on the other hand he never publicly apologized for his original support for the regime that exterminated millions of Jews and other "subhumans." In the late 1940s, in an extremely rare reference to the Holocaust, he said: "Agriculture is now a motorized food industry, essentially the same thing as the fabrication of cadavers in the gas chambers of the extermination camps, the same thing as the blockades and the reduction of countries to famine, the same thing as the fabrication of hydrogen bombs."[31] In this astonishing statement, Heidegger glided over the fact that the Holocaust was a *German* phenomenon involving the slaughter of millions of *Jews*. Instead, he chose to view the Holocaust as a *typical* episode in the technological era afflicting the entire West.

Nevertheless, in speaking of the Holocaust in the same breath with the hydrogen bomb, Heidegger was making an important point. Mass extermination in the Nazi camps was possible only because of developments within industrial technology. Moreover, the Nazis spoke of the Jews as if they were little more than industrial "waste" to be disposed of as efficiently as possible. Officials in charge of planning strategic use of nuclear weapons must be trained to conceive of the enemy populace in wholly abstract terms. Heidegger argued in several places that the hydrogen bomb—an instrument of mass extermination—was not the real problem facing us. Instead, the problem is the perversion and constriction of humanity's understanding of being itself in the technological era. Extermination camps and hydrogen bombs, from Heidegger's viewpoint, were both symptoms of humanity's conception of itself and everything else as resources to be produced and consumed, created and destroyed, at will.

There is, however, something problematic but also typical of Heidegger's tendency to explain specific events and deeds in Germany as if they were typical of the entire Western world. Hence, during the 1940s he spoke of Hitler as if he were the inevitable manifestation of certain trends within the history of productionist metaphysics:

> One believes that the leaders had presumed everything of their own accord in the blind rage of a selfish egotism and arranged everything in accordance with their own will. In truth, however, they are the necessary consequence of the fact that entities have gone over into the way of erring in which the vacuum [resulting from the self-concealment of being] expands which requires a single order and guaran-

tee of entities. Therein is demanded the necessity of "leading," i.e., the planning calculation of the securing of the whole of the entity. For this such men who serve the leading must be directed and armed. The "leaders" are the authoritative mobilization-workers [*Rustungsarbeiter*] who oversee all sectors of the securing of the consumption of the entity, because they see through the whole of the surrounding [sectors] and thus dominate erring in its calculability. [VA I: 85–86/105]

Hitler's National Socialism, we are told, resulted *not* from peculiar historical conditions, such as widespread German anti-Semitism and hostility toward Enlightenment political values, but instead from a metaphysical process that determined events throughout the West. Surely such an analysis is open to question. The same kind of exculpatory metaphysical explanation can be discerned in Heidegger's account of Nietzsche's racism: "Nietzsche's racial thought has a metaphysical, not a biological sense." [N II: 309] Here, Heidegger sought both to protect Nietzsche from abuse at the hands of various Nazi ideologues, and to clarify that his own conception of the German *Volk* was a metaphysical, not a racist, one. But, as Derrida asks, "Is a metaphysics of race more serious or less serious than a naturalism or biologism of race?"[32] Nevertheless, Derrida also cautions that in thinking Nazism and anti-Semitism, we must not focus on Heidegger alone, for "Nazism could have developed only with the differentiated but decisive complicity of other countries, of 'democratic' states, and of university and religious institutions."[33]

During the 1920s and 1930s, German universities—professors and students alike—were mostly either conservative or reactionary in outlook.[34] Hence, students and faculty helped both to make Hitler's accession to power possible and to consolidate that power. After the war, reflecting on his Rektor's address, Heidegger argued that it had been intended to defend the university against political co-optation. But colleagues who may have read his ambiguous words in this manner in 1933 were soon expressing outrage at his accumulation of university authority, at his allowing himself to be called *Führer* of the university, and above all at his eagerness (expressed in a telegram to Hitler!) to cooperate with the *Gleichschaltung*, i.e., with the "coordination" of the university into the totalitarian National Socialist state.[35] Shortly before resigning as Rektor, Heidegger spelled out his official goals:

> Since the beginning of my installation, the initial principle and the authentic aim [of my rectorate] . . . reside in the radical transformation of intellectual education into a function of the forces and demands of the National Socialist state. . . . One cannot presume [to know] what will remain of our transitory works. . . . The only certainty is that our fierce will, inclined toward the future, gives a meaning and brings support to our most simple effort. *The individual by himself counts for nothing. It is the destiny of our nation incarnated by its state that matters.*[36] [My emphasis]

In light of *Being and Time*'s emphasis on individuation as an essential ingredient in authentic existence, we may be amazed at the assertion that the individual "counts for nothing" in the Nazi state. We should recall, however, that in the late chapters of *Being and Time* Heidegger argued that authentic individuation could occur only within the context of an entire generation willing to submit to its common destiny. The explicitly political Heidegger of 1933–34 sought to achieve his own authentic individuation by surrendering himself to what he believed were the "powers of being" at work in National Socialism.

Two provocative passages shed light on Heidegger's political engagement. In the first passage, from 1935, Heidegger spoke of the "violent one" who tries to overpower the overpowering being: "The more towering the summit of historical *Dasein*, the deeper will be the abyss for the sudden fall into the unhistorical, which merely thrashes around in issueless and placeless confusion." [GA, 40: 170/135] Apparently, Heidegger was aware of the risk he was taking by supporting National Socialism. In his own defense, he insisted that it was better for him to become engaged than to sit on the sidelines like so many others did. He did not escape unscathed by his decision. So disturbed was he by his de-Nazification hearings, and by related threats by French occupation authorities to deprive him of his home and his personal library, that he had a nervous breakdown in the spring of 1946 and spent three weeks under psychiatric care. In light of the global havoc and personal disaster which followed upon Heidegger's decision to lend his philosophy in support of Hitler, we may find particularly ironic Heidegger's remark from 1929–30:

> What can [today's philosopher] not report with the most modern slogans about the world situation, spirit, and the future of Europe, the coming age of the world and the new Middle Ages! How he can speak with unsurpassable earnestness about the situation of the university and its concerns, ask what man is, whether he is a transition to or [a matter of] boredom to the gods. Perhaps he is a comedian— who can know that? . . . If he is one who philosophizes, why does he relinquish his solitude and loiter about as a public professor in the market? But above all, what a dangerous beginning is this ambiguous behavior! [GA, 29/30: 18–19]

While Heidegger's maverick version of National Socialism was incomparably more sophisticated than the primitive writings of many Nazi ideologues, nevertheless he shared with other Nazis a deep mistrust for the concept of individual civil, political, and economic liberties. Regarding such liberties as invitations to socially corrosive egoism, he proclaimed that only by surrendering to a higher power could Germans achieve genuine freedom. This conception of freedom was shared not only by Heidegger and Hitler, but by Jünger as well. Heidegger's relationship to National Socialism cannot be understood unless we see the extent to which Heidegger believed that it offered an alternative to the technological nihilism predicted by Jünger. Jünger called on Germans to submit to that nihilism, while Hitler—so Heidegger at first believed—called on Germans to submit to the dangerous venture leading *beyond* such nihilism. In the following chapter, we shall examine Jünger's conception of the nihilism of modern technology.

Chapter 4. Jünger and the *Gestalt* of the Worker

> And so I see a new, leading race rising to the surface in
> Europe, a race fearless and fable-like, without blood-
> shyness, without looking back; accustomed to endure the
> fearsome and to do the fearsome and to posit the highest
> in their goals. A race which builds and defies machines,
> for whom machines are not dead iron, but organs of
> power which it governs with cold understanding and hot
> blood. This gives the world a new look.
>
> Ernst Jünger, *Das Waldchen 125: Eine Chronik
> aus den Grabenkampfen 1918* (1925)

Although for his conception of modern technology Heidegger owed more to
Ernst Jünger than to anyone else, the views of both men were consistent with
those of certain other members of the "conservative revolution." They agreed
that technology could not be understood in terms of social, political, or economic categories; instead, the socio-economic structures of modernity, including great industries and mechanical warfare, were the empirical manifestations
of that which transcends the causal-material realm. Under Nietzsche's influence, Jünger conceived of modern technology as the latest manifestation of
the eternal but hidden Will to Power. The essence of technology, then, was
nothing mechanical or technical; rather, the world was mobilized in technological terms because humanity itself had become the primary instrument
required to carry out the latest historical phase of the Will to Power. Likewise,
Heidegger concluded that the essence of technology was the disclosure of all
entities as standing-reserve for enhancing the sheer Will to Will.

Both Jünger and Heidegger regarded themselves as chosen to peer beneath the material surface of events into the elemental-spiritual domain.
Jünger portrayed the titanic process of modern technology as an aesthetic
phenomenon, as a terrifying but sublime spectacle that was "beyond good and
evil." Refracting technology through the prism of literary-aesthetic categories,
he explained socio-economic reality in terms of mythical symbols of a super-
natural and irrational power. Heidegger also claimed to discover an intimate
relationship between technology and art: technology, like art, was a way of
disclosing the being of entities. Moreover, like the work of art, modern technology could not be conceived in pragmatic or purposive terms; rather, technology was that mode of disclosure driven by the sheer Will to Will, by a Will

that willed no goal except its own expansion. In the late 1930s, Heidegger began to conclude that Jünger's predictions about the technological future were accurate and unavoidable. Paradoxically, Heidegger believed, moving beyond the nihilism and violence brought by modern technology was possible only on the condition that humanity first submit to the claim of modern technology.

Despite the extent of his debt to Jünger's vision of technology, however, and despite his sympathy with Jünger's political vocabulary, Heidegger viewed him as a philosophical primitive, whose account of technology had to be reinterpreted in light of Heidegger's ontological categories. In what follows, I provide an outline of those early writings of Jünger that were influential on Heidegger's thinking. In chapters five and six, I explain how Heidegger appropriated Jünger's ideas.

A. REACTIONARY MODERNISM

As we have seen, many Germans yearned to replace the rootlessness of modern urban-industrial society with an organicist, authoritarian *Volksgemeinschaft* grounded in German "blood and soil." Unlike those reactionaries who rejected industry and technology as inherently bad for the *Volksgeist*, however, other members of the conservative revolution concluded that Germany must adopt modern technology, while at the same time establishing a socio-economic system that would properly contain it. The task, then, was manifold: (1) to restore Germany's national power by appropriating in a suitably German way the fruits of modern industry and technology; (2) to restore the social fabric by instituting an authoritarian-nationalistic socialism that would be a "third way" between the *Volk*-destroying evils of boundless egoism ("the spirit of Manchester") and blind collectivism ("godless communism"); (3) to replace the Enlightenment spirit of calculating rationality with the self-sacrificing but danger-loving individual. These particular conservative revolutionaries envisioned a future in which technological power would be wielded by Germans who incarnated the ancient virtues of manliness, courage, resoluteness, hardness, discipline, and honor. Above all, as Pierre Bourdieu has noted, they sought a revolution that would preserve those distinctions which separated them from the anonymous industrial masses, the proletariat.[1] Their "solution" to social problems amounted to a magical reconciliation of intractably opposed social forces.

Jeffrey Herf has used the term "reactionary modernists" to describe this technological-romantic branch of the conservative revolution.

> In the country of romantic counterrevolution against the Enlightenment, [the reactionary modernists] succeeded in incorporating technology *into* the symbolism and language of *Kultur*—community, blood, will, self, form, productivity, and finally race—by taking it *out of* the realm of *Zivilization*—reason, intellect, internationalism, materialism, and finance. The integration of technology into the world view of German nationalism provided a cultural matrix that seemed to restore order into what these thinkers viewed as a chaotic postwar reality.[2]

Herf argues that National Socialism can be understood as the political fulfillment of reactionary modernism. This thesis is controversial because it holds that Hitler's "ideology," so far as it went, was internally consistent. Some historians have argued that Hitler cynically used *völkisch* rhetoric both to conceal and to justify his commitment to the very industrialization which was for many *völkisch* writers incompatible with *Volksgemeinschaft*.[3] There is no doubt that Hitler spoke in terms of the blood-and-soil mythology of late nineteenth-century *völkisch* ideologues. As Fritz Stern explains, "The promise of the Third Reich, of the unity of the racial *Volk*, of its aggrandizement, of the resolution of its internal conflicts, the invocation of heroism, of individual exertion and national will—all of these seemed timely echoes of an ideology first disseminated [by *völkisch* ideologues] in the days of Bismarck."[4]

But Hitler was a fiery nationalist who, especially after 1936, committed Germany to an intense program of industrialization in order to rebuild the armed forces. Herf contends, however, that the Nazis were sincere in reconciling romantic mythology with modern technology, irrationalism with the products of rationality. In fact, there was nothing unique about the Nazi position; it was, in effect, one version of the reactionary modernism which sought to reconcile romantic-*völkisch* symbols and modern technological practices. Such reactionary modernism enabled Hitler to justify his distinction between *liberal* capitalism, which fostered the war of all against all, and *German* capitalism, which promoted *Volk* unity and strength.

Following Horkheimer and Adorno, Herf at first conceived of the relation between modern technology and National Socialism as a German product of the "dialetic of the Enlightenment." While it had supposedly dis-enchanted the world, the Enlightenment had simply removed all traditional and religious obstacles to humanity's secret longing to gain total control over nature. Behind the emancipatory rhetoric of the Enlightenment, then, lay a mythical element: the myth of the unconstrained human Will to Power that would enslave nature by scientific-technological means. According to Horkheimer and Adorno, National Socialism was a manifestation of the technological Will to Power that afflicted the entire West. Herf eventually concluded, however, that this hypothesis not only was misguided but included a curious kind of apology for Germany's misadventure: National Socialism was, allegedly, merely the German version of what was happening throughout the West.[5] Heidegger, too, interpreted National Socialism as a German version of the consequences of Enlightenment metaphysics. Indeed, the confluence of the Heidegger and the Horkheimer-Adorno interpretations of Western history and modern technology is remarkable, especially considering Adorno's famous attack on Heidegger in *The Jargon of Authenticity*.

I agree with Herf's argument that Germany's history was unique and thus not really comparable to how the English, French, or Americans dealt with the challenge of modern technology. Far from being the consequence of the Enlightenment, National Socialism resulted from insufficient Enlightenment: Germany had not successfully incorporated the liberal-democratic political values that helped other Western countries to pass through the crisis of industrialization without becoming totalitarian states. Many Germans could not believe that there was an authentic emancipatory impulse at work in the Enlightenment call for constitutionally guaranteed political and economic liberties. Supposedly, the only "liberty" guaranteed by Enlightenment values was the liberty to use calculating rationality to enslave one's fellow man in soulless factories and lifeless cities.

Ignoring Max Weber's warning that politics should have limited and pragmatic aims, reactionary modernists claimed that only an apocalyptic socio-political transformation could save the German soul.[6] Paradoxically, then, they rejected economic and scientific rationalism, but affirmed the technological powers arising from it. As we saw in the first chapter, even German engineers argued that the rationality needed to design machines should be subordinated to an authoritarian state guided by mythical-symbolic principles. Herf maintains that the "central legend" of reactionary modernism was

> the free creative spirit at war with the bourgeoisie who refuses to accept any limits and who advocates what Daniel Bell has called the "megalomania of self-infinitization," the impulse to reach "beyond: beyond morality, tragedy, culture." From Nietzsche to Jünger and then Goebbels, the modernist credo was the triumph of spirit and will over reason and the subsequent fusion of this will to an aesthetic mode. If aesthetic experience alone justifies life, morality is suspended and desire has no limits. . . . As aesthetic standards replaced moral norms, modernism indulged a fascination for horror and violence as a welcome relief to bourgeois boredom and decadence. . . . When modernists turned to politics, they sought engagement, commitment, and authenticity, experiences the Fascists and Nazis promised to provide.[7]

While Oswald Spengler and Werner Sombart were important proponents of reactionary modernism, Ernst Jünger was its leading figure. Before we can discuss Heidegger's debt to him, we must examine Jünger's notion that the *Gestalt* of the worker mobilizes the earth in the technological era.

B. JÜNGER'S WAR EXPERIENCE: ENCOUNTER WITH THE PRIMAL WILL

Jünger's views on technology were deeply influenced by his experience of modern warfare. Some critics argue, in fact, that his war experience proved so shocking to his sensibilities that he became a detached observer who protected himself from the horrors of modern technology by elevating it into a superior power against which humankind was ultimately powerless. On the field of battle, he experienced himself at times as a cog in a gigantic technological movement. Yet, unexpectedly, by surrendering himself to this enormous process, he experienced an unparalleled personal elevation and intensity which he regarded as authentic individuation. Generalizing from this experience, he concluded that the best way for humanity to cope with the onslaught of technology was to embrace it wholeheartedly. Jünger proclaimed the technological Overman, the "organic construction," who combined human passion with technical precision, hot flesh with cold steel. In 1930, he declared himself a "historical real-ist," who tried to comprehend "that substance, that layer of an absolute reality of which ideas as well as rational deductions are mere expressions. This stance is thus also a symbolic one, in so far as it comprehends every act, every thought and every feeling as the symbol of a unified and unchangeable being which cannot escape its own inherent laws."[8]

Born in 1895 near Hanover, Jünger was an adventurous lad who ran away to join the French Foreign Legion at the age of sixteen. Bored with the limitations and constraints of Wilhelmine Germany, he was overjoyed at the onset of World War I. He started as a volunteer infantryman, but his extraordi-nary valor soon led him to be promoted to a troop commander. Wounded fourteen times, he was one of Germany's most decorated war heroes. His first book was a vivid account of his experience in the trenches. Its very title, *The Storm of Steel (In Stahlgewittern)*, already contains the core of his later doctrine that modern technology is a natural phenomenon, like a storm. He was not alone in embracing the war as a liberation from the tedium of everyday life in a rigid society. But whereas the initial enthusiasm and nationalistic fervor of the volunteers were ground down by the relentless and horrifying character of trench warfare, Jünger maintained his war lust long after 1918. His early books reflect his attempt to discover an inner truth that would somehow "justify" the horrors of war and Germany's defeat. He concluded that war was an aesthetic phenomenon. As one critic has observed:

> The fight, stripped of any remaining moral motivation, could thus be carried on *for its own sake*, as the expression and correlate of inner experience. The mon-strously senseless battles and their total challenge to the subjective, boundless ability to hold one's ground could be grasped irrationally as a "volcanic process," a "well of life." . . . By mystical submission one could achieve the most painful, yet most heroic experience.[9]

In books such as *War as Inner Experience* (1922), Jünger maintained that technological warfare was a manifestation of elemental, mythical, irrational forces that transcended "bourgeois" economic concepts and political ideolo-gies. British soldiers, too, reverted to mythical thinking in order to cope with the unbelievable scale of the Great War. Paul Fussell has remarked:

> That such a myth-ridden world could take shape in the midst of a war represent-ing a triumph of modern industrialism, materialism, and mechanism is an anom-aly worth considering. The result of inexpressible terror long and inexplicably endured is not merely what Northrop Frye would call "displaced" Christianity. The result is also a plethora of very un-modern superstitions, talismans, wonders, miracles, relics, legends, and rumors.[10]

Jünger's mythologizing was shaped by contemporary artistic movements, including modernism, futurism, art for art's sake, and the aesthetic of horror. Through the prism of these movements, the dreadful events of the Great War were refracted as a hauntingly beautiful spectacle which symbolized the primal

Will at work beneath everything. Jünger's symbolic interpretation of historical reality was influenced by Spengler. Using popularized notions of *Lebensphilosophie*, Spengler viewed history "morphologically," i.e., in terms of the spatio-temporal shapes (*Gestalten*) which, as invisible but governing symbols or metaphors, organized empirical socio-political reality. Spengler was interested in what empirical reality *signified*. For him, it was the manifestation of a hidden and animating possibility: "not 'I' realizes the possible; rather, 'it' realizes itself through me as an empirical person."[11] For Jünger and Spengler, world history was a symbolic and aesthetic phenomenon, a spectacle configured by the governing *Gestalt* which was itself a manifestation of the eternal Will. Marxist critics such as Lukács have complained that Spengler and Jünger made these world-shaping *Gestalten*, not human agents, "the leading actors in the drama of history."[12] Rejecting the Enlightenment view of human "progress," Spengler concluded that each historical epoch was both unique and inexplicable in causal terms. Like other reactionaries, he was trying to discover metaphysical principles of history that were more basic than Marx's materialistic view of history. Heidegger's own "history of being" parallels, but is more sophisticated than, Spengler's attempt to find an unsurpassable, non-rational source for historical reality.

Jünger believed that his war experience gave him a magical perspective on the hidden *Gestalten* that governed history.[13] This perspective led him to conclude that the soldiers he once faced across the trenches were not the enemy, but instead were comrades joined in a gigantic adventure: the transformation of the earth into a totally administered technological marvel. Despite writing in militantly nationalist journals during the 1920s, Jünger regarded nationalism primarily as a source of friction that would lead to wars, which were for him "the father of all things." Warfare furthered the planetary technological *Gestalt* by hastening the industrialization process in which workers in factories and soldiers on the field became virtually indistinguishable. Soldiers consumed what the workers produced. This constantly expanding cycle is now called "the permanent war economy."

The masculinist Jünger called for a new man, hard and courageous, the worker-soldier, who would embrace technology as the latest manifestation of the same vital Will that animated more primitive men. Unlike "civilized" man, the worker-soldier would surrender to the atavastic Will. Passionate yet icy cold, he would do whatever was needed to fulfill the technological imperative of planetary domination. On the battlefield, Jünger saw the prototype of this new breed:

> The strangest thing of all was not the horror of the [battle-scarred] landscape in itself, but the fact that these scenes, such as the world had never known before, were fashioned by men who intended them to be a decisive end to war. Thus all the frightfulness that the mind of man could devise was brought into the field; and there, where lately there had been the idyllic picture of rural peace, there was as faithful a picture of the soul of scientific war. . . . [E]ven in this fantastic desert there was the sameness of the machine-made article. . . . And it seemed that

man, on this landscape he had himself created, became different, more mysterious and hardy and callous than in any previous battle. . . . After this battle [the Somme] the German soldier wore the steel helmet, and in his features there were chiselled the lines of an energy stretched to the utmost pitch, lines that future generations will perhaps find as fascinating and imposing as those of many heads of classical or Renaissance times. [SS: 109]

Jünger's literary idealization of his war experience appealed to those veterans who in war had experienced a life intensity that was wholly missing in postwar life. Some veterans, who became members of the infamous *Freikorps* that helped Hitler's rise to power, were disillusioned by the treachery allegedly involved in Germany' surrender, by the subsequent catastrophic inflation, and by the atomistic individualism and nihilism fostered by the Weimar Republic. These men provided a receptive audience for Jünger's proclamation of the coming of a new *Typus* of mankind, the worker-soldier who would sacrifice himself for the incomparable experience of being on the front lines of the technological *Gestalt*.[14] One critic has observed:

> For [Jünger], war is the most dangerous work; he sees in the soldier the primordial form of the modern worker and in the battle of material the authentic vision of modern times. Thus, the trench-fighters of the Western front appeared as "a new race, self-formed by the hard breeding of war . . . , raised in the school of battle and entrusted with the hard-working tools with which the *deadly work* gets performed." In the battle of material, he expressed the famous unity of war-frenzy and rational precision: "here speaks a more severe and a drier seriousness, a marching-beat which awakens the representation of vast industrial realms, masters of machines, battalions of workers, and cool men of power."[15]

The *Freikorps* credo of manliness and erotic hardness is reflected in Jünger's rapture about the splendor of battle as depicted in *War as Inner Experience* (*Der Kampf als inneres Erlebnis*): "The baptism of fire! There the air was so laden with overflowing manliness that every draw of breath intoxicated, that one would have to weep without knowing why. Oh, hearts of men, who can feel that!" [KIE: 22] *Materialschlacht*, the "material battle" fought with thousand-pound cannon shells, produced men who were "more ruthless, wilder, and more brutal than any other begotten men that the world had ever before seen." [KIE: 40] Jünger proclaimed that the "holy" fighting man was "a wholly new race, incarnated energy and laden with the highest burden. Supple, lean, sinewy bodies, striking faces, eyes in a thousand horrors turned to stone under helmets. . . . Jugglers of death, masters of explosives and flames, magnificent beasts of prey, they moved quickly through the trenches. . . . " [Ibid.]

But not everyone was one of the new type of man: "Certainly there were only a few chosen ones in whom, thus driven, the war gathered itself, yet the spirit of an age gets borne only by individuals." [KIE: 40–41] Only the elite were able to look directly into Gorgo's face, the gruesome mask of modern war, thereby discovering an unexpected sublimity. For Jünger, the horrifying was unsurpassably thrilling. On the battlefield: "The first dead [man], an unforgettable moment of truth [*Augenblick*] which froze the heart's blood to thickened crystals of ice." [KIE: 22] For him, courage in war was an ecstatic, erotic experience: "That is a frenzy beyond all frenzies, an unchaining that leaps all bonds. It is a fury without consideration and limits, comparable only to the violences of nature. There man is like the raging storm, the churning sea, and the roaring thunder. . . . " [KIE: 52]

In *The Aesthetic of Horror*, Karl Heinz Bohrer has argued that Jünger was one of the last representatives of an aesthetic tradition initiated by Edgar Allan Poe and furthered by late nineteenth-century aesthetes, decadents, and dandies such as Baudelaire, Wilde, and Beardsley.[16] Poe's aesthetic of the thrill of horror, which made him appear to be promoting amoral attitudes, was altered by Wilde and others to evoke a calculated sensational effect in readers bored by Victorian society. For Wilde, art revealed the dark mystery of the soul, full of secret longings and dreadful passions. Encounters with the horrifying brought one before the spiritual abyss, before the "terrifying eternity" of things. Wilde helped develop the idea of the non-ordinary temporality that became so important for twentieth-century literature: the "epiphany" that enables one to see through the veil of everyday life into the unfathomable, eternal, arational, and amoral mystery lurking beneath. Many German writers, including Paul Tillich and Heidegger, spoke of the epiphany as the "moment of truth" (*Augenblick*) in which the banality of everyday life was transformed by a profound shift of temporality that disclosed the situation in a radically new way.[17] Such a radical transformation called for a "decision" which could itself not be justified by existing political or social structures, since the epiphany revealed the groundlessness of those structures.[18]

Between 1880 and 1930, even the best representatives of German intellectual life—Dilthey, Simmel, Scheler, Weber—had questioned the progressive and rationalistic interpretation of world history. Increasingly, the world seemed to have lost its sense of direction, so that people were looking for a new worldview that would restore "weight" to things. Dilthey observed that "human substance fills us all with a feeling of fragility, of the might of the dark impulses, of being afflicted with obscure visions and illusions, of the finiteness in everything that constitutes life, even when these things give rise to the highest constructions of communal life."[19]

For Nietzsche, the task for humanity after the death of God was to look into the abyss of its "dark impulses" in order to discover and to affirm the eternal Will to Power at work in all things. But glimpses into the horrifying would undermine life, unless the glance into the abyss was accomplished under the guise of art. In *The Birth of Tragedy*, Nietzsche argued that the Apollinian function of Greek tragedy was to provide a healing balm for people who had seen into the darkness of life during Dionysian frenzy: "action requires the veils of illusion. . . . [A]n insight into the horrible truth outweighs any motive for action, both in Hamlet and in Dionysian man."[20] Only art could

turn these nauseous thoughts about the horror or absurdity of existence into notions with which one can live. . . . Here, when the danger to [man's] will is greatest, *art* approaches as a saving sorceress, expert at healing. She alone knows how to turn these nauseous thoughts about the horror or absurdity of existence into notions with which one can live: these are the *sublime* as the artistic taming of the horrible, and the *comic* as the artistic discharge of the nausea of absurdity.[21]

Like Nietzsche, Jünger concluded that life could be "justified" only as an aesthetic phenomenon. His oddly detached descriptions of the horrors of war turned reality into a picture which ignored the social, political, and economic dimensions of the twentieth century. His "magical realism" and aestheticism ran the risk of ignoring the reality of the human suffering involved in the "spectacle" of warfare. But he believed that his "inner experience" of war revealed its essence to him. As he proclaimed in *Fire and Movement* (*Feuer und Bewegung*), "As war brings to expression not a part of life, but life in its full violent power, so is this life itself at bottom wholly of a warlike nature." [FB: 112] Pacifists failed to see that war is the very "magnetic center" of a people, that the will to destruction is essential to man, and that all great historical creation necessarily involves destruction. While technological war was "the most fearsome form [*Form*] in which the world spirit has until now shaped [*gestaltet*] life" [FB: 106], the terror of mechanical warfare could be assuaged if one apprehended it as a picture: "There is hidden a beauty therein that we are already capable of surmising, in these battles on land, on the water, and in the air, in which the hot will of blood restrains and expresses itself through the mastering of the technological wonder-works of man." [KIE: 106] Like proponents of "art for art's sake," Jünger conceived of war as a sublime spectacle that was intrinsically important. At the end of *War as Inner Experience*, Jünger voiced his vision of war as an aesthetic phenomenon:

> All goals are past, only movement is eternal, and it brings forth unceasingly magnificent and merciless spectacles. To sink into their lofty goallessness as into an artwork or as into the starry sky, that is granted only to the few. But who experiences in this war only negation, only inherent suffering and not affirmation, the higher movement, he has experienced it as a slave. He has no inner, but only an external experience. [KIE: 107]

C. TOTAL MOBILIZATION: TECHNOLOGY AS AN AESTHETIC PHENOMENON

Jünger was hardly alone in noticing that the early twentieth century was the age of the revolutionary worker. Communists and Nazis alike proclaimed themselves representatives of the real interests of working people. Although National Socialists called for personal surrender to a higher destiny, and while they also viewed historical events as aesthetic phenomena, their "revolution" was too crude and too bourgeois for his tastes. Hence, when invited to join the regime in 1933, Jünger declined.

The elitist Jünger asserted that in the nihilistic technological era, the ordinary worker either would learn to participate willingly as a mere cog in the technological order—or would perish. Only the higher types, the heroic worker-soldiers, would be capable of appreciating fully the world-creating, world-destroying technological-industrial firestorm. He coined the term "total mobilization" to describe the totalizing process of modern technology. Already in World War I, soldiers realized that they were in a "war of attrition." The country which could produce the most material would win. The war was a "storm of steel" because of the truly massive scale of war material involved. The war was, in effect, a gigantic process of labor, involving constant production and consumption. In his essay "Total Mobilization" (*Die totale Mobilmachung*) (1930), Jünger stated:

> Next to the troops which encounter each other on the battlefield, there arises a new kind of troops of commerce, of provisions, of the arms industry—the army of work in general. In the final phase, already indicated toward the end of this war, no more movement occurs—even if it is a homemaker at her sewing machine—which does not dwell within a performance that is at least mediated in terms of war. In this absolute seizing of potential energy, which transforms the war-leading industrial states into Vulcan-like forges, there is indicated in perhaps the most obvious way the beginning of the age of the worker. . . . [TM: 130]

Germany lost the war, in Jünger's view, because it had failed to mobilize totally. Too attached to the values of comfort, security, pleasure, individuality, personal liberty, rationality, investment, and progress, Germans had been unwilling to risk everything on behalf of a daring mission. But in future wars, *total wars*, no one would be safe. Anticipating the air raids of World War II, Jünger prophesied: "We have already left behind us the age of aimed bullets. The squadron leader . . . no longer knows the difference between combatants and non-combatants, and the deadly cloud of gas draws in as an element over every living thing. The possibility of such threats, however, presupposes neither a partial nor a general, but a *total* mobilization which extends itself even to the child in the cradle." [TM: 132]

Across the planet, the technological *Gestalt* transformed primitive and ancient societies; venerable traditions were turned into "half-grotesque, half-barbaric" cults of machine technology: "The abstractness, and thus also the ferocity of all human relations, grows in an unbroken fashion." [TM: 145] This *Gestalt* compelled humanity either to surrender or to be destroyed: "in war and peace, [the *Gestalt*] is the expression of the mysterious and compelling claim to which this life in the age of masses and machines subjugates us." [TM: 132] For Jünger, then, the essence of modern technology was nothing technical; instead, the essential was the fact that humanity had been gripped by an irresistible will to dominate, which expressed itself in the guise of machine technology. He believed that humanity would be both saved and elevated only if it submitted itself to the nihilistic claim of the technological Will to Power.

Jünger's belief was widely shared. In England, D. H. Lawrence described workingmen as believing that they were "exalted by belonging to this great and superhuman [industrial] system which was beyond feeling or reason, something really godlike. . . . It was pure organic disintegration and pure mechanical organisation."[22] In prewar Italy, F. T. Marinetti also spoke of the new alliance between man and machine. The founder of futurism, he hated bourgeois values because they impeded the new technological aesthetic. He proclaimed that "we must prepare for the imminent, inevitable identification of man with motor, facilitating and perfecting a constant interchange of intuition, rhythm, instinct, and metallic discipline of which the majority are wholly ignorant, which is guessed at by the most lucid spirits."[23] In his *Futurist Manifesto* (1909), he sounded like a less compassionate Walt Whitman:

> We will sing of great crowds excited by work, by pleasure, and by riot; we will sing of the multicolored, polyphonic tides of revolution in the modern capitals; we will sing of the vibrant nightly fervor of arsenals and shipyards blazing with violent electric moons; greedy railway stations that devour smoke-plumed serpents; factories hung on clouds by the crooked lines of their smoke; bridges that stride the rivers like giant gymnasts, flashing in the sun with a glitter of knives; adventurous steamers that sniff the horizon; deep-chested locomotives whose wheels paw the tracks like the hooves of enormous steel horses bridled by tubing; and the sleek flight of planes whose propellers chatter in the wind like banners and seem to cheer like an enthusiastic crowd.[24]

Like Marinetti, Jünger frequently used metaphors to describe man as the "organic construction" whose nerves were electric wire, and whose muscles were pistons. The idea of the "man of steel" continues to have widespread appeal: consider, for example, such popular recent films as *Robocop* and *The Terminator*, which depict "supermen" whose mechanical interior is housed in soft flesh. Jünger's vision of the man of steel involved a sado-masochistic element. In "On Pain" (*Über den Schmerz*), he argued that bourgeois civilization tried to alleviate pain by abolishing torture and slavery, by developing smallpox vaccination, and by inventing the whole world of technical and political "comforts." Nietzsche's "last man" was a denizen of this pain-free and fully insured bourgeois world. Scorning this world, Jünger proclaimed that when one sees something *comfortable*, one ought to ask immediately where the *burden* is. [US: 160] Only the man willing to shoulder the burden and pain of the technological era was truly a man. The commando, the artist, the hero, and all who know the value of self-discipline—they understand that pain is a direct manifestation of life's elementary power. Jünger claimed that the body must no longer be conceived as valuable in itself, but instead only as an object or as an instrument in the service of attaining the higher values demanded by the technological impulse: man must become a machine. The state must inculcate discipline in youth. A discipline is "the form through which man maintains contact with pain." [US: 171] For the bourgeoisie, a "good" face was considered "sensitive"—nervous, moving, changing, open to various influences. "By way of contrast, the disciplined face is closed; it possesses a strong point of view and is one-sided, objective, and rigid. With each kind of correct training one soon remarks how the operation sets down firm and impersonal rules and procedures into the hardening of the face." [Ibid.]

Above all, one had to learn the value of *sacrificing* oneself to something higher. If one could only attain the aesthetic perspective, then one could come to terms with the complete objectification of one's own body. Such self-objectification could occur only in a world in which space and time had essentially changed. Jünger believed that the art of photography represented this transformed temporality. He published three books of photographs depicting the marvels of the technological era, including towering dams and speeding airplanes. Photography was a technical method consistent with the curious epiphany-like temporality of the technological epoch. The photograph "holds fast the bird in flight just as much [as it does] the man in the moment-of-truth [*Augenblick*] in which he gets torn apart by an explosion. That is . . . the mode of seeing peculiar to us; and photography is nothing other than a tool for this our peculiarity." [US: 188] Photography makes it possible to objectify pain, for example, in news photographs of athletes who have made their bodies objects of measurement and discipline. The chiseled face of the athlete "possesses a pure relation to photography. It is one of the faces in which the *Typus* or the race of the worker brings itself to expression." [US: 193]

D. JÜNGER'S CONCEPT OF THE *GESTALT* OF THE WORKER

In *The Worker*, Jünger defined this new kind of humanity, which was compelled by the *Gestalt* of the worker to produce ever more powerful technological devices in the service of planetary domination. What Jünger meant by *Gestalt* is not entirely clear. In psychological terms, it means a shape or form that is more than the sum of the perceptual elements organized by that shape. Jünger often let the word be defined negatively, by that with which it stood in contrast: the spineless, fragmented bourgeois world. Jünger's *Gestalt* promised a totalizing, unifying movement that would heal the class conflicts and social disorders plaguing the Weimar Republic. Following Spengler, Jünger contrasted *Gestalt* with reason: the latter was abstract, lifeless, repetitive, and derivative; the former was concrete, lively, novel, and creative. The *Gestalt* was associated with rooted *Kultur*, not with rootless *Zivilization*. *Gestalt* makes history and cannot be explained in terms of the heartless rationalism of bourgeois *Zivilization*. [DA: 88ff.] *Gestalt* "stamps" everyone with its character; leaders and followers alike are shaped by it. [DA: 75–76, 89–90] The *Gestalt* of the worker, then, names the metaphysical "stamping" [*Prägung*] and "imprinting" [*Stempeln*] that organizes the worker's experience and behavior. "By *Gestalt* we refer to the highest meaningful reality. Its appearances are meaningful as symbols, representations and impressions of this reality. The *Gestalt* is the whole which embraces more than the sum of its parts. This 'more' we call totality." [DA: 323]

Moreover, "from the moment in which one begins to experience things in terms of a *Gestalt*, everything becomes *Gestalt*. The *Gestalt* is not a new magnitude which would be disclosable in proximity to the already known, but instead by a new stroke of the eye the world appears as a stage [*Schauplatz*] of *Gestalten* and their relations." [DA: 39] The *Gestalt* is like a metaphysical magnet that organizes things and people by invisible lines of force. However, "one ought never forget that here it is a question not about cause and effect, but about simultaneity. There is no purely mechanical law; the changes in mechanical and organic conditions [occur] through the superordinate realm [of the *Gestalt*], from which is determined the causality of individual processes." [DA: 137] The *Gestalt*, then, cannot be conceived as an entity, but instead as the very "being [*Sein*]" of the worker. By "being," Jünger sometimes meant the "deepest reality," which for him amounted to "life" or to the Will to Power that animates all things. At other times, he spoke about being in a way that was close to what Heidegger meant by it. Jünger did not conceive of his new breed of men in biological-racist terms, but instead in metaphysical terms, as the manifestation of a new organization of all entities. [DA: 160] Ontologically conceived, the *Gestalt* is a kind of prism through which one perceives the "light" that constitutes reality in the technological era. [DA: 92] This prism changes unpredictably, such that in different historical epochs men experience themselves and the world in different ways. "Values" thus are not eternal but perspectival; they serve the Will to Power that manifests itself now in one way, now in another. Far deeper, more refined, more meaningful than man is the hidden and eternal power that works through him. The industrial world of factories and cannons, then, amounts to a "projection" in the spatio-temporal domain of natural laws of an eternal being (*Sein*) that lies beyond human comprehension. [DA: 147–148]

The *Gestalt* imprints all entities so that they appear as raw material in the process of total mobilization: "The *Gestalt* of the worker mobilizes the whole standing-reserve [*Bestand*] without distinction." [DA: 160] Such imprinting changes the character of activities which might otherwise be regarded as "timeless." For example, "the farm worked with machines and fertilized with artificial nitrogen from factories is no longer the same farm." [DA: 176]

Jünger argued that in the 1930s humanity was in the "between" period, the period of anarchism and nihilism, which had to be worked through before the *Gestalt* of the worker could be fully projected. This period is marked by war, confusion, and horror. Nevertheless, he announced, people must press on like soldiers on a battlefield who somehow know they are part of something grander than themselves. In the technological era, we are all expressions of the *Gestalt* of the worker-soldier: "Among all uses which are to be carried out in the realm of work, armament is the most significant. This is explained by the fact that the most mysterious sense of the *Typus* and his means is directed to domination. Here there is no—be it ever so special—means that is not at the same time a power-means, i.e., an expression of the total character of work." [DA: 313]

Emerging in the place of collapsed Christian and liberal values, the *Gestalt* of the worker organizes all human experience symbolically in terms of participation in work, which for Jünger is the human version of cosmic energy. In this mythological view, the worker cannot be conceived as a member of a mere "class," for the concept of "class" belongs to the bygone bourgeois world. The worker has, "from the ground of his being [*Sein*]," a relation to "elementary powers" of which the bourgeois world has no suspicion. [DA: 23]

The heroic worker, like the Unknown Soldier who died on Flanders' fields, surrenders totally to the higher claims of the *Gestalt*. [DA: 153, 162ff.] Devoid of bourgeois individuality, he becomes a *Typus* ("type") who achieves "freedom" by surrendering to the higher power working through him. At first, Jünger spoke of the *Typus* as having a steely, chiseled face, but later he spoke of the omnipresence of the mask, whether it be the gas mask of the soldier, the mask of the industrial welder, the face mask of the hockey player, or even the make-up of the modern woman. The masked, depersonalized *Typus* goes everywhere in uniform. [DA: 153, 241] Having moved beyond the dualism of subject and object, the worker becomes "standing-reserve" to be "stamped" by the *Gestalt* of the worker. [DA: 155ff.] The *Typus*, more disciplined even than the Prussian officer corps or the Jesuits [DA: 222], would achieve a measure of serenity in his submission:

> There is produced the possibility of a serene anarchy that simultaneously goes together with the strictest ordering, a spectacle—as is already shown in the great battles and in the enormous cities—whose image stands at the beginning of our century. In this sense, the motor is not the master but the symbol of our time, the image of a power for which explosion and precision are not opposites. . . .
>
> . . . As *Gestalt* the individual encompasses more than the sum of his forces and capacities; he is deeper than he can achieve in his deepest thought and more powerful than he can bring to experience in his most powerful deed.
>
> . . . [T]he unlosable inheritance of the individual is that he belongs to eternity, and in the highest and most indubitable moments is aware of that. It is his task, that he bring this [eternity] to expression in time. In this sense his life becomes a likeness of the *Gestalt*. [DA: 41–43]

Jünger maintained that many young men were willing to let themselves become the "stuff of nature" to be "shaped according to the demands of the *Gestalt*. . . . " [DA: 221] He proclaimed that "no spirit can be deeper and more knowing than those beloved soldiers who fell somewhere on the Somme or in Flanders." [Ibid.] Those soldiers ushered in the age of technology, that is,

> the ways and means by which the *Gestalt* of the worker mobilizes the world. The degree to which man stands decisively in relation to technology and is not destroyed by it depends on the degree to which he represents the *Gestalt* of the worker. In this sense technology is the mastery of language that is valid in the realm of work. This language is no less important or profound than any other since it possesses not only a grammar but also a metaphysics. In this context the machine plays as much a secondary role as man; it is only one of the organs through which this language is spoken. [DA: 165]

Jünger's own language produces a striking but difficult-to-describe effect upon the reader. Despite his talk of the hot-bloodedness of "inner experience," he described things in a strangely detached way, as if he were an observer in a nightmarish world from which there was no awakening. Yet because his writing involves a flow of icy hard depictions of the overwhelming horrors and uncheckable advances of the technological impulse, Jünger's language comes close to realizing his view that technological language is the "clatter of machine guns at [the World War I battle of] Langemarck." [DA: 145] Not identical with the language normally associated with instrumental rationality, he believed, this new language is the symbolic manifestation of the ultimately meaningless, eternal, non-historical Will to Power. No words are needed to understand this new language, which gives voice to the "secret mythical law." The power at work in modern technology "stands in separable binding with a firm and determinate life-unity, an unquestionable being [Sein]—the expression of such a being it is which appears as power. . . . " [DA: 79] The rationality at work in modern technology is merely an expedient means to an inherently non-rational end: production for production's sake. Hence, the Gestalt of the worker does not "mean" anything apart from the restless activity of the cosmos at work; the new world forged by the worker is simply a more straightforward manifestation of the Will to Power. The Gestalt cannot be explained in human terms; rather, human experience must be organized in terms of the Gestalt which in itself "possesses no qualities." [DA: 90] Just so, technology cannot be conceived in merely technical terms, but instead as the manifestation of elemental powers. [DA: 22–23, and passim]. About the quality-lessness and meaninglessness of the Gestalt of the worker, one commentator has observed:

> This, in effect, generates a rather amazing paradox. Technology means the domination of a language whose symbolism is derived from the rationalistic attributes of work. But that domination is maintained by a totality deprived of any signifying quality whatsoever, namely the Gestalt. This being the case, language in its ordinary sense is redundant. Bereft of will and bereft of values, the worker relies in his work-performance on his purely "existential being." What characterizes this existential being? The fact that his performance or achievement is "an achievement without any conceptual attributes, so that in a very cogent sense, the worker is a supporter of a revolution *sans phrase*."[25]

Although at times Jünger spoke as if the age of the worker is totally dedicated to means, at other times he acknowledged that this age contains its own goal: world domination by the master race of workers. Attaining this goal, however, required a period of instability and nihilism:

> Our technological world is not an area of unlimited possibilities; rather, it possesses an embryonic character which drives toward a predetermined maturity. So it is that our world resembles a monstrous foundry. . . . [I]ts means have a provisionary, workshop character, destined for temporary use.
> To this condition it corresponds that our environment has a transitional nature. There is no stability of forms; all forms are constantly molded by dynamic unrest. There is no stability of means; nothing is stable outside of the rise of production curves. [DA: 181–182]

The bourgeoisie viewed the changes wrought by industrialization as part of the progress toward "a rational and virtuous perfection. It is therefore bound up with the evaluation of knowledge, morality, humanity, economics, and comfort." [DA:171–172] But Jünger insisted that the idea of progress was rooted in illusory beliefs about human mastery of history. In fact, he argued, history is the manifestation of forces stronger than merely human ones. In surrendering to these higher forces, so Jünger wrote, the worker develops cultic symbols of technology that replace traditional religions.

> Reserved for [the *Typus*] is the rediscovery of the strong way in which life and sacred ceremonies are identical—a fact which, except in certain borderlands and mountain valleys, is lost upon men of our continent.
> This interpretation allows one to risk the suggestion that a deeper piety may be observed today among a movie or auto-racing audience than beneath the pulpits or before the altars. . . . It would be a good guess that *other games, more sacrifices, other revolts are imminent.* [DA: 171; my emphasis]

The suppliant attitude on the part of workers toward modern technology reveals their understanding that technology is an expression of a transcendent and "hidden center," which has no purpose external to the completion of the goal of the *totality.* [DA: 153] Technology is a manifestation not so much of the Will to Power, but of what Heidegger spoke of as the sheer Will to Will. Yet Jünger also spoke as if the nihilistic period that grips contemporary humanity would eventually lead to a period of repose, characterized by total organization, total planning, total control. "The goal at which the efforts are aimed consists in planetary domination as the highest symbol of the new *Gestalt.* Here alone rests the measure of a superordinate security, which reaches beyond all warlike and peaceful work-processes." [DA: 321] Still, this totally controlled world would not have a "purpose" beyond functioning in its deterministic way. Eventually, this era would give way to another, equally astonishing "spectacle" in the cosmic drama to which the elite can bear witness.

Before the culmination of the present *Gestalt* can occur, however, existing states must become authoritarian. Hence, today

> one experiences the spectacle of the dictators which the people impose upon themselves, in order that the necessity can be ordered; [the spectacle] of dictators in whose appearances there is achieved a strict and sober style of work. In this appearance is embodied the attack of the *Typus* against the values of the masses and of the individual—an attack which immediately shows itself as against the organs (already moving toward collapse) of the bourgeois concepts of freedom, as against the parties, the parliament, the liberal press, the free economy. [DA: 283]

While the bourgeois revolutions took place within absolute regimes and made use of spiritual-conceptual weapons, the workers' revolution will take place within bourgeois democracies and will make use of objective-technical weapons. All instruments of bourgeois culture—from radio to newspapers to schools to barracks to factories—will become objective means for a revolutionary process that will culminate in planetary domination. [DA: 296–299] The power of these technical instruments is not infinite in scope, but only totalizing. The technological age, in Jünger's view, is neither the last, nor the best, but only yet another manifestation of the eternal Will to Power.

Unlike many German reactionaries, Jünger admired the Soviet Union, which had submitted itself to the "Prussian Leninism" required to transform Russia into a tautly organized, highly productive, technological society. Influenced by Ernst Niekisch and other members of the German "National Bolsheviks," Jünger became increasingly disenchanted with the halfway measures promised by National Socialism and more attracted to the ruthless, whole-hearted centralized planning practiced in the Soviet Union.[26] The possibility for a true workers' state appeared to live under communist rule, where leaders did not hesitate to liquidate whole populations for the sake of forging the state into a living machine.[27] Jünger regarded Marxist communism as a mask for Russian nationalism, which itself was only a phenomenon in the worldwide process of revolution through the *Gestalt* of the worker.[28] From the Marxist viewpoint, Jünger was a technological determinist, i.e., he ascribed to technology (as the *Gestalt* of the worker) intrinsic powers that shaped history, and thereby failed to see that the destructive, worker-enslaving features of modern industrial technology were functions of capitalist social relations. Likewise, the heirs of Enlightenment liberalism would find shocking and reprehensible his irrationalism, his appeal to mystery, his call for human surrender to a higher power, his denial of the reality of human progress, and his scorn for the importance of individual liberty and purpose.

We must recall, however, that Jünger read events in aesthetic, not in moral or political, terms. Reading his writings, one is reminded of old newsreels in which the booming voice of the announcer describes how the industrial armies are being mobilized to complete another titanic dam, ship, airplane, or skyscraper. This frenzied activity was strangely beautiful to him:

> One begins to acquire a disposition for higher temperatures, the icy geometry of light, and the white glare of superheated metals. The environment becomes more constructive and more dangerous, colder and more luminous; there disappear from it the last remains of *Gemütlichkeit*. . . . One avoids secondary goals such as taste; one elevates the formulation of technical questions to the decisive position; and one does well thereby, since more than the technological is concealed behind these questions. [DA: 183]

Concealed behind technology is the eternal artistic Will that shapes reality now this way, now that: "It is evident that art, which the *Gestalt* of the worker has to represent, is to be sought in strict connection with work." [DA: 299] The high-level planning of the modern nation-state is a manifestation of "a will to *Gestaltung*, which seeks to grasp life in its totality and to bring it to form." [DA: 231] Clearly, "for a will which conceives the planet Earth as its elementary material, tasks cannot be lacking. . . . [T]here is no domain of life which is not to be considered also as material for art." [DA: 232] The *Gestalt* of the worker made room once again for the "adventurous heart," for the one who wanted to escape the confines of a stultified world. He concluded *Der*

Arbeiter by saying: "Here to take part and to serve: that is the task which is expected of us." [DA: 322]

Like many in the fascist movement, with which Jünger cannot unambiguously be identified, he "spiritualized" labor by depicting it as a manifestation of the same cosmic force that moves the planets. Although much of what he said helped to pave the way for the coming of National Socialism, Jünger himself became increasingly aloof from the movement. Indeed, his best-selling novel *On the Marble Cliffs* (*Auf den Marmorklippen*, 1939) offered a thinly veiled attack on National Socialism.[29]

By the 1940s, Jünger was conceiving of technology in terms of the same dualisms that had been expressed by critics at the turn of the century: cultural world vs. technological civilization; personal productivity vs. the mechanical performance of work; the true and natural vs. the technological and perverted life.[30] In an entry in his journal from September 12, 1945, he concluded, "Man as technologist, as spiritual-abstract being [*Wesen*], is necessarily the enemy and exploiter of . . . men of culture."[31] The fearsome character of our times, he alleged, can be attributed to goal-free, value-neutral modern science. Hence, in Hitler he saw "the incarnation of the scientific theories of the nineteenth century in their full loathsomeness."[32] But Jünger never abandoned the idea that modern humanity is in the grip of the technological *Gestalt*. In 1981, for example, he stated that

> [In 1932, I saw] that a new type of planetary man was putting itself in place and that he bore within himself a new metaphysics of which, on the empirical plane, he is not yet really conscious. Today still his visage remains partially hidden. . . . But this [technological] process, I repeat, one must avoid representing it solely from an historical, economic, social, or even ideological angle. The phenomenon transpires at a much greater profundity. [. . .]
>
> Today, we live in a transitional stage between two immense moments of history, as it was the case in the time of Heraclitus. The latter found himself between two dimensions: on the one side, there was myth, on the other, history. And we, we find ourselves between history and the appearance of something completely different. And our transitional era is characterized by a phase of Titanism which the modern world expresses at all levels.[33]

Before we consider Heidegger's appropriation of Jünger's thought, some critical observations are in order. Hans-Peter Schwarz claims that Jünger simplified and explained in mythical terms a many-sided and confusing manifold of conflicting phenomena (the rise of modern technology and its relation to political-cultural changes in Europe). Ignoring historical differences in the technological and cultural developments of the major world powers, Jünger found "the universal formula which made it possible to explain all modern phenomena. Movement, the consuming of every substance in favor of speeding up, transposing of life-stuff into energy—with this basic law of motion as the 'universal key,' Jünger has posited the horoscope of his age."[34] According to Schwarz, numerous non-Nazi contemporaries of Jünger took a dim view of his talk about the *Gestalt* of the worker; indeed, they regarded it as a kind of fancy fascism. Proponents of reactionary modernism, including Heidegger, however, believed that Jünger was telling it like it was.

Karl Heinz Bohrer maintains, by way of contrast, that Jünger—far from providing a factual description of events in the early 1930s—was writing what amounted to a dystopian novel akin to Huxley's Brave New World, or analogous to Fritz Lang's 1926 film Metropolis.[35] According to Bohrer, everything said in The Worker follows a literary concept of style. This emphasis on style, appearance, form, phenomenon is also apparent in Jünger's effort to describe the worker not in economic terms, but instead as the governing "style" in comparison to which actual economic circumstances were unimportant. The mask of the worker symbolizes that man is eternally alienated from himself and always acts out the impulses of the recurrent Will to Power. Jünger was not interested, then, in concrete historical, social, or political events, but instead was concerned to provide a descriptive "phenomenology" of the worker as the latest aesthetic manifestation of the Will to Power. Jünger's "report" about the technological era was not social analysis, but instead science fiction of the utopian (or dystopian) style. His writings call to mind contemporary cinema in the film noir genre, which depicts the future as a technological totalitarianism with no room for human uniqueness or feelings. Bohrer remarks:

> One best convinces oneself of the utopian character if one recognizes its analogies to contemporary fantastic space-utopia and science fiction: in Kubin's novel The Other Side, as also in modern technological science fiction, the super-terrestrial hero is characterized by the traits of being masked, lacking in feeling, and the claim to immortality, properties which make him superior to the ordinary earth-dweller. Thus Jünger also distinguished as unique the "Gestalt" of the Typus from the "abstract masses."[36]

Jünger's frequent references to the "mask" of the worker and to the "planetary" scale of his ambitions eliminate all socially relevant distinctions that would enable the reader to locate the worker in Germany's actual historical situation in 1932. Jünger's worker, then, was an imaginary figure searching for new worlds to conquer, just like a character in science fiction novels.[37] Jünger's worker not only renounces the ordinary human quest for "happiness," but even welcomes pain in objectifying his own body. These characteristics make him completely alien to all humanistic and Enlightenment traditions.[38] Jünger's rejection of the aim of "happiness" "distinguishes him above all from atavistic perversion of the real-utopian projects in the National Socialist workers' state. . . . In the rejection of happiness and the fantastic representation of the 'mask' and of 'planetary domination' there appears, namely, a counter-Enlightenment radicality, which comes to meet not the fascist program, but the pessimist introversion of modern art and literature."[39]

Bohrer's charge that Jünger refracted modern technology through the aesthetic prism of a novelist may give us pause. One might ask: did Heidegger base his understanding of modern technology on fictional writings? In reply, we should note that Jünger's "fiction" was not the product merely of fantasy but of his own encounter with the extraordinary events of the twentieth century. His writings do offer a different and sometimes compelling way of seeing the technological world. Who in this century has not wondered at times whether human beings have become not the masters but instead the instruments of technological power? Nevertheless, Jünger's *affirmation* of world wars, technological totalitarianism, elitism, and authoritarianism is an affront, though perhaps not a surprise, to those (including me) who regard his amoral aestheticism as reprehensible.

I would also like to take issue with Bohrer's conclusion that Jünger's vision was finally incompatible with National Socialism. It is true that in historical practice, the Nazis often adhered to bourgeois concerns about comfort and security. But, as I have argued, National Socialism also followed to some extent the vision of reactionary modernism. The Nazis celebrated violence, danger, adventure, and beauty; they called for a new breed of man who would combine spiritual and technological power. Joseph Goebbels gave expression to this vision in his "steely romanticism." Consider the following passage from the February, 1939, issue of *Deutsche Technik*:

> We live in an era of technology. The racing tempo of our century affects all areas of our life. There is scarcely an endeavor that can escape its powerful influence. Therefore, the danger unquestionably arises that modern technology will make men soulless. National Socialism never rejected or struggled against technology. Rather, one of its main tasks was to consciously affirm it, to fill it inwardly with soul, to discipline it and to place it in the service of our people and their cultural level. National Socialist public statements used to refer to the steely romanticism of our century. Today this phrase has attained its full meaning. We live in an age that is both romantic and steellike, that has not lost its depth of feeling. On the contrary, it has discovered a new romanticism in the results of modern inventions and technology. While bourgeois reaction was alien to and filled with incomprehension, if not outright hostility to technology, and while modern skeptics believed the deepest roots of the collapse of European culture lay in it, National Socialism understood how to take the soulless framework of technology and fill it with the rhythm and hot impulses of our time.[40]

Goebbels's "steely romanticism" was in some ways reproduced in Heidegger's language of hardness and ruthlessness in 1933–34. Nevertheless, despite his fascination with the masculinist rhetoric of Goebbels, Göring, and Jünger, Heidegger resisted their impulse to transform the German people into an "organic construction" in the service of the Will to Power. Although he perceived modernity, industrial technology, and National Socialism primarily through the literary optic provided by Jünger, Heidegger longed for an alternative to Jünger's vision of the future. In the next two chapters, we examine Heidegger's attempt to use Jünger to think against Jünger.

Chapter 5. Heidegger's Appropriation of Jünger's Thought, 1933–34

Anyone familiar with Heidegger's thought will recognize in Jünger's writings striking parallels with Heidegger's concept of modern technology. Like Jünger, Heidegger was a member of the "generation of 1914." Although his military service during the Great War was restricted primarily to rear echelon assignments, he—like many other soldiers home from the gruesome war— glorified the bravery exhibited at the front and the solidarity formed in the trenches by men of all social classes. Like Jünger, he believed that the old bourgeois world had been destroyed by that war; hence, postwar Germany was in between worlds, in a condition verging on nihilism.[1] Jünger claimed to see the new world arising: the technological age governed by the *Gestalt* of the worker. But while Heidegger also saw the possibility of a new world of work in 1933–34, his vision of that world was different from Jünger's.

Heidegger's relationship to Jünger's writings is both complex and ambiguous. Attracted to Jünger's critique of bourgeois decadence, to his affirmation of an elite future humanity, and to his longing for an authoritarian *Gemeinschaft*, Heidegger was also put off by Jünger's affirmation of the technological future. Nevertheless, Heidegger used Jünger's voluntaristic, martial, and manly rhetoric in calling for a National Socialist revolution to forestall the technological destiny forecast by Jünger. Gradually, he concluded that there was no way for humanity to avoid being "stamped" by the *Gestalt* of the worker. Increasingly, Heidegger called for humanity to submit to that destiny, much as Jünger had done at the close of *Der Arbeiter*. Paradoxically, only by such submission could there occur the "turning" needed for humanity to gain a proper relationship to modern technology.

In this chapter, we first see how Heidegger's Nazi addresses of 1933–34 blended his own philosophical vocabulary with Jünger's vocabulary of steely romanticism. During this period, Heidegger claimed that the university would be a major battlefield in which the revolutionary forces of change would fight those who stood in the way of Germany's new *Volksgemeinschaft*. Only the courageous, the hard, the manly, those willing to risk everything to further a cause higher than themselves, were welcome in the new university; only such higher types understood the real meaning of "science" (*Wissenschaft*); only such men could steer Germany away from the physical and spiritual wasteland produced by modern technology. Heidegger acknowledged that the revolutionary world would be a *worker's* world, but he sought to define both work and worker in a way other than how Jünger had defined them. While adopting Jünger's masculinist rhetoric, Heidegger fought against Jünger's vision of technological nihilism. Heidegger, in effect, used Jünger *against* Jünger.

A. HEIDEGGER'S ADDRESS AS REKTOR IN 1933

In 1933, unlike those who assumed that Hitler's accession to power meant that the "revolution" had already achieved its aim, Heidegger believed that the "real" revolution would succeed only if he became its spiritual *Führer*. Recognizing the importance of the step taken by Hitler, however, Heidegger willingly furthered the National Socialist *Gleichschaltung*, whereby formerly democratic institutions were transformed into those suitable for the totalitarian regime. Yet in so doing he used Jünger's masculinist rhetoric of hardness, courage, self-sacrifice, and manliness.

Heidegger first publicly articulated his vision of the new world of work in his Rektor's address, "The Self-Assertion of the German University" (May, 1933).[2] After 1945, Heidegger was to say that by "self-assertion" he meant the university's assertion of its independence in the face of the political intrusions of National Socialism into university life. In 1933, however, he evidently had something else in mind by "self-assertion," namely, that the German university should assert its proper role of leadership in the revolutionary movement. In terms reminiscent of Jünger, Heidegger began his address by saying that leaders themselves must be "led by that unyielding spiritual mission that forces the fate of the German people to bear the stamp of its history." [SDU: 10/471] Further,

> The will to the essence of the German university is the will to science as the will to the historical mission of the German people as a people that knows itself in its state. *Together*, science and German fate [*Schicksal*] must come to power in this will to essence. And they will do so if, and *only* if, we—this body of teachers and students—*on the one hand* expose science to its innermost necessity and, *on the other hand*, are equal to the German fate in its most extreme distress. [Ibid.]

Heidegger defined science in terms of its origins in Greek philosophy. The Greek philosopher first "stands up to the *totality of what is*, which he questions and conceives as the entity that it is." [SDU: 11/471–472] Hence, "All science is philosophy. . . . " [Ibid.] True science was not the busy activity of normal researchers, but instead the philosopher's act of uncovering Germany's future which lay hidden in the destiny bestowed by the Greeks. The beginning, Heidegger maintained, "stands *before* us. . . . There it awaits us a distant command bidding us to catch up with its greatness." [SDU: 12–13/473] Authentic science (philosophy) was "the highest mode of *energeia*, of man's 'being-at-work.' " [Ibid.] Philosophy's resolute working, i.e., authentic questioning, could open up a truly spiritual world, the power of which "most deeply preserves the people's strengths, which are tied to earth and blood. . . . " [SDU: 14/475]

If the philosopher's struggle is the highest form of being-at-work, and if the revolution is to inaugurate a world of authentic work, then—so Heidegger concluded—the philosopher should lead the *Volk* into that new world. In light of the primacy of philosophical work, Heidegger divided German "work service" into three "equiprimordial" groups that correspond to the divisions of the ideal city in Plato's *Republic*: armed-service (*Wehrdienst*), work-service (*Arbeitsdienst*), and knowledge-service (*Wissensdienst*). Note that the first two forms of work—those of the soldier and the worker—were covered by Jünger's idea of the *Gestalt* of the worker-soldier. At least during his term as Rektor, Heidegger not only supported paramilitary training for all students, but would also show up unexpectedly at drills to exhort the student-soldiers. According to Hugo Ott, Heidegger was enraptured by Jünger's accounts of how the "front experience" transformed men into the steely, resolute, ruthless, and courageous types necessary to carry out the revolution.³ Winfried Franzen has also described Heidegger's "yearning for severity and arduousness" during 1933–34.⁴ Heidegger apparently believed the prevailing propaganda which proclaimed Hermann Göring, the daring World War I aviator who had risen to power with Hitler, to be the new Nazi man, because Heidegger gave to a friend a copy of Göring's biography as a birthday present in 1933!⁵ Heidegger maintained that risking one's life on the battlefield was a decisive way in which to achieve the manhood which Germany required for the great and difficult tasks lying ahead. Consider the martial pathos at work in the following passage taken from Heidegger's last public political speech, given at Tübingen in November, 1933:

> We of today stand in the struggle for a new reality. We are only a transition [*Übergang*], only a sacrifice. As fighters of this fight [*Kampfes*] we must have a severe stock, which no longer depends on anything individual, [but] which establishes itself on the ground of the *Volk*. The fight concerns not persons and colleagues, also not empty formalities and general measures. Every pure fight bears the enduring traits of the image of fighting and its work. Only war deploys the true laws for the actualizing of things, war is that which we will: we fight heart by heart, man by man.⁶

Also in November, 1933, Heidegger gave an impassioned speech urging his students to support Hitler's plebiscite calling for Germany's withdrawal from the United Nations. Moreover, he called for the annexation of Austria, a German-speaking country that was clearly a part of the new German *Volk*. Despite his martial rhetoric regarding neighboring countries, however, Heidegger did not support Jünger's idea that wars of conquest were a way of furthering the inevitable arrival of the age of the worker-soldier. In 1937 he defined the authentic struggle between peoples not as armed warfare, but instead as that spiritual struggle in which neighbors returned each other to their own unique ways of being.⁷

In his Rektor's address, Heidegger stated that work-service involved the communal tasks that must be carried out by all "estates" or "classes"

(*Staende*) of the *Volk*. Elsewhere, he maintained that the National Socialist revolution meant the end of the division of the *Volk* into classes: "There is only a single German 'living class' [*Lebensstand*]. That is rooted in the supporting ground of the *Volk* and in the *worker's class* freely articulated in the historical will of the state, [a class] whose stamping [*Prägung*] becomes pre-formed in the movement of the National Socialist German *Workers' Party*."[8] The idea of a heroic, one-class workers' state was one of the guiding principles of the "revolution." As one National Socialist writer put it:

> And when another saying of the *Führer* goes: "in the future there will be only one more nobility, the nobility of labor," this shows that the proletarian colouring of the concept "laborer" and the fighting attitude toward another rank has been extinguished. Accordingly, all literary attempts to construct a proletarian culture have become pointless and forgotten. There is only one culture and one life form, that of the German people. It is clear that from all the efforts to transform the [industrial] plant into a cell of community life, a life style of the German worker must emerge.[9]

Neither Hitler nor Heidegger envisioned the "one-class" workers' state as an undifferentiated mass society like the one they attributed to Russia or America. Instead, they held forth an organic ideal in which each worker contributed in his own way to the maintenance of the community. Heidegger specified that those engaged in "knowledge-service" (*Wissensdienst*), which corresponded to the rank of guardians in Plato's *Republic*, were the leaders of the one-class community. These guardians, including "the statesman and the teacher, the doctor and the judge, the minister and the architect," served the knowledge entrusted to them by the *Volk*. Such knowledge is "the most severe endangerment of *Dasein* in the midst of the overwhelming power of entities. The very questionableness of being, indeed, compels the people to work and fight and forces it into its state, to which the professions belong." [SDU: 16/477]

B. THE POLITICAL SPEECHES ON THE NATURE OF WORK AND THE WORKERS' STATE

Taking seriously the Nazi slogan "Work Makes Free" (*Arbeit Macht Frei*), Heidegger insisted that Freiburg University students avoid the dangers of elitist intellectualism by learning the value of manual labor. Such work-service would create "the basic experience of the hardness, nearness to the soil and tools, the lawfulness and strictness of the most simple bodily and thereby essential work in the group."[10] Rejecting the "cultured man's" effete concept that only "spiritual creators" do real work, Heidegger insisted that "every work is *spiritual as work*."[11] He sought to forge a bridge between the work of the "fist" and the work of the "brow."[12] This bridge was now possible because the revolution ended the misuse of "science" as an instrument misused by the privileged in order to "exploit" (*ausbeuten*) working people. Essentially, he argued, "the knowing of pure science is absolutely not distinguished from the knowing of the peasant, the lumberjack, the earth- and mine-worker, the hand-worker."[13] Heidegger's proximity to National Socialist views becomes crystal clear when we compare the preceding sentence to one from Hitler's *Mein Kampf*: "The blacksmith stands again at his anvil, the farmer walks behind his plough, and the scholar sits in his study, all with the same effort and the same devotion to their duty."[14] One critic has suggested that Heidegger, by emphasizing the proximity of ditch-digging and philosophy, was protecting his own academic position from those Nazis who criticized intellectual "parasites" in the new "workers' state."[15] Nevertheless, Heidegger was almost certainly sincere in his affirmation of the spiritual character of all work.

His criticism of the degradation of work and worker in the industrial plants of Germany of 1933 was in some ways consistent with Nazi propaganda, which almost demonized technology and which affirmed the importance of the human-scale workshops that promoted the well-being of the *Volk*. In 1933, this emphasis on small industry and agriculture may have retained at least a superficial measure of plausibility, since even six years later one-third of all workers labored in shops of ten employees or less.[16] Yet in 1933–34 Heidegger did not make a careful analysis of the actual character of work in Germany. Emphasizing the non-industrial "worker," the farmer, day laborer, and artisan, he did not explain how the industrial worker could be brought into the domain of authentic work. He had no clear program for the economic transformation of Germany, for such matters were presumably left up to political administrators who would enact the vision of spiritual leaders such as Heidegger.

In a speech to six hundred unemployed workers gathered together in an auditorium at the University of Freiburg in February, 1934, Rektor Heidegger proclaimed that he knew "whither urbanization [*Verstädterung*] has brought the German man" and that only his return "to soil and land" would renew the German community. Cities were the home of industry, exploitation, dehumanization, and class conflict.[17] Apparently for Heidegger, only the hard stock of the peasantry, grounded in their homeland, could provide the model necessary for German "manhood" to save itself from the evils of modernity. It is worth recalling here that Heidegger's grandfather was a shoemaker, his father a part-time cooper, and his mother a peasant. His overt concern about the fate of these simple and noble ways of life suggests to some critics that Heidegger's political views were linked with those of the conservative agrarian movement of the early twentieth century.[18]

Nowhere are Heidegger's appreciation of down-home life and his contempt for urban life clearer than in his radio address of March 7, 1934, "Why Do We Stay in the Provinces?", in which he explained his decision to turn down yet a second invitation to accept the chair of philosophy at Berlin. Speaking from his mountain hut near Todtnauberg, he compared his own philosophical work both to the interplay of natural forces (mountain brooks, primeval rock, winter snowstorms) and to the stern simplicity of the peasantry:

The effort of molding something into language [*sprachlichen Prägung*] is like the resistance of the towering firs against the storm. [. . .] This philosophical work does not take its course like the aloof studies of some eccentric. It belongs right here in the midst of the peasants' work. When the young farmboy drags his heavy sled up the slope and guides it, piled high with beech logs, down the dangerous descent to his house, when the herdsman, lost in thought and slow of step, drives his cattle up the slope, when the farmer in his shed gets the countless shingles ready for his roof, *my work is of the same sort. It is intimately rooted in and related to the life of the peasants.*

. . . The inner relationship of my own work to the Black Forest and its people comes from a centuries-long and irreplaceable rootedness in the Alemannian-Swabian soil.[19]

This romantic depiction of the nature of authentic work and social reality was quite consistent with Nazi propaganda about the need to restore the primal values of the peasants rooted in the land. Indeed, the Nazi "revolution" seemed to promise an alternative to the *Volk*-destroying industrial labor practices. Heidegger claimed that authentic work, which could take place only far from the newspapers and idle chatter of the big city, involved simplicity, devotedness, endurance, courage, and a capacity for disclosing appropriately the thing being worked upon. In his radio address, as elsewhere, however, Heidegger distinguished between his view of "rootedness" and that of incompetent, city-dwelling Nazi ideologues. Their talk of "*Volk*-character" and "rootedness in the soil" was out of tune with peasant existence. The best thing city people could do would be to stay far away from the peasant country, instead of invading it for skiing holidays that "destroy more in one evening than centuries of scholarly teaching about *Volk*-character and folklore could ever hope to promote."[20]

Heidegger retained his antipathy toward urban life long after the 1930s. In 1961, for example, in a lecture celebrating the seven hundredth anniversary of his hometown, Messkirch, he mused that only the small town could preserve a sense of "homeland" in a world suffering from the alienating power of modern technology:

> Homeland is most possible and effective where the powers of nature around us and the remnants of historical tradition remain together, where an origin and an ancient, nourished style of human existence holds sway. Today for this decisive task perhaps only the rural counties and small towns are competent—if they recognize anew their unusual qualities, if they know how to draw the boundaries between themselves and life in the large cities and the gigantic areas of modern industrial complexes. . . . [MHZ: 38–39/234]

C. HEIDEGGER'S APPROPRIATION OF JÜNGER'S MASCULINIST RHETORIC

Heidegger's belief that the New Reich would restore the German *Volk* to its homeland was diametrically opposed to Jünger's belief that such nationalism would provoke the world wars needed to erase all homelands and to turn the entire planet into a unitary technological plant. Despite rejecting Jünger's vision of the workers' state, Heidegger nevertheless drank deeply from Jünger's intoxicating literary and political cup, as we can see in the following typical passage from one of Heidegger's lectures (November, 1933) in support of National Socialism:

> [The new German student] enrolls himself consciously in the worker front. Follow-ership wins comradeship, which educates those nameless and unofficial leaders, who do more because they bear and sacrifice more; they carry the individuals out beyond themselves and *stamp on them the imprint of a wholly proper stamp of young manhood [Jungmannschaft]* [my emphasis]. With the new reality, the essence of work and of the workers has also changed. *The essence of work now determines the Dasein of man from the ground up* [my emphasis]. The state is the self-forming articulation [*das . . . sich gestaltende Gefüge*], in work and as work, of the *völkisch* Dasein. *The National Socialist state is the state of work.* And because the new student knows himself joined for the carrying out of the *völkisch* claim to knowing, accord-ingly he is the worker; he studies, because he is the worker; and he marches into the new order of the national Dasein and its *völkisch* knowing such that he himself must co-shape [*mitgestalten*] on his part this new ordering. The immatriculation is no longer the mere admission into a present-at-hand corporation; it becomes decision [*Entscheidung*], and every pure decision displaces itself into acting within a determinate situation and environment.[21]

In this passage, we encounter Jünger's vocabulary of decision, will, work, sacrifice, youth, radical change, *Gestalt*, and stamping. Let us examine briefly how Heidegger appropriated this vocabulary in his attempt to find a "third way" between the unacceptable alternatives of industrial communism and capitalism.

Will and Submission: Jünger claimed that there was no way for twentieth-century humanity to avoid being "stamped" by the *Gestalt* of the worker, but Heidegger for a time believed that a resolute German *Volk* could "co-*gestalt*" itself in a way that provided a radical alternative to modernism and industrial-ism. In his Rektor's address, he proclaimed that unless the German people willed to create a spiritual world for itself, then the joints of the world would finally collapse, suffocating everything in madness. [RA: 19/479–480] What was required of Germany was to make "the constant decision between the will to greatness and a letting things happen that means decline. . . . " [RA: 14/475] The martial flavor of the speech can be discerned in Heidegger's use of a phrase from the military theorist von Clausewitz: "I take leave of the frivolous hope of salvation by the hand of accident." [RA: 18/479] Heidegger's "decisionism" resonated not only with Jünger's writings, but also with those of the leading Nazi jurist, Carl Schmitt, and the ideologue Alfred Rosenberg. Like Heidegger, Rosenberg scorned the Weimar weaklings who claimed pitifully

> that they wanted the best, only circumstances were too difficult to achieve it. But they forget or do not want to admit that they did not really *will* it. Perhaps they

wished, perhaps they whispered something to each other in private; but in the moment in which they did not kindle the will to power in order to replace this ostensible willing with action, weak phrases were all that remained. There was no glorification of the creative power of a man or of a people, which in the desire to create something new storms against all obstacles, but there were only rootless abstract conceptions like humanity, the brotherhood of man, and other nice things.[22]

While Jünger urged Germans to will their inevitable technological destiny, Heidegger urged them to choose an alternative. This "choosing," however, involved submitting to Hitler's decision: "Not doctrinal propositions and 'ideas' are the rule of your being. The *Führer* himself and only he is the contemporary and future German reality and your law."[23] Heidegger insisted, of course, that Hitler's will was not merely arbitrary, but instead attuned to "the distant command [*Verfügung*] of the incipience [*Anfang*] of our spiritual-historical *Dasein*." [RA: 16/477] Heidegger tempered his call for submission by arguing, "All leading must grant the body of followers its own strength. All following, however, bears resistance within itself. This essential opposition of leading and following must not be obscured, let alone eliminated." [RA: 18/479] To be authentic, such "resistance" had to be rooted in a profound vision of the genuine possibilities of the German *Volk*. Such a vision evidently could not be found in the works of most intellectuals, who were presumably blind to what Heidegger had been "chosen" to see. In emphasizing the contribution to be made by followers, he was presumably staking out his own claim to insight into Germany's higher mission, which was being overlooked by those who purported to be National Socialist "philosophers."

Despite his occasionally melancholy reflections about the destruction of the old world, Jünger urged not only Germans but all European peoples to submit to the inevitability of what was facing them. For Jünger, submission to the inexorable historical process of total industrialization could give that process meaning, just as submission to the gigantic process of material warfare gave meaning to the Great War for soldiers such as Jünger. In such submission, whether on the battlefield or in the factory, the authentic worker-soldier could see beneath the horrifying spectacle of industrial warfare, into the eternal process of the Will to Power lurking beneath. In calling for an alternative to Jünger's vision, Heidegger used much of Jünger's language, in part because he was appealing to the sensibilities of the same two generations: those who had fought the Great War, and those disaffected youths who had come of age right after it. These people appreciated the martial language of hardness, violence, resolve, danger, victory—and defeat. Note the extraordinary parallels to Jünger's language of sacrifice in the following "Call to Work Service" issued by Heidegger on January 23, 1934:

> The lame, the lazy, and the half-hearted will "go" into work service, because staying away might threaten their prospects for examinations and positions. The strong and unbroken, who carry out their *Dasein* from the exciting mystery of a new future of our *Volk*, are proud of the fact that hardness gets demanded

[*abverlangt*] of them; for that is the moment of truth [*Augenblick*] in which they raise themselves up to the hardest tasks, for which there is neither reward nor praise, but only the "blessedness" ["*Beglückung*"] or readiness for sacrifice and service in the sphere of the innermost necessities of German being [*Sein*].[24]

Like Jünger, Heidegger also appealed to wartime experience in the trenches to explain authentic *Gemeinschaft* in the coming age. Genuine community was formed only when each individual bound himself to what is higher than either individual or community. Calling on *Being and Time*'s vocabulary of authentic being-toward-death, Heidegger claimed that the comradeship of the front soldiers arose because

> the nearness of death as a sacrifice stood before each one in the same nullity, so that this could become the source of unconditioned belonging-together. Precisely death, which each individual man must die for himself, which individualizes each individual in the uttermost way, precisely death and the readiness for its sacrifice creates first the space of community, from which comradeship arises. [GA, 39: 73]

The Revolutionary Generation: Like Jünger, Heidegger had high expectations of the generation of the Great War. Born in the same year as Hitler (1889), Heidegger believed that they were chosen to erect new standards to replace the materialism of their parents. Some historians argue that National Socialism can be understood as a generational phenomenon.[25] Hitler had an extraordinary appeal among young people, especially those in universities, where National Socialism enjoyed great support. Heidegger was much influenced by the generational conflict, despite saying that *Being and Time* was not the product of the revolutionary mood of today's youth. [GA, 30: 101–105] While Rektor, Heidegger sought to further the National Socialist transformation of the university by involving students and younger faculty members in university administration. He also organized what he called a "*Todtnauberg Lager*," an outdoor seminar to which he marched his students and lecturers for the purpose of discussing how to change higher education in light of the new Heideggerean attitude toward science. [DR: 35ff./495ff.] By 1941, and perhaps long before, Heidegger had become disillusioned with contemporary youth. In lectures from that year, he stated, "The possession of knowledge which today's youth brings with it, corresponds neither to the greatness nor to the seriousness of the tasks." [GA, 51: 13] Before he became disillusioned, however, he called the young generation to sacrifice itself for the new reality that was dawning. On November 27, 1933, for example, he is reported to have said that in the new German reality, "the German student [enters] into a new, although unbloody sacrifice under the image of Langemarck [a bloody World War I battle that had become a symbol of the German's manly willingness to sacrifice himself for the *Volk*]. The new German student now goes through *work-service*; he stands by the S.A."[26]

Nationalism and Heroism: During the 1920s, Jünger wrote essays for right-wing journals, in which he praised the heroic deeds of those who sacrificed themselves in the Great War in order to renew the German spirit. Heidegger had his own version of this martial and nationalistic spirit. On May 27, 1933, he spoke in commemoration of the tenth anniversary of the death of Leo Schlageter, a decorated young veteran and later a member of the *Freikorps*. The highly patriotic Schlageter, who as a martyr became a symbol for German resistance to the treaty of Versailles, was executed by the French for trying to blow up railway tracks in French-occupied Rhineland. Heidegger noted that Schlageter was forced to face the firing squad alone; to endure this blow, he had to place "before his soul the image of the future outbreak of the *Volk* to its nobility and greatness, and die in belief in that. Whence this hardness of will [*Haerte des Willens*], to endure that which is most difficult? . . . Freiburg student! Let the strength of the homeland mountains [in the *Schwarzwald*] of this hero stream in your will!"[27]

Schlageter, a hero from Heidegger's own Swabian homeland, embodied the manly virtues of courage, clarity, and will required in Germany's hour of need. Schlageter derived his courage from being rooted in the mountains and sun, streams and valleys of his homeland. *Bodenständigkeit*, rootedness in the soil, was widely prized by reactionaries of all stripes and nationalities during the early twentieth century. Rektor Heidegger challenged the rootless, urbaninfluenced, bookish German student to rise above mere scholarship and grade-grubbing. He proclaimed that studying must once again become a "venture," not a "shelter" for the timid. The struggle required to carry out the German revolution "brings about the *complete transformation of our German Dasein*."[28] Such a transformation implies the emergence of a new kind of humanity, a new "breed," akin to the *Typus* of the worker-soldier spoken of by Jünger. Heidegger proclaimed on July 1, 1933,

> Whoever does not endure the struggle, remains left behind. The new courage must accustom itself to steadiness, because the struggle for the educational places of the leaders will last a long time. It gets fought from the forces of the New Reich, which the *Volk*-Chancellor Hitler will bring to reality. A hard stock [*Geschlecht*] without the thought for itself must be equal to it; this stock lives from constant testing and for the goal which it assigns for itself.[29]

It is disturbing to read what Heidegger says about the need to leave behind the weak, the halfhearted, the lame, and others who stood in the way of the revolution. Even more disturbing is how he used the vocabulary of hardness, resoluteness, and courage to describe those higher types capable of walking ahead while others fall (or are pushed) to the side. The same vocabulary was used later by Goebbels and the S.S. to describe the character of the "heroic" men who were called on to perform the "terribly hard" tasks necessary for the completion of the German revolution. Of course, these tasks included rounding up into concentration camps, and later killing, millions of Jews and other "un-German" types. Heidegger apparently did not directly support such measures, but he was certainly aware of the concentration camps for political prisoners (including Social Democratic legislators and officials) that were set up near Freiburg shortly after Hitler's accession to power in 1933.

Although his political speeches were profoundly shaped by Jünger's masculinist and violent language, Heidegger hoped that National Socialism would provide an alternative to the world envisioned by Jünger. From 1934 on, Heidegger became increasingly concerned that the historical movement of National Socialism was headed in the wrong direction, toward a strengthening, not a surmounting, of the technological era. Moreover, he also began questioning whether his own vocabulary of hardness and resoluteness was influenced more than externally by the technological Will to Power. In his lectures on Hölderlin in 1934–35, for example, he spoke of the need for the German *Volk* to be opened up by the mood of holy mourning, to become receptive to the affliction of living in a world from which the gods had fled. Heidegger's meditations on Hölderlin eventually led to the idea that only through "releasement," not through resolute will, could humanity be freed from its technological destiny. Throughout the 1930s, Heidegger's reflections on technology involved a constant exploratory movement back and forth— from Jünger and Nietzsche, on the one hand, to Hölderlin, on the other. Jünger described modern technology better than anyone else, but took his ideas about technology as an aesthetic phenomenon from Nietzsche. Nietzsche's doctrine of art as form-giving activity that restores weight and meaning to life resonated with Heidegger's conviction that only a great work of art could save Germany from the leveling effects of the one-dimensional technological mode of working and producing. Hölderlin, in contrast with the voluntaristic Jünger and Nietzsche, spoke of art not so much as the activity of giving shape, but instead as the drawing out of forms that are somehow already there. We shall postpone consideration of Hölderlin, however, until we complete our study of how Heidegger's mature conception of technology was shaped by his meditation on the writings of Jünger and Nietzsche.

Chapter 6. Jünger's Thought in Heidegger's Mature Concept of Technology

As we have seen, in the first phase of his confrontation with Jünger, Heidegger appropriated Jünger's language in order to support a revolutionary movement which heralded an alternative to the technological future forecast by Jünger. After the Röhm purge on June 30, 1934, Heidegger began the long process of distancing himself from the "political reality" of National Socialism, but continued to meditate on its "inner truth and greatness." This meditation, which involved a turn toward art, was carried out in his lectures on Nietzsche and on Hölderlin. Nietzsche's views on the world-shaping powers of art offered a way of understanding the metaphysical basis for Jünger's doctrine of the *Gestalt* of the worker, while Hölderlin's poetry seemed to offer a saving alternative to Jünger's technological future. In this chapter, we begin to study Heidegger's artistic "turn" in connection with his confrontation with Jünger's "aesthetic" view of modern technology. Here, aiming to disclose how Jünger's writings influenced Heidegger's mature concept of technology, we shall move through the 1930s and 1940s into Heidegger's postwar writings on technology. In chapters seven and eight, we shall move back to 1934–35 to consider in more detail how Heidegger's reflections on Nietzsche and Hölderlin led to new insights about the relationship between art and *techne*. I divide the analysis in this way to facilitate exposition of material which is in fact highly interrelated and tangled. The reader should keep in mind that in his search for a new, non-technological world, Heidegger was constantly playing off Jünger's views on art and technology against those of Nietzsche and Hölderlin.

A. CHANNELING THE PRIMAL SURGE

Before 1934, Heidegger mentioned art infrequently, although even in 1919 he spoke of the poet as a *Gestalter*. More important than the artist, he thought at first, was the thinker who renewed a people's understanding of the being of entities. Beginning in 1934, however, Heidegger elevated the artist—particularly the poet—to a level perhaps even higher than the thinker. A great thinker had to work in connection with a great artist who pointed beyond the present age. Hence, he remarked in 1937 that the basic behavior and moods of a historical people "win their governing *Gestalt* and their beguiling force in great poetry, the imaging arts, and in the essential thinking (philosophy) of a *Volk*."[1] Only great poetry could stem the tide of the technological nihilism, in the face of which *Kultur* and Christendom were impotent: "Christendom is just as nihilistic as Bolshevism and thus even as mere socialism." [GA, 43: 31] Heidegger's idea of the world-saving power of art arose partly in response to Jünger's aesthetic interpretation of the world-shaping Will to Power.

We recall that for Jünger aesthetic experience enabled one to encounter the horrifying dimension of the eternal Will to Power at work in all things. At times, Heidegger spoke of his own work as an encounter with the horrifying, an encounter that required a "bearing" reminiscent of the aesthetic attitude. In the 1930s, he said:

> The basic mood of *the* philosophy, i.e., the *future* philosophy, we call . . . *restraint* [*Verhaltenheit*]. Primordially unified and belonging-together in it are: the *horror* [*Erschrecken*] before this closest and most obtrusive [fact], that the entity is, and at the same time the dread [*Scheu*] before the furthest, that in the entity and before every entity being [*Seyn*] presences. Restraint is that mood in which that horror is not overcome and put aside, but instead is preserved and guarded through dread. [GA, 45: 2]

Just as Jünger glimpsed the eternal Will in the gruesome character of technological warfare, so Heidegger apprehended the self-concealing being of entities in the horrifying meaninglessness of entities in the technological era. Total mobilization had eliminated rank-ordering and distinction; everything was reduced to the same undifferentiated raw material for industrial production. Germany, Heidegger believed, was about to be swept into oblivion by the flood-tide of technological beasts from East and West. In *Male Fantasies*, Klaus Theweleit has inquired into the psychological implications of the reactionary image of the communist tide threatening to flood Germany. A typical expression of this fear is found in the statement made in 1918 by W. von Oertzen: "The wave of Bolshevism surged onward, threatening not only to swallow up the republics of Estonia and Latvia, neither of which had yet awakened to a life of its own, but also to inundate the eastern border of Germany."[2] Theweleit claims that terror of being drowned in the "Red tide of communism" was linked to the fears of reactionary German men that the menstrual tide of the unclean, indeterminate, and boundless female would overwhelm their rigid, masculinist ego-boundaries. In other words, such men projected onto communism their own fears of the female.

To compensate for such fears, many postwar Germans—especially members of the renegade *Freikorps*—adopted Jünger's language of hardness, courage, and manliness. Paradoxically, however, the image of the all-destroying flood was also attractive to some reactionaries because they longed for the burial of the bankrupt bourgeois world. Jünger's book *The Storm of Steel* expressed both the positive and the negative dimensions of the elementary natural floods and currents at work in the Great War and in world history. Such floods may overwhelm the individual, but they also wipe away the old order and make way for a new, "higher" individual. For reactionary thinkers such as Jünger, then, the key was not to get rid of the flood, but instead to channel it. Theweleit states, "[Our soldiers] want to stand with both feet and every root firmly anchored in the soil. They want whatever floods may come to rebound against them; they want to stop, and dam up, those floods."[3]

Heidegger used very similar vocabulary in his discussion of the founding involved in bringing forth a work of art. Consider these remarks from 1935 about the struggle to found a world: "The battle is then sustained by the creators, poets, thinkers, statesmen. Against the overwhelming force [*Walten*] they set the barrier of their work to capture in this the world opened up thereby. It is with these works that the elemental power, *physis* [presencing, being] comes to stand. Only now does the entity become entity as such." [GA, 40: 66/51]

Humanity is capable of apprehending (*vernehmen*) entities *as* entities only because humanity can delimit and thus bring to a stand their overpowering presencing. The *martial* element of this account of human existence is evident from the following passage: "To apprehend in this twofold sense [of both accepting and determining] means to let something come to one, not merely accepting it, however, but taking a receptive attitude toward that which shows itself. *When troops prepare to receive the enemy, it is in the hope of stopping him at the very least, of bringing him to stand* [*zum Stand bringen*]." [GA, 40: 146–147/ 116; my emphasis]

While Jünger conceived of this power as the primal Will, Heidegger conceived of it as the overwhelming being or presencing of entities. Humanity, so he believed, was *compelled* by being to bring being to a stand, to limit and bound being so that entities could manifest themselves in determinate ways. Jünger took from Nietzsche the idea that the world is a self-producing work of art, a dramatic spectacle which is brought forth through and for the Will to Power. While Heidegger criticized Jünger's claim that this spectacle was a "symbol" of an eternal "absolute reality," Heidegger nevertheless spoke as if that which was of ultimate importance was self-concealing being, not the entities which appeared by virtue of that being. Becoming fixated on entities meant that one had become oblivious to the transcendent. To become open once more to the overwhelming power of transcendent being was the great risk taken by world-founding poets and thinkers, crucial players in the vast ontological "game" in which being both revealed and concealed itself. Through the poet spoke the founding word of being: "The word, the name, restores the emerging entity from the immediate, overpowering surge [*Andrang*] to its being and maintains it in this openness, delimitation, and permanence. . . . Pristine speech opens up the being of the entity in the structure of its collectedness." [GA, 40: 180–181/144]

Only a great artist could bring forth the work needed to delimit the flood or the surge of being. Like Nietzsche, Heidegger regarded art as being worth more than truth, if we understood "truth" in its ordinary meaning as a body of correct propositions, while defining "art" as the event of ontological disclosiveness (*aletheia*). While the work of art opened up new ways for entities to manifest themselves, "truth" petrified that manifesting into something fixed and useful for humans. Sounding like Jünger, Heidegger claimed that the work of art "stamped" the emerging world with a new *Gestalt*, one which disclosed entities in a new way. Heidegger's later notion that in the technological era humanity is "challenged forth" by *Gestell*, i.e., compelled to treat itself and all other entities as "standing-reserve" (*Bestand*) for total mobilization, was analogous to Jünger's notion that the *Gestalt* of the worker compelled humanity to mobilize itself and the earth as standing-reserve for enhancing the technological project of total control. In 1933–34, Heidegger agreed that the new world would be one of work, but by "work" he meant not industrial labor, but instead the activity of "letting things be" according to the authentic possibilities both of the *Volk* and of the things to be disclosed. Increasingly, however, Heidegger came to see that the National Socialist "revolution" would not open a new world, but instead would pave the way for Jünger's technological one.

B. HEIDEGGER'S TRANSFORMATION OF JÜNGER'S CONCEPT OF THE *GESTALT* OF THE WORKER

Heidegger's doctrine of the "history of being" has certain similarities to Jünger's idea that world history is constituted by various manifestations of the Will to Power. Heidegger argued that during the course of Western history being "gave" itself in such a way that it also concealed itself. It gave itself in an increasingly fixed and thing-like way, as "beingness." Metaphysics, oblivious to the sheer event of presencing, focused on beingness, the permanently present ground or foundation for things. Metaphysicians gave various names to this foundation, such as *eidos, ousia, energeia, actualitas, actus purus,* Creator, self-certain subject, and Will to Power. Beingness "stamps" the people and things of a particular era in a certain way, i.e., it determines the horizon in which they can show themselves.

Neither Heidegger's "beingness" nor Jünger's *Gestalt* is an entity, but instead they constitute the ontological "field" that discloses and gathers entities in particular ways. For example, people living *within* the world mobilized by what Jünger called the *Gestalt* of the worker can conceive of "cause and effect" only as principles of explanation. But both Heidegger and Jünger claimed that cause-and-effect explanations were inadequate for understanding either the emergence of a historical *Gestalten* or the way in which they held sway in a particular epoch. Heidegger argued that technological humanity's fascination with cause-and-effect explanations demonstrates that people are now governed by the *Gestalt* of *ratio*—calculating rationality—which commands that everything give a reason for itself to the tribunal of the subject's self-certain consciousness. Gripped by the technological *Gestalt*, humanity overlooks the originary event (*Anfang*), the ontological "movement" of "jointing and fateful dispensing" (*Fügung und Schickung*) which imprint each epoch of Western history with its own mode of beingness. Later Heidegger concluded that the task for thinking is to recall the event of appropriation (*Ereignis*) which "gives" both beingness and human existence. Above all, he argued, we must conceive of joining and dispensing not in causal terms but as the disclosive events which first make it possible for things to be conceivable in cause-and-effect terms. [GA, 52: 100]

Heidegger maintained that something can "be" without being the product of a cause. In the technological era, of course, "to be" means "to be produced" according to natural or technical causes understandable by reason. For Heidegger, however, "to be" meant to be manifest—a disclosive event that was prior to and more basic than the entities disclosed through it. Required for such disclosure is both the self-manifesting (presencing) of entities and the temporal clearing or horizon (human existence) in which such self-manifesting could occur. Neither presencing nor clearing can be understood in terms of categories applicable only to entities. That which does not appear is more primordial than what does appear. Presencing and clearing articulate, order, shape, or "joint" the spatio-temporal-causal world of a particular epoch. In English, we say that "things are out of joint" when the organizing principle of the world begins to give way. For things to be "in joint," then, means that the organizing principle—the work of art, the world axis, the totem—maintains itself and articulates a world in which things can display themselves according to appropriate rank, order, and limits.

At first glance, it would appear that Heidegger agreed with Jünger's view that the *Gestalt* is a transcendental, non-empirical ordering principle which "stamps" or "joints" each particular epoch in a certain way. Heidegger argued, however, that Jünger's *Gestalt* never escaped from the limits of the metaphysical understanding of being:

> Ernst Jünger's work *Der Arbeiter* is important because it, in another way than Spengler, achieves what all the Nietzsche literature was up to now unable to achieve, namely, to communicate an experience of the entity and of how it is, in the light of Nietzsche's project of the entity as Will to Power. To be sure, Nietzsche's metaphysics is by no means conceived in a thoughtful way [*denkerisch begriffen*]; on the contrary, instead of being questionable, in the true sense, this metaphysics becomes self-evident and apparently superfluous. [ZSF: 42–45]

Trapped within Nietzsche's metaphysics, Jünger conceived of the *Gestalt* of the worker in terms of a certain kind of humanity. He spoke as if the *Gestalt* of the worker forged together in humanity the calculating, steely powers of the machine and the atavistic, passionate energies of the Will to Power at work in all life. As we have seen, however, Heidegger believed that this view of humanity as half-animal, half-rational was the final stage of the decline of Aristotle's doctrine of the "rational animal" in Nietzsche's "blond beast" who would dominate the earth with modern technology.

While thinking in such anthropocentric and subjectivistic terms, however, Jünger also began to think of technology in terms of something even more radical: subjecthood (*Subjektität*). According to Heidegger, "subjecthood" describes human existence which discloses things in such a way that the subject-object distinction is virtually overcome in the uniformity of standing-reserve. While Jünger was moving in the right direction, he was not able to clarify the relation among *Gestalt*, work, and worker. He defined "work" as the "representation" of the *Gestalt* of the worker, but in what sense is the *Gestalt* a "representation" of anything? Does it itself have the characteristic of work? If not, how does it gain specificity as the *Gestalt* of the worker? Heidegger argued that Jünger would do well to understand "work" as a "phase of the appearance of the being of the entity." [ZSF: 63] The *Gestalt* of the worker is a misnomer, if we understand thereby that worker-humanity owns or gives rise to itself as such a *Gestalt*. To describe the technological era, Heidegger implied that it would be more accurate to speak of the *Gestalt* of *work*, i.e., the final phase of the productionist-oriented history of being, a history which began with Plato's doctrine of the *eidos*, the permanently present ontological model or blueprint which makes an "imprint" on things. In 1955 Heidegger wrote to Jünger: "You also think of the relationship of *Gestalt* to that which it 'forms' as the relationship of stamp and impression [*Stempel und Prägung*]. To be sure, you understand the stamping in the modern sense as bestowing 'meaning' on the meaning-less. *Gestalt* is the 'source of meaning-bestowing.' . . . " [ZSF: 53]

Gestalt, then, turns out to be yet another "master-name" in the history of metaphysics, another name not for being but for the "beingness" that grounds entities. From Heidegger's viewpoint, Jünger spoke as if technological humanity transformed itself into the meaning-giver and ground-bestower to all things. Heidegger maintained, by way of contrast, that humanity has been transformed into the worker because today "to be" means "to be worked upon and transformed in accordance with the imperative of production for its own sake." Instead of being an expression of humanity, then, the *Gestalt* of the worker names the constricted "presencing" of entities (including humanity) as standing-reserve. Heidegger asked: "Does the essence of *Gestalt* arise in the area of origin of what I call *Ge-stell*? . . . Or is the *Ge-stell* only a function of the human *Gestalt*? If this were the case, then the *presencing* [*Wesen*] of being and the being of the entity would be completely the making of human conception." [ZSF: 62/63] Heidegger, of course, rejected the idea that humanity is the foundation for being, since that idea itself is the ultimate, though inverted, version of the metaphysical notion that being is the basis (ontological, creative, causal, technological) for entities. Heidegger agreed with Jünger that the history of these historical epochs is bound up with the history of the Will to Power. Missing from Jünger's scheme, in Heidegger's view, was insight into the primordial ontological character of the fluid "motion" of the synchronic event of presencing (*Anwesen*), which "hardens" itself diachronically into particular historical modes of being present (*Anwesenheit*, beingness).

Heidegger was not entirely fair to Jünger. While Jünger did not understand the ontological difference, and had thus been forced to think within the limitations of Nietzsche's metaphysics, Jünger did not conceive of the *Gestalt* of the worker as a product or function of humanity, but instead conceived of humanity more or less as a pawn in the eternal playing of the Will to Power. Whatever Jünger meant by the Will to Power, and his conception of it was certainly obscure, he did not conceive of it as on the same plane as the causal relations which obtain at the level of empirical reality. The *Gestalt* of the worker was not itself a "cause," but instead the illuminating principle in light of which entities could appear as nothing but standing-reserve for total mobilization. Jünger's concept of the Will to Power and its historical *Gestalten* cannot be so easily inserted into Heidegger's conception of the history of being. Nor can Jünger's conception of Nietzsche's thinking be so readily fitted into Heidegger's scheme.

The fact is that Heidegger's conception of modern technology—not only in terms of its empirical-historical symptoms, but also in terms of its ontological dimension—amounts almost to an appropriation of Jünger's thinking. In regard to many thinkers to whom Heidegger was greatly indebted, he had the tendency either to discount their influence upon him or to show that he had thought more deeply than they had about a given topic. His treatment of Jünger was no exception in this respect.

While Heidegger spoke of the history of being, and Jünger of the history of the Will to Power, both believed that the "multifarious transformations" assumed by being or the Will to Power in different epochs presented "the heroic spirit with an engrossing drama." Both thinkers believed that they were called on to bear witness to the "play" of those transformations, while ordinary people were blind to that play and were thus "stamped" to act according to the imperatives demanded by the mode of being governing the age of the worker. Jünger and Heidegger appreciated Nietzsche's doctrine of the importance of rank and ordering among people. At first, Heidegger said that great artists, thinkers, and statesmen are the vessels through which these world-governing modes of being first "stamp" themselves on humanity. By the late 1930s, however, he no longer included politicians (such as Hitler) in the world-founding class. Instead, a people's basic mode of comportment is to be determined by the poet and the thinker. The thinker's task is "to conceive the metaphysical essence of technology and thereby to carry it out—as one form of the directing of entities—into its possible *Gestalten*."[4]

C. HEIDEGGER'S CHANGING ATTITUDE TOWARD JÜNGER'S VISION OF THE FUTURE

Increasingly convinced that Jünger's vision would be fulfilled after all, Heidegger could not avert his eyes from the horrific "presencing" of entities in the technological era. Like the great Greek thinkers he so admired, Heidegger believed that he too had to keep his gaze fixed on the horror of the sheer fact that entities *are*. Most men avert their gaze from the overpowering presencing of things, which reminds them of their own ultimate groundlessness, and which also threatens constantly to overturn every historical project or community. Preferring security to the courageous venture with truth, most people focus on how to gain some control over entities that appear *within* the given historical world. Such worlds, however, are constituted by those daring individuals who are summoned and claimed by presencing (being) itself as the site for a new disclosure of entities as entities. Such a disclosure calls into question the previous world and inevitably involves destruction and enormous risk. Heidegger quoted Nietzsche's phrase: "Philosophy . . . is a voluntary living amid ice and mountain heights." [GA, 40: 15/11]

In 1934, he waxed eloquent about the importance of philosophy: "A historical *Volk* without philosophy is like an eagle without the high expanse of the glowing ether, wherein its soaring [*Schwingen*] attains its purest flight [*Schwung*]." [GA, 45: 2] Philosophy was not a merely practical affair which was supposed to "assist" the revolution, but instead was to *guide* it. In response to the propaganda minister who proclaimed that Germans needed no more poets and thinkers, but instead more "grain and oil" [GA, 54: 179], Heidegger replied,

> For the correct remedying of economic difficulties, for the correct improving and securing of the health of the *Volk*, our [ontological] reflections upon correctness and truth will indeed achieve nothing, not even for the correct augmentation of the velocity of airplane engines, not even for the correct improvement of radio receivers and just as little for the correct basis for the plan of instruction in the schools. Over against all these pressing issues of everyday "living" philosophy disappoints. . . . It is the *immediately useless* knowing—and yet still something else: the *dominant* knowing. [GA, 45: 30]

Heidegger claimed that only authentic thinking and poetry could save Germany in its hour of crisis. By "thinking," he did not mean rational calculation, but instead that mode of comportment which opened one up to the awesome and dreadful *presencing* of things, to the fact *that things are* at all. In 1933 he had emphasized that only the philosopher could hold himself out in the midst of the overpowering presencing of entities, so as "not to close [him]self off from the horror [*Schrecken*] of the unrestrained and the confusion of that which is dark."[5] Unfortunately, what the Nazis meant by the "unrestrained" and the "dark" was not the being of entities, but instead blood and instinct, frenzy and violence, domination of humanity and nature. As reactionary modernists, the Nazis united instinct with technology in a way which led to unparalleled devastation. During World War II, Heidegger remarked:

> Before being can occur [*ereignen*] in its original truth, being as the will must be broken, the world must be forced to collapse and the earth must be driven to desolation, and man to mere labor. Only after this decline does the abrupt dwelling of the origin take place for a long span of time. . . .
> The decline has already occurred. . . .
> The still hidden truth of being is withheld from metaphysical humanity. The laboring animal is left to the giddy whirl of its products so that it may tear itself to pieces and annihilate itself in empty nothingness. [VA I: 65/86–87]

Jünger, too, had argued that after humanity had surrendered to the *Gestalt* of the worker, there would occur a period of titanic destructiveness and productivity, which would be followed by a period of repose—a technologically perfect world which would ultimately give way to yet another form of the Will to Power. Despite his obvious antipathy toward what the technological era was doing to the world, and in opposition to his description of it as a period of "decline," Heidegger also argued that humanity's "abandonment" by being

ought to be construed not as something negative but instead as the final stage
in a historical process over which humanity could exercise little control. Not
without conviction, although somewhat in opposition to things he said elsewhere, Heidegger also insisted that he was no "enemy" of technology and that
it was not the "work of the devil." The technological epoch was both exceed-
ingly dangerous and full of hope, since the new beginning longed for by
Heidegger could arise only from out of a period of darkness and despair.

His conception of how long that period would last changed over the
years. In 1933, he believed that the new dawn had been initiated by Hitler;
perhaps things could be transformed in a few generations. In the middle of
World War II, he was still unsure whether the space-and-time-grabbing process of planetary domination served to prepare the *Lebensraum* appropriate to
technological humanity, or whether it would simply drive on to develop ever
newer possibilities for transforming space and time. [GA, 53: 59] If the former
were the case, the time of redemption might be nearer; if the latter, that time
might be put off for centuries. At times, Heidegger concluded that modern
humanity had become a "planetary adventurer" whose colonization of other
peoples and places was but a symptom of total uprootedness. [GA, 52: 180]
The time of transformation seemed to recede ever further into the distance.
Perhaps the technological age, he concluded, would last for several centuries.

As he realized that the "revolution" was not at hand, Heidegger sought to
achieve—like Jünger—a detached appreciation of the technological *Gestalt*,
the transitional stage to the new beginning. In this mood of acceptance, he
could assert defiantly that "only a dreamer and a fantasizer would be able to
deny that in the age now broken forth upon the whole earth, man as the
worker and soldier experiences the authentic being and provides what alone
should be valid as being." [GA, 51: 38] In 1941, he added,

> [Today], "*the worker*" and "*the soldier*" determine the view of reality in a thorough-
> going way. These two names are not meant as names for a class of people and an
> occupational guild; they characterize in a unique fusion the kind of humanity which is
> authoritatively taken into claim for carrying out the current world-convulsion, and
> gives direction and instruction for relationship to the entity. The names "worker" and
> "soldier" are thus metaphysical titles and name the human form of the carrying out of
> the being of the entity which has become manifest, which being *Nietzsche* has con-
> ceived pre-thinkingly as the "*Will to Power*." [GA, 51: 18][6]

Heidegger went on to say that in the age in which workers and soldiers
are "imprinting the basic structure of human making/creating" [GA, 51: 37–
38], "work (cf. Ernst Jünger, *Der Arbeiter*, 1932) is now reaching the metaphysical rank of the unconditional objectification of everything present which
is present in the Will to Will." [VA I: 64/85] "Stamped" by the *Gestalt* of the
worker-soldier, technological humanity worked differently than did previous
stages of humanity. "The field that the peasant formerly cultivated and set in
order appears different from how it did when to set in order still meant to take
care of and to maintain." [VA I: 14/14–15] Further, "The forester who measures the felled timber in the woods and who to all appearances walks the

forest path in the same way his grandfather did is today ordered by the industry that produces commercial woods, whether he knows it or not. He is made subordinate to the orderability of cellulose. . . . " [VA I: 17/18] Spengler had already said the same thing in 1932. In that same year, Jünger, too, asserted, "The farm worked with machines and fertilized with artificial nitrogen from factories is no longer the same farm." [DA: 176]

Jünger and Heidegger agreed that this transformation of work was closely associated with a radical alteration of language. Heidegger argued that the "instrumentalizing" of language in the technological era had turned it into an armament for the human domination of the world. This process did not have to be spoken of merely negatively. Jünger, for example, described this instrumental use of language in positive terms. Heidegger stated that Jünger had called forth, from within the same metaphysical sphere determined by Nietzsche's interpretation of being as the Will to Power, a "counter-movement which strives for a new 'instrumentizing' of language, in order to make of it the highest 'accuracy of fire.' " [GA, 52: 35] Such instrumentalized language could take the form of the various media of mass culture, including motion pictures, which "projected" the world as a "picture" to be dominated by the subject. According to Heidegger, Jünger likened language "to the 'film,' a form of armament through which the *Gestalt* of the worker dominates the world." [Ibid.]

It may seem odd, at first glance, to regard motion picture films as armament for dominating the world. For Heidegger, however, film and photography were instruments in the technological drive to make everything wholly present, unconcealed, available for use. Jünger himself was fascinated by photography. He recognized that it was not just a new technique of reproduction, but was an essential manifestation of the *Gestalt* of the worker, which sought to objectify everything by reducing all space and time to the immediate present. It was no accident that Heidegger entitled one of his essays on modern science and technology "The Age of the World-Picture" (*Die Zeit des Weltbildes*, 1938). This essay describes how scientific explanation "projects" a mathematical realm in terms of which a thing must be encountered if it is to be re-presentable as a genuine "object" (*Gegenstand*). What, then, is the modern "world-picture"? Ordinarily, by "picture" we mean a representation of something else; hence, the modern world-picture would be a representation of the world. Heidegger argued that for modern humanity, such a picture would be radically anthropocentric. The term "picture" must be understood as it works in the German expression, which may be translated literally as "We are in the picture," which may be rendered more freely as "We get the picture." To put oneself in the picture regarding something means setting the thing in place before oneself and fixing it before oneself precisely as set up in this way. In short, to "picture" something means to establish it as an object over against the viewing subject. Yet "We are in the picture" about something does not mean only that we set it before us or represent it to us, "but that what is stands before us—in all that belongs to it and all that stands together in it—as a system." [HW: 82/129] Being "in the picture" also means being acquainted with and ready for the things before us.

Where the world becomes picture, what is, in its entirety, is juxtaposed as that for which man is prepared and which, correspondingly, he therefore intends to bring before himself and have before himself, and consequently intends in a decisive sense to set in place before himself. . . . Hence world-picture, when understood essentially, does not mean a picture of the world but the world conceived and grasped as a picture. What is, in its entirety, is now taken in such a way that it first is in being and only is in being to the extent that it is set up by man, who represents and sets forth. [HW: 82–83/129–30]

There was no medieval "world-picture," because medieval humanity did not conceive of the world as a human representation, but instead as a creature of God. And the ancient Greeks did not think of themselves as subjects gazing upon projected objects, but instead as being gazed upon or looked at by the entities themselves. For technological humanity, however, to re-present something, to turn it into a picture, means forcing it back into the relationship to oneself as the normative realm: "Wherever this happens, man 'gets the picture' in precedence over whatever is. But if that man puts himself into the picture in this way, he puts himself into the scene, i.e., into the open sphere of that which is generally and publicly represented." [HW: 84/131–132] Assuming the role of the "representative" of everything that is, technological man for the first time in human history takes a "position." This position taken on things for the purpose of making them reveal themselves in ways satisfactory to the one doing the re-presenting is called, in the motion picture industry, the "point of view" of the camera. For Heidegger, television was an inevitable invention of the technological subject, for television makes it possible for virtually the entire human species to witness the same event from the same point of view at the same moment.

According to Heidegger, the publishing business is another instance of the technological will to make everything present. The publishing business, especially in its marketing dimension, develops plans and procedures that will best "bring the world into the picture for the public and confirm it publicly." [HW: 90–91/139] The proliferation of collections and sets of books, as well as pocket editions, coincides perfectly with the aim of researchers to make everything available to everyone. Mechanical reproduction of art works makes the achievements of the great masters available to everyone, but—as Walter Benjamin made clear in a famous essay—in such a way that the "aura" surrounding the work disappears entirely.[7] Heidegger described this "aura" as the capacity of a work of art to "organize" a world as the clearing in which human life can transpire and through which the earth can manifest itself. Nothing like this occurs either when looking at pictures in a museum or when browsing through a book of photographic reproductions of famous paintings.

The phenomenon of mass culture, which some have described as the "society of the spectacle," was for Jünger and Heidegger an essential feature of technological civilization.[8] Jünger had predicted that the masses would turn pilots and racecar drivers into cult figures who conquered space and time by means of the machinery of industrial technology. And he forecast that people would worship more willingly at the altar of electronic gadgets than at traditional temples. Heidegger regarded mass culture in more negative terms, asking in 1935, "What can it mean?" when millions gather together for torchlight parades or listen in rapt attention to what a boxer (Max Schmeling) has to say about the world situation. [GA, 40: 41/31] For Heidegger, mass culture was the result of public fascination with entities, a fascination that grew as the being of entities increasingly hid itself from humanity.

In their evaluations of mass culture, Heidegger and Jünger were both deeply affected by Nietzsche's doctrine of master and slave morality. Mass culture, i.e., the bourgeois world, was associated with slave morality, which yearned for comfort, security, anonymity, averageness, equality, and protection. As elitists, Heidegger and Jünger believed that the technological era could be carried to its completion only by an elite corps of humanity who scorned the cheerful optimism of mass culture. Jünger's "worker" was by no means equivalent to Marx's "proletariat"! Heidegger and Jünger looked to Nietzsche for insight into the remarkable men, the "overmen," needed to complete the process of nihilism. Jünger had spoken of his interpretation of such men in his collection of essays *The Adventurous Heart*. [AH] These men were willing to use violence in order to pursue the uncharted paths being opened up by modern technology.

Regarding his attitude toward Jünger's technological Overman, Heidegger's attitude was ambivalent. On the one hand, the Overman as technological "adventurer" and destroyer was a manifestation of the nihilism resulting from the total self-concealment of being. On the other hand, the Overman as the one who went "over and beyond" existing humanity could be, paradoxically, a harbinger of the new, post-nihilistic world heralded by Hölderlin. Speaking of the former kind of Overman, Heidegger remarked, "The *Gestalt* of the adventurer is possible only within the realm of destiny of modern humanity and its 'subjectivity.' Odysseus was not yet an adventurer. The seafarers meant, i.e., poetized, by Hölderlin are no longer [adventurers]. The 'adventurous' heart belongs in the sphere of the metaphysics of the Will to Power." [GA, 52: 180]

Heidegger observed that the modern "adventurous heart" often "takes flight into some drunken frenzy and this may be only blood-frenzy. . . ." [Ibid.], a remark intended as a rebuke to Jünger's celebration of murder and warfare. For Jünger, the adventurous heart was one who appreciated the experience of beauty and horror associated with plunging into the fiery abyss of the Will at work beneath ordinary events. Heidegger agreed that witnessing the invisible "play" at work behind empirical events was a decisive insight, but he distanced himself from Jünger's interpretation of this play:

> Who wants to deny that since all human planning and working announces with special clarity the character of a great "game," in which no individuals and also not all [people] together are able to raise the stakes for which the "world-game" gets played? Who might wonder that in such a time, since the world up until now has gone out of joint, the thought awakens: now only the lust for danger, the "adventurer" can be the way in which man becomes certain of reality? [GA, 51: 36]

In 1942, Heidegger observed that the "fascination" with the play of industrialization and its "disciplines" could "shroud the misery [*Elend*] in which technolization thrusts man. Perhaps for the wholly technicized man there is no more of this 'misery.'" [GA, 53: 54] In "On Pain," Jünger argued that the crucial feature of the bourgeois age was the quest to eliminate pain. The new technological man, by way of contrast, defies pain by turning his body into an object. Jünger proclaimed that "man becomes capable of defying the attack of pain to the extent to which he is able to set himself outside himself. This setting-outside, this impartialization [*Versachlichung*] and objectification of life increases unceasingly." [US: 196] The masks and uniforms worn by the technological worker-soldier are instances of the armor required by the new age. Heidegger agreed that "a man without a uni-form today gives the impression of being something unreal which no longer really belongs." [VA I: 89/108] Moreover, "It almost seems as if the nature [*Wesen*] of pain is not disclosed to man under the domination of the Will. . . . " [VA I: 91/110] By seeking to dominate pain, technological humanity made itself ever more available to the Will to Will, for pain is able to disclose that which is *other than* what is made present by the Will. The Will makes entities present only as standing-reserve, but pain reveals "presencing" as such, and thus the ontological difference between presencing and entities. As Parvis Emad has observed,

> Pain's appearance means the emergence of something which escapes Will's total control and planning. . . . Pain confronts the Will with a difference, one between its calculative arrangements and a state of being which resides outside of this arrangement. . . . Curiously enough, pain confronts the Will with the presence of something which defies it. Thus pain proves capable of manifesting a difference between the constant presence of the Will and the presence of something else.[9]

In speaking of the importance of experiencing pain, loss, grief, however, Heidegger would seem to be contradicting his claim that humanity must submit to the deadening *Gestalt* of the worker: "Technology becomes mastered only when from the first and without reservation 'yes' is said unconditionally to it. This means that the practical mastering of technology in its unconditional unfolding already presupposes the metaphysical submission to technology." [GA, 51: 17; cf. also GA, 48: 205]

Especially in 1933–34, Heidegger used Jünger's rhetoric of hardness, but increasingly he distanced himself from such language. He concluded that the steely resoluteness celebrated by Jünger and many National Socialists was incapable of carrying humanity beyond the technological era. Jünger was right, so Heidegger believed, in saying that humanity must submit to the technological *Gestalt*, but such submission did not mean *objectifying* oneself in the sense of hardening oneself to pain or of turning one's body into an insensitive instrument. Heidegger seems to have believed that submission to the technological *Gestalt* meant being open to the bodily pain and spiritual grief associated with the technological domination of humanity and the earth. Only such submission would make possible the fulfillment of the technological destiny, so that thereafter a new start could be made. For Heidegger, Jünger's technological man had *not gone far enough*. By objectifying his body, Jünger's worker sought to attain a detached perspective to it, but such detachment would block the profound pain from which could arise the possibility of an alternative. Humanity's "new beginning" was possible only at the furthest extent of humanity's greatest danger.

This idea has remarkable parallels to Marx's doctrine that the proletariat could become the revolutionary renewer of humanity only by attaining to human universality. But the condition necessary for such universality was that the proletarian class would have to be reduced to sheer abstract humanity emptied of specific contents apart from the creative capacity definitive of humanity. This curious parallel between Heidegger's vision of the preconditions needed for change and Marx's vision was not entirely accidental. Like Jünger, Heidegger believed that the Soviet Union may have been the most advanced manifestation of the *Gestalt* of the worker. If the spiritual dimension of the Russian people could encounter the full misery imposed by that *Gestalt*, perhaps something interesting could come out of the Russian experience with Bolshevism. Here, and in other respects as well, Heidegger's views may be compared with those of another fierce critic of communism, Alexander Solzhenitsyn.[10]

In an extraordinary passage written while the German army was being defeated at Stalingrad, Heidegger asserted:

> Who has ears to hear . . . can already for two decades hear the word of Lenin: *Bolshevism is Soviet power* + *electrification*. This means: Bolshevism is the "organic," i.e., calculatively organized (and as +) thrusting together of the unconditioned power of the party with fully realized technologization. The bourgeois world has not seen and today in part still does not want to see that in "Leninism," as Stalin calls this metaphysics, has been realized a *metaphysical leap forward*, from which in a certain way first becomes comprehensible the metaphysical passion of today's Russian for technology, from which he brings to power the technological world. [GA, 54: 127]

While sometimes charging the Soviets with the highest form of the Will to Will, at other times Heidegger accused America of being its home: "Bolshevism is only a variety of Americanism. The latter is the genuinely dangerous form of the measureless, because it arises in the form of bourgeois democracy and is mixed with Christendom, and all of this in an atmosphere of decisive history-lessness." [GA, 53: 86–87]

D. HEIDEGGER'S POSTWAR ATTITUDE TOWARD THE TECHNOLOGICAL ERA

Because he was convinced that American democratic capitalism and Russian communism were metaphysically the same, and since these powers were victorious in World War II, Heidegger concluded that the war had not solved anything! [WHD: 66/65] Such a view is considered astonishing to those who regard German fascism as having been an evil of such magnitude that the Soviet Union and the United States temporarily set aside basic differences in order to combat it. By expressing such views in 1951–52, Heidegger inflamed critics, who concluded that he remained unpenitent regarding Germany's war guilt.

In reply, Heidegger pointed out that the onslaught of the Cold War almost immediately after the end of World War II demonstrated the correctness of his own view of world affairs. Having crushed Germany, Russia and America—the technological giants—could begin grappling for total control of the earth. Heidegger agreed with Jünger that in the age of technology, the distinction between war and peace had disappeared: "War has become a distortion of the consumption of entities which is continued in peace." [VA I: 87/106] Postwar Europe still demanded a decision which the current political system was incapable of making. Heidegger feared that "these matters . . . will once again be forced into politico-social and moral categories that are in all respects too narrow and faint-hearted, and thus will be deprived of a possible befitting consideration and reflection." [WHD: 64/66–67]

Heidegger declared that what was needed in our destitute time was not political ideology, but instead a proper relationship with being. Only then would humanity enter into a relation with the essential nature of technology. Postwar civilization was still gripped by the technological Will to Power which started both world wars. Such a civilization, still conceiving of technology as an instrument to ensure human security and happiness, could not be open to the difficult truth which humanity had to face in order to move beyond the technological era. In his version of Nietzsche's Overman, Heidegger saw the kind of man needed to take on the high responsibility of assuming dominion over the earth. Nietzsche claimed that man must pass beyond himself to his essential truth: the Overman. Heidegger insisted that the Overman did not mean a slope-shouldered, ape-like brute, someone who made "sheer caprice the law and titanic rage the rule." Instead,

> the Over-man [*Über-mensch*] is the one who first leads [*überführt*] the essential nature of existing man over into its truth, and so assumes that truth [*übernimmt*]. Man up to now, thus established in his being [*Wesen*], is to be brought into the condition of being the future master of the earth, i.e., of wielding in a lofty direction the possibilities of power which fall to the future man from the being of the technological transformation [*Umgestaltung*] of the earth and of human activity. . . . We may assume . . . that here and there, still invisible to the public, the Overman already exists. But we must never look for the Overman's essential *Gestalt* in those figures who by a shallow and misconceived Will to Power are pushed to the top as the chief functionaries of its various organizational forms. Nor is the Overman a wizard who will lead mankind toward a paradisiacal happiness on earth. [WHD: 26/59–60]

For Heidegger, the Overman was a Janus-faced figure: Jünger's man of steel, on the one hand, and Hölderlin's god-seeking poet, on the other. Quoting Nietzsche, Heidegger observed that the Overman was "*Caesar with the soul of Christ.*" [WHD: 67/69] As we shall see in the next chapter, Heidegger longed for the arrival of new gods to restore meaning to the nihilistic technological world. Much of his vocabulary about a destitute humanity wandering in the technological desert was drawn from the Biblical tradition.[11] For him, the Overman was in effect a transitional but salvific figure. Such an extraordinary half-divine man would discard the ideal of "boundless, purely quantitative nonstop progress," but he would nevertheless be "poorer, simpler, tenderer and tougher, quieter and more self-sacrificing and slower of decision, and more economical of speech." [Ibid.] The Overman depicted here was much like Jünger's "organic construction," tougher and more self-sacrificing than herd-like bourgeois man. Such an elite and extraordinary man could appear only in a world where rank-ordering had been carried out: "By rank order in its essential meaning—not merely in· the sense of an arrangement of existing conditions according to this or that scale—Nietzsche understands the standard that all men are not equal, that not everybody has aptitude and claim to everything, that not everybody may set up his everyman's tribunal to judge everything." [WHD: 67/69]

In the 1950s, then, unchastened by the political consequences of rejecting Enlightenment guarantees of individual human rights, and apparently regarding as contemptible the widespread human longing for material well-being and political liberty, Heidegger did not hesitate to call upon humanity to prepare the way for an Overman who would offer a higher possibility to humanity. Tough but tender, the Overman was concerned not about technological "progress," but instead with bringing the technological age to its high point, so that by completing the possibilities of that age, a new age might begin. The Overman would fulfill the ultimate possibility of technological humanity. In effect, he would be the sacrificial lamb, innocent but hard, tender but tough, Caesar with the soul of Christ. Hölderlin had depicted Christ as the secret brother of Heracles and Dionysius, a trio who represented a "still unspoken gathering of the whole of Western fate, the gathering from which alone the West ['evening-land'] can go forth to meet the coming decisions—to become, perhaps in a wholly other mode, a land of morning." [WHD: 67/69–70] In taking upon himself the full "stamp" of the technological epoch, in experiencing its nihilism, its horror, its absence of god, its lack of meaning, its supreme need, the Overman would simultaneously be the demigod who would announce the arrival of the new gods needed to open up a new world. This new world would be post-technological not because it got rid of everything technical, but instead because it made possible a new way for humanity to relate to those devices without being "stamped" by the *Gestalt* of the worker. Post-technological humanity would make use of its technological powers for "high purposes," which cannot be known to those of us who remain trapped in the technological epoch.

In 1959, Heidegger expanded upon the theme of the destiny of a Europe stamped by the technological *Gestalt*. For such a Europe, the poet's "earth" and "sky" have disappeared. Europe's "technological-industrial sphere already covers the whole earth. And this is already interpreted as a planet in the interstellar cosmic realm, which is posited for the planned realm of human action." [GA, 4: 176] Citing Paul Valéry, Heidegger noted that Europe may have become a "mere stub" of the Asian continent, but also could be the "brain" of Earth, in the sense of the technological-industrial calculation sweeping over the planet. But was there not a third possibility? Might not the West make a new beginning? Even postwar Heidegger believed that the National Socialists had taken some steps toward such a new beginning. Democracy, socialism, and communism were not the answer, for they were the late consequences of the history of the first beginning in ancient Greece. Humanity could never escape the first beginning in Greece, but could make a new beginning which would transform the Greek technological heritage. Such a new beginning, however, could occur only if humanity became attuned to the uncanny destiny governing the whole planet: the destiny of modern technology. Somehow in becoming attuned to the radical nihilism of modern technology, in surrendering to the affliction associated with it, there would simultaneously awaken the possibility of a new disposing (*Verfügen*) of human existence. Jünger was incapable of such an awakening, because he was still governed by representational-calculative thinking. Hence, he was not sufficiently attuned to the profound affliction of the technological age. In willing that it go forward, he willed in despair of any alternative. But Jünger was not alone. All humanity, in Heidegger's view, desperately awaits the attunement that would show the way out of the technological desert. Such attunement, Heidegger argued, was possible only through the work of art. Let us now consider his meditations on Nietzsche and Hölderlin's views about the world-altering power of the work of art.

Chapter 7. National Socialism, Nietzsche, and the Work of Art

In part because of his confrontation with Jünger's "aesthetic" interpretation of the *Gestalt* of the worker, Heidegger was led to analyze the conception of art to which Jünger owed so much—that of Nietzsche. By 1934–35, Heidegger had ceased to praise the importance of "scientific philosophy" as he had in his Rektor's address, and had begun to emphasize the need for a work of art to found the "new" German world. It was in this period that he came to see more clearly the inner relationship between the Greek concept of art as *techne* and modern technology. In his lectures on Nietzsche, Heidegger analyzed the idea that the great work of art is *techne* in that it provides the *Gestalt* that gives measure, limit, boundary, and form to things. Modern technology is a degenerate form of *techne*, in that it imposes a highly constricting measure upon things, so that they can show themselves only instrumentally. Reflecting upon the inner relation between *techne* and modern technology, Heidegger amplified his insight into the history of productionist metaphysics. The Greek conception of art or *techne* as measure-giving disclosure had ended in the technological era. Now what demanded attention was a new understanding of art that would make possible an authentic alternative to industrial modes of production.

In this chapter, we first examine the fact that around 1934 Heidegger began thinking of German (and Western) history in terms of artistic categories drawn from Greek tragedy. Next, we shall explain Hitler's claim that National-Socialism would provide the German people with the new work of art, the new myth, necessary to lead them out of the wasteland of modernity and industrial technology. Heidegger's interpretation of the saving power of art was at least in part an attempt to provide National Socialism with a proper understanding of the role of art in the "new" Germany. Finally, we shall consider Heidegger's creative appropriation of Nietzsche's views about the ontologically formative power of the work of art. Heidegger's lectures on Nietzsche attempted to provide both the metaphysical basis for, and a saving alternative to, Jünger's technological vision of the future. It is important to keep in mind that the poetry of Hölderlin was the foil for Heidegger's analyses of the works of Nietzsche and Junger. In chapter eight, we shall examine Hölderlin's importance for Heidegger's concept of modern technology.

A. HEIDEGGER AND "THE TRAGIC VIEW OF LIFE"

Influenced by Nietzsche, Heidegger began interpreting the decline of the West as a kind of tragedy, resulting from the fact that "man had fallen out of being without knowing it." [GA, 40: 40/30] He believed that ontological blindness led to *hubris* in the form of arrogant optimism about humanity's prospects for gaining total control of its own destiny. After the great age of the Greeks, "the light of [the human] clearing was dimmed by the blazing fire of arrogance [*Vermessenheit*], which only calculated the measure from the entity." [GA, 55: 327] Humanity's "insurrection" against being invited *nemesis*, in the form of the technological nihilism. While capable of making everything present and disposable as standing-reserve, technological humanity had become completely blind to the presencing (being) that enables entities to be disclosed as entities. Events such as world wars, destruction of traditional values, global uniformity, and industrial pollution clouded the smug optimism of those whom Nietzsche called "last men."

While depicting the course of Western history in tragic terms, Heidegger portrayed himself as a tragic hero within that larger tragic drama. Like the king Oedipus, the philosopher Heidegger wanted to stand in the light of truth so much that he was willing to risk everything in the process. The philosopher's task, as he conceived it, was to restore "to things, the entity, their weight (being)." [GA, 40: 13/9] This task involved challenging the people to confront the truth about their desperate situation. Like Sophocles' Oedipus, Heidegger wanted to discover the source of the "spiritual decline" plaguing his country. Symptoms of that decline included "the darkening of the world, the flight of the gods, the destruction of the earth, the transformation of men into a mass, the hatred and suspicion of everything free and creative. . . . " [GA, 40: 41/31] Oedipus, disregarding the warnings of the seer, sought the source of his kingdom's troubles by returning to his own obscure origins. So, too, Heidegger sought to resolve the "spiritual decline" of the West by encountering being in an originary way, analogous to what he believed the ancient Greeks had achieved when they initiated Western history. Initiating a new beginning would require attacking those historical figures blocking the way to a new experience of the incipient event (*Anfang*) of being: "It is the privilege—in the history of everything essential—but also the responsibility of all descendants that they must become the *murderers* of the ancestors and themselves stand under the fate of a necessary murder!" [GA, 31: 36]. In 1935, he added:

> With the passion of a man who stands in the manifestness of glory and is a Greek, Oedipus unveils that which is concealed. Step by step, he must move into unconcealment, which in the end he can bear only by putting out his own eyes, i.e., by removing himself from all light, by letting the cloak of night fall round him, and, blind, crying out to the people to open all doors that a man may be made manifest to them as that which he *is*.
>
> But we cannot regard Oedipus only as the man who meets his downfall; we

must conceive of him as the *Gestalt* of Greek *Dasein*, in which ventures forth its basic passion in the broadest and wildest way, the passion for unveiling being, i.e., the struggle for being itself. . . . Hölderlin wrote keen-sightedly: "Perhaps King Oedipus has an eye too many." This eye too many is the basic condition for all great questioning and knowledge and also their unique metaphysical ground. [GA, 40: 114/90–91]

In the mid-1930s, Heidegger believed that he himself, standing at the historical crossroads as Germany's self-proclaimed spiritual leader, was blessed and cursed with Oedipus's "eye too many." Peter L. Rudnysky has pointed out that Heidegger's account of the philosopher's clarity of vision at the crossroads "precisely registers the evolution of Oedipus from *Oedipus the King* to *Oedipus at Colonus.*"[1] Heidegger interpreted this evolution as follows: the philosopher must seek to be touched anew by being, so that entities can disclose themselves other than as undifferentiated standing-reserve; the philosopher may "shatter" in this process, but his sacrifice may usher in a new age. The older Oedipus, blind to entities but capable of "seeing" being, represents the broken but ontologically healed philosopher.

In 1954, however, an older Heidegger offered a different interpretation of the Oedipus sequence:

When a man goes blind, there always remains the question whether his blindness derives from some defect and loss or lies in an abundance and excess. In the same poem that meditates on the measure and measuring, Hölderlin says: "King Oedipus has perhaps one eye too many." Thus it might be that our unpoetic [technological] dwelling, its incapacity to take the measure, derives from a singular excess of frantic measuring and calculating. [VA II: 77/273][2]

Now, the young Oedipus represented those previous metaphysicians whose ontological "night vision" was blinded by the shining allure of entities available for domination. The arrogant young Oedipus had taken the only road he could see, the road which made his kingdom a wasteland. The older Oedipus, by way of contrast, may be compared with the later Heidegger. In spite of his having been shattered by ontical events (National Socialism, Germany's defeat, personal humiliation), Heidegger remained convinced that he was still gifted with the ontological vision needed for insight into the self-concealing "event" (*Ereignis*) which "gives" the various modes of presencing that constitute the history of being. Only such vision could save the West.

Sophocles' theme of blindness and vision in the Oedipus trilogy provides a common motif in both Jünger and Heidegger—that the "seer" or the "visionary" must be able to keep his gaze upon the horrifying and the awesome. This capacity for apprehending the horrifying was, for Jünger and Heidegger, the highest form of the artistic experience. A common source for their attitude in this matter was Nietzsche's book *The Birth of Tragedy*. For Nietzsche the role of tragedy is to garb the horror of mortality in the "healing balm" of moving and beautiful language that somehow reconciles us to our dreadful destiny. For Jünger, holding one's gaze steady while a man is torn apart by a shrapnel shell, or while another man is transformed into a half-mechanical beast by industrial technology, enables one to penetrate into the secret force at work in the world, the eternal Will to Power that links men across historical epochs. Through a method that he called "historical realism," Jünger claimed that he could disclose that hidden layer of "absolute reality" of which all thought and feeling are but symbolic expressions. The "heroic spirit," so he wrote, sees war not as a human event but as a sequence of "multifarious transformations and disguises" of war which "present the heroic spirit with an engrossing drama."[3] War, in other words, is a titanic work of art. Years earlier, Nietzsche argued that the artist is

> the medium through which the one truly existent subject celebrates his release in appearance. For to our humiliation *and* exaltation, one thing above all must be clear to us. The entire comedy of art is neither performed for our betterment or education, nor are we the true authors of this art world. On the contrary, we may assume that we are merely images and artistic projections of the true author [the Will], and that we have our highest dignity in our significance as works of art—for it is only as an *aesthetic phenomenon that existence and the* world are eternally *justified*—while of course our consciousness of our own significance hardly differs from that which soldiers painted on canvas have of the battle represented on it.[4]

The Greeks spoke as if human life was not an end in itself, but was justified only insofar as it provided a pleasing aesthetic display for the gods. Humans, then, were like shadowy projections with little autonomy, puppets of the gods. But to some extent, even the gods were incapable of altering or resisting a still higher power, that of fate. The passage cited above suggests that, for Nietzsche at least, the depersonalized Will to Power had replaced both the gods and fate. Like Nietzsche, Jünger believed that world history was an aesthetic phenomenon, a spectacle which provided a sublime experience. But who was capable of apprehending this experience? Ultimately, Jünger believed, only those warrior-artists who were themselves masks through which the Will to Power could experience itself at work in the world. The Will, in other words, was both the origin and the appreciator of its spectacle. The technological era was merely the latest manifestation of the cosmic "play" of the spectacle-producing Will to Power.

Heidegger developed in his own way the theme that humanity is a participant in a drama or play that is not of human origin. He claimed that human existence was the clearing (the mask) through which entities could display themselves. Discarding the usual notion that humans are active subjects who gaze upon and understand entities, Heidegger maintained that they look at themselves through the ontological "openness" which has appropriated humanity. "Looking" (*Blicken*), we are told, is best understood in terms of the Greek verb *theo*. Usually this verb has been regarded only in its medial form, *theaomai*, which may be translated as "to look at." From this verb, we have the word *theatron*, the showplace, ordinarily understood as the theater, the site where people go to look at something. Conceived more radically, however, *theaomai* means for something to bring *itself* to view, in the sense of the view which a thing offers from and gives of itself. *Thea*, viewing, thus does not at all mean "looking" in the sense of the re-presenting viewing through which the ego-subject directs itself toward and grasps the object. Rightly understood, human existence is in effect "grasped" by the being of entities so that they may show themselves and thus "be." We do not encounter things; rather, *things encounter us*. [GA, 54: 152ff.] In effect, then, Western humanity has been unwittingly playing the role of mask or clearing or theater by virtue of which the being of entities may display itself as a cosmic spectacle. We must be careful, however, in saying that "the being of entities displays itself," for "being" names no entity at all and can never itself "appear." It may, then, be better to say that human existence is the clearing through which entities may display themselves (and thus "be") as a cosmic spectacle. This spectacle occurs not for any "purpose," especially not for any merely *human* end, but instead is ultimately purposeless, the cosmic-ontological version of a work of art.[5]

Having become blind to its supporting role in this event of ontological disclosedness, man has arrogantly presumed that he is the lead actor, author, producer, witness, and beneficiary of the drama called "Western history." Gripped by *hubris*, man races toward the ever-receding goal of controlling all things: by rewriting the script of history, by recreating the world in his own image. The essence of this quest for control, Heidegger maintained, is the craving for more power as an end in itself. Hence, the technological era is essentially a Will to Will, a self-empowering process that has no "purpose" beyond its own expansion. If human *Dasein* could only awaken to the ultimate purposelessness of history, if *Dasein* could only see that people are players in a cosmic game which transcends human ends, so Heidegger believed, that awakening in and of itself would make possible a new, non-instrumental, non-domineering era of history.

The unfortunate change of humanity to the self-centered, power-hungry subject was not really a human decision, but instead occurred because being concealed itself from humanity. Just as Jünger saw history as a series of beautiful spectacles generated by and for the Will to Power, so too Heidegger came to see history as a series of crystallizations of being in the mode of beingness: *eidos, energeia, actualitas, actus purus, res cogitans*, Will to Power, Will to Will. The historical formations of beingness "stamped" things, or provided the measure or form according to which all things and all human behavior could show themselves. Hence, the history of being was in effect a history of world-forming, artistic disclosures. In apprehending these various disclosive events, these forms of "beingness," the great thinker could also catch an elusive glimpse of that which could never appear: being as such. Self-concealing being, according to Heidegger, was more sublime, simpler, more extraordinary, more awesome, and more horrifying than any mere "entity." Just as for Jünger the artist was the "mask" through which being as the Will to Power could manifest itself indirectly in the historical world, so too for Heidegger the thinker and the artists were "masks" through which the being of entities could manifest itself in its various stages.

B. NATIONAL SOCIALISM AS "NATIONAL AESTHETICISM"

While in his Rektor's address in 1933, Heidegger spoke of the German revolution in terms of authentic science (*Wissenschaft*), within two years he had made art the basis of the revolution. One reason for this shift may have been that Hitler celebrated not science but art as the source for saving Western culture. Adolf Hitler, himself a frustrated artist, conceived of his mission as saving the West from spiritual degeneration. In 1938 he remarked that "the world will come to Germany and convince itself that Germany has become the guardian of European culture and civilization."[6] Many National Socialist ideologues conceived of their political project in artistic terms: they would give a new *Gestalt* to Germany. In 1933, for example, Joseph Goebbels wrote the following to Wilhelm Furtwängler:

> Politics, it too, is an art, perhaps the most elevated art and the greatest that exists, and we—who give form to modern German politics—we feel ourselves like artists to whom have been conferred the high responsibility of forming, beginning with the brute masses, the solid and complete image of the people. The mission of the artist is not only to unify, but goes much further. He is obliged to create, to give form to eliminate what is sick, to open the way to what is healthy.[7]

In his 1935 Nuremburg address "Art and Politics," Hitler specifically called for the "revival and resurrection of German art."[8] Such a revival meant reproducing the "eternal" works of the ancient Greeks in a way compatible with contemporary German social needs. There was to be no place in Germany for the corrupt practices or the Dadaists and Cubists, whose works were held up to public ridicule in the infamous exhibit of "degenerate art" (*Entartete Kunst*) held in Munich in 1938.[9] Hitler maintained that "art belongs to the whole complex of the racial values and gifts of a people."[10] Where, he asked, "would the Egyptians be without the pyramids and temples, without the adornments of their human life; what would the Greeks be without Athens and the Acropolis; what Rome without its buildings; what the lives of our German Emperors without their cathedrals and their palaces; what our Middle Ages without their . . . churches or cathedrals?"[11]

In spite of the economic depression, Hitler insisted that Germany must use its resources to build great monuments that would express the renewed German *Geist*: "We shall discover and encourage the artists who are able to *impress upon the state of the German people the cultural stamp [Stempel] of the German race which will be valid for all time.*"[12]

It was no accident that Heidegger read the first version of his essay "The Origin of the Work of Art" in 1935, not long after Hitler's Nuremberg speech about art and architecture. As a primary example of a great work of art, Heidegger used a Greek temple. The art of the temple opened up the world of the *polis*, in which entities could first manifest themselves in their own specific shapes and forms, and in which Greek humanity could make the decisions that would determine its destiny. Insofar as he agreed that art and *polis* had a distinct relationship Heidegger agreed with Hitler. But he also disagreed with

Hitler's view of this relationship. For genuine art to "work," in Heidegger's view, it would have to reveal the fragility of human existence, its finitude and mortality. Hitler, by comparison, with his talk of the "thousand-year Reich" and with his view that great art was the manifestation of the eternal, called for totalitarian "temples," the scale of which was designed to give the impression not of history but of eternity. This vision of Germany indicated to Heidegger that Hitler was still under the sway of foundational metaphysics. For Heidegger, if the new Germany was to constitute itself in a way which was somehow comparable in force and novelty to the Greek *polis*, it would have to forswear any illusions about basing itself on "eternal" foundations.

Yet historical evidence suggests that Hitler was aware that even his Reich would end, though this fact still did not detract from his foundationalist fantasies. He ordered his chief architect, Albert Speer, to construct gigantic buildings which would be impressive even as ruins, like the Coliseum in Rome or the Great Pyramid in Egypt.[13] One building in Berlin was to include a dome almost one thousand feet high. In addition to this role as future testimony to the greatness of the Third Reich, Speer's extraordinary monuments had two other goals: (1) they would serve as public-works projects necessary to stimulate the economy; and (2) they would encourage the people to submit to the Third Reich in the face of such larger-than-life embodiments of its will. It is important to see that National Socialist art was an instrument of domination which worked hand in hand with the more overt forms of terror and violence used to subdue resistance.

Albert Speer also served as the director of the "Bureau of Beauty of Labor" (*Amt Schönheit der Arbeit*), which sought to transform the workplace into an aesthetically pleasing, clean environment, for the purpose of restoring to the depressed working man "the feeling for the worth and importance of his labor."[14] At first, this movement to "aestheticize" work and to restore dignity to it was consistent with the pre-1936 Nazi ideology which emphasized the need to restore the vitality of the working class by rejuvenating ancient folkways and customs. The Nazis commissioned artists, writers, musicians, theater directors, and folklore scholars to celebrate the values of the *Volk* in factories, public exhibitions, displays, and meetings of all types. These public displays were extraordinary. As one critic has noted, "The greatest aesthetic accomplishments of German fascism were the stagings of the Party conventions" in Nuremberg.[15] Architect Speer also designed and constructed these truly massive human "buildings." On the great parade grounds (thirty square kilometers), surrounded by skyward-trained searchlights forming a "cathedral of ice," a million people would be assembled both to witness and to serve as the constituent elements of totalitarian "art" at its finest. Hitler understood the political power of great spectacles, and the use to which they could be put on film. Indeed, it has been argued that the rallies were designed with films, such as Riefenstahl's *The Triumph of the Will*, in mind.

Hitler's political-artistic ambitions went far beyond public buildings and spectacles to encompass the entire German landscape. Hans Jürgen Syberberg has argued that Hitler viewed the Third Reich as a total work of art (*Gesamt-*

kunstwerk), analogous on a national scale to the total works of art attempted by Wagner at Bayreuth.[16] Hitler conceived of the *Autobahnen* he built throughout Germany, for example, as akin to the paths through gardens of French chateaux of the medieval era, or through the municipal parks of the bourgeois era. Syberberg, himself a noted German filmmaker, acknowledges that we ordinarily think of Hitler's interest in film pejoratively, but have failed to consider the astonishing possibility that World War II was "directed like a big-budget war film" to be viewed from Hitler's bunker. The gruesome image of Germany in flames in 1945 may be regarded as the final scene in a historical spectacle generated by a suicidal impulse. The drive to create an amoral state defined by art and violence could only end in self-destruction.

Most people could not foresee the dreadful conclusion of Germany's "revolutionary" drama. Hence, National Socialism at first gained the support of many leading writers and artists, including the famous poet Gottfried Benn. In an address given in November 1933, Benn defended his former expressionist style against the Nazi view that expressionism was degenerate art. Here, he made clear the importance of art—including expressionism—for the Third Reich:

> The leadership of the new Germany is extraordinarily interested in questions about art. As a matter of fact, their outstanding intellectuals . . . make art almost daily a matter of extreme urgency for the state and the public in general. The enormous biological instinct for racial perfection . . . tells them that *art is the center of gravity, the focal point for the entire historical movement*: art in Germany, art not as an aesthetic achievement but *as a fundamental fact of metaphysical existence*, which will decide the future, which is the German Reich and more. . . . [My emphasis][17]

In view of the ideological importance attributed to art by National Socialism, Philippe Lacoue-Labarthe has described the whole movement as "national aestheticism."[18] There is much evidence to support this view of National Socialism. Hitler proclaimed that art had the capacity of transforming and renewing the *Volksgeist*. This elevated conception of art had its roots far earlier than Hitler, and in more sophisticated German thought. Kant's *Critique of Judgment*, Schiller's *Letters on the Aesthetic Education of Mankind* (1794-95), and Hegel's lectures on art had already idealized art's ability to ennoble human existence and to disclose aspects of reality hidden from calculating rationality. During the latter part of the eighteenth century, Germans developed the odd belief that they embodied in a unique way the spirit of the ancient Greeks in a way like no other people. The Greeks invented both art and politics, and saw the internal relationship between them. Hence, so a number of eighteenth-century Germans concluded, Germany could attain to its own greatness only if it defined itself in terms of the spiritual power of art.[19]

Many years later, although in a cruder fashion, National Socialists also conceived of the new Reich as a collective, harmonious work of art, a totalitarian *Gemeinschaft*. Like the Greeks, according to Lacoue-Labarthe, the Germans under Hitler were both the subjects who witnessed the new *polis* and the objects who constituted it. Of course, the Germans were not the first to use classical antiquity as a model for political renewal, to aestheticize politics, and to worship the nation-state in place of the deposed Judeo-Christian God. The French initiated these practices during the French Revolution, especially during its Jacobin phase. Apparently ignorant of the extent to which the French Revolution anticipated their own totalitarian practices, most National Socialists condemned it for promoting the nihilism of modernity. The French Revolution failed in part, according to Nazi ideologues, because it imitated *Roman* antiquity, which was a degenerate derivation from Greek culture. If one wanted an authentic model for the radical renewal of the West, so argued Hitler and his colleagues, then one ought to choose the Greeks. Thus, many National Socialists, like many German romanticists before them, concluded that to save Germany, the *Volk* had to revive the Greek past, its architecture, its cult of athleticism and bodily beauty, its art, its social cohesion ("organicism").

But Heidegger wanted to surpass such empty repetition. He sought to imitate only this: the creative leap made by the Greeks. In Heidegger's view, this leap was discontinuous with preceding human history. The Greek *polis* and *ethos* were not the products of human intentions or plans, but instead erupted as a primal response to the initial event in the history of being itself. What Burckhardt said of Venice, Heidegger applied to Athens: "Venice recognized itself as a strange and mysterious creation [*Schöpfung*]—in which something other than human ingenuity had been at work from long ago."[20] Heidegger was not alone in believing that the same *creativity* which gave birth to Athens was not at work sculpting the new Germany. Long in search of their own national "identity," many Nazis concluded that Germany had to invent itself by formulating a myth which would transform and unify the German *Volk* in a way analogous to, but not identical in content with, how Homer bound together the Greeks.

This extraordinary attempt by Germany to create itself through self-generated mythification, combined with its rejection of the traditional Enlightenment political ideals, clearly involved a self-destructive "logic." Rejecting any existing standard against which to judge and to evaluate their behavior, apart from its likeness to the uniquely creative power of the Greeks, the Germans ended up creating something horribly new: a totalitarian state which overreached all acceptable standards by building gas chambers for mass extermination. Heidegger certainly did not favor such extermination, but by proclaiming that Germany was infused with the creative power to initiate a radically new beginning for the West, he contributed to Germany's delusional and destructive self-mythification. Moreover, by holding up the ideal of tragedy and self-sacrifice as the highest possibility for an authentic people, he contributed to National Socialism's horrifying and suicidal climax. In his own way, he shared in the Nazi fascination with the relationship between art and violence. As Fritz Stern has pointed out, the Nazi emphasis on sacrifice and victims led to a scheme "in which *value is commensurate with the catastrophic nature of a man's existential project.*"[21]

National Socialism utilized art in the service of its political ambitions, rather than allowing great art to serve in its utopian capacity of pointing beyond the limitations of existing society.[22] Walter Benjamin argued that fascists believed the unruly proletarian masses could gain (non-revolutionary) satisfaction by being provided with new ways of "expressing" themselves, e.g., in enormous parades and spectacles, with planes and zeppelins flying over them and three-hundred-foot swastikas standing before them. Such larger-than-life spectacles, like great public buildings, gave the masses a sense of "belonging" to something far larger and more important than themselves. In fascism, then, politics used modern technology to transform politics into an artistic event, one which would mystify and conceal the fact that property relations went unchanged despite "revolutionary" events. Benjamin concluded:

> "*Fiat ars—pereat mundus*," says Fascism, and as Marinetti admits, expects war to supply the artistic gratification of a sense perception that has been changed by technology. This is evidently the consummation of "*l'art pour l'art*." Mankind, which in Homer's time was an object of contemplation for the Olympian gods, now is one for itself. Its self-alienation has reached such a degree that it can experience its own destruction as an aesthetic pleasure of the first order. This is the situation of politics which Fascism is rendering aesthetic. Communism responds by politicizing art.[23]

Whereas Kant and Schiller had conceived of art as "disinterested" or "playful," in contrast with the utilitarian concerns of everyday life, the National Socialists subordinated art to political interests. Sharing Benjamin's insight, Rainer Stollman has remarked that "German fascism knew how to transfer all energies, wishes, yearnings, psychic drives and fantasies into an aesthetic, socialistic illusion which worked to cover up the real causes of economic and psychic misery; indeed, it could even push for their continuation."[24] Most attracted to the aesthetic illusions of National Socialism were precisely those most threatened by industrialization: artisans, small shopkeepers, the petit bourgeoisie of all varieties. Resentful of and anxious about capitalist industrialism, these people willingly embraced the extravagant myths and irrational schemes offered by National Socialism. Only in 1936 did the extent of the Nazi commitment to industrialization become clear. In that year, the first Four Year Plan was instituted, which emphasized full employment for the sake of maximizing Germany's industrial and military power.

Some National Socialists, including Goebbels, were influenced early on by the achievements of industrial technology, and by the futurist Marinetti.[25] In 1934, a major exhibit on "Italian Futurist Air[plane] Pictures" was held in Germany. While many Nazi ideologues criticized futurism because it conflicted with the pre-1936 emphasis on pre-industrial folkways and values, it soon found a place in what was to become the steely romanticism of post-1936 Nazi Germany. Futurism, as we noted earlier, praised war as an aesthetic movement, which brought forth new forms of architecture, shiny tanks, the geometrical flights of soaring airplanes, and "the smoke spirals from burning villages."[26] Hitler increasingly embodied this fascination with the shiny machinery of modern technology as he sped around Germany in his Mercedes convertible, or flew from place to place in his airplane—descending like a god from the skies.

After 1936, earlier veneration of pre-industrial ways of life was suppressed in favor of the exaltation of the precision, efficiency, order, rationality, power, cleanliness, and beauty of industrial technology. This pro-industrial attitude was in some respects an intensification of pre-1936 efforts to restore dignity to labor, but differed in a crucial respect: earlier Nazi ideology castigated industrial technology as an enemy of the *Volksgeist*, whereas post-1936 ideology praised it as the very expression of the *Volksgeist*.

Historians have pointed out that the pro-industrial Nazi "turn" was in part a rebirth of the functionalist movement in industrial design known as *Die Neue Sachlichkeit* ("The New Objectivity or New Sobriety"), which had flourished in the 1920s. With its newfound emphasis on industrial efficiency and order, National Socialism appeared to be moving ever closer to Jünger's aesthetic vision of modern technology. Regarding the shift in Nazi ideology, one critic has remarked:

> The romantic image of the handicraft shop, venerated in the early days of the Nazi movement, was scrutinized and purged of pre-industrial characteristics: old tables, rotten from wood worms, had to be replaced so that handwork could "understand the needs and demands" of the "epoch of the machine." Technology was aestheticized as the extension of handicraft production. A series of photographs displaying the aesthetic qualities of hand motions in both mechanized and unmechanized production illustrated the point that "handicraft work is not eliminated but transformed." For Beauty of Labour the enormous gears of modern industry became the objects of aesthetic contemplation, and rows of shiny oil cans became a symbol for "the hand tools of the machine masters." The mistrust among German artisans provoked by the technocratic revival was condemned as *Maschinenstürmerei* [attacking machines].[27]

In this effort to persuade workers that efficiency and productivity were aesthetic phenomena, i.e., ends in themselves apart from any merely "commercial" values, the Nazis drew a mystifying veil over the real economic interests at work in the Third Reich. Proclamation of the idea that handicraft skill could be expressed in the industrial plant meant that "reactionary modernism" as an aesthetic program had now become explicit Nazi ideology. Anson G. Rabinbach argues that

> Beauty of Labour not only integrated aesthetics into the world of production, but derived from production a technocratic aesthetic which dissolved the *völkisch* and pre-industrial imagery of pre-1933 Nazism into a new legitimation based on the autonomy of technical rationality. If Nazism did not display the veneration of machinery that characterized Italian fascism in the early 1920s, or the Soviet Union in the early 1930s, this was true *only before 1936*, when Germany's conditions could be attributed to the ills of modernity, and the support of the *Mittelstand* could be secured by the image of its dissolution.[28]

The public phase of Heidegger's political involvement with National Socialism coincided with the height of its critique of industrialism and modernity. His attempt to found an emerging social order had a genuine utopian element that was missing from the cynical "revolution" promised by Hitler. The National Socialists revealed their hand in 1936, when they affirmed their attachment to the nihilistic technological power drives which, in Heidegger's view, were at work in America and the Soviet Union. His increasing disenchantment with the political *reality* of National Socialism, as opposed to its "inner truth and greatness," may be attributed to his discovery that it used art and philosophy instrumentally, as veneers concealing its *real* agenda: planetary domination. Far from enabling Germany to avoid being stamped by the *Gestalt* of the worker, the National Socialists pushed Germany ever more rapidly into the sphere of technological nihilism.

Heidegger, viewing life tragically through the eyes of his favorite poets and philosophers, maintained that all great ventures are threatened by disaster. This attitude may be readily discerned in his interpretative remarks in 1935 about a famous strophe from Sophocles' *Antigone*. In Heidegger's reading, Sophocles in this strophe speaks of man as the one who

> sails into the very middle of the dominant order [*Fug*], inscribes [*reisst*] (in the "rift" [*Riss*]) being into the entity, and yet can never master the overpowering. Hence, he is thrown back and forth between dominant order and dis-order, between the evil and the noble. Every violent curbing of the powerful is either victory or defeat. . . . In venturing the mastering of being [*Seins*], he [the violent one] must risk the pressure of that which is not [*Un-seienden*], *me kalon*, must risk dispersion, instability, lack of structure, and disorder. The more towering the summit of historical *Dasein*, the deeper will be the abyss for the sudden fall into the unhistorical, which merely thrashes around in issueless and placeless confusion. [GA, 40: 170/135]

Heidegger's postwar remarks were sufficiently ambiguous that one cannot decide whether he believed that World War II was precisely such a plunge from the "towering summit" into "confusion." Certainly in 1942, Heidegger clung to his belief in the uniqueness of the West's two unique beginnings: that of Greece and that of Germany. In his lectures on Hölderlin, he noted that one can scarcely read anything about the Greeks without being assured that the Greeks were already somehow National Socialists. But, he remarked, "one does not serve the knowledge and evaluation of the historical uniqueness of National Socialism if one now interprets the Greeks such that one could suppose the Greeks were already National Socialists." [GA, 53: 106 and 98] The task for the Germans was not to repeat what the Greeks had done, but to begin something new which would nevertheless make possible for the Germans a community and homeland as great and as novel as that of the Greeks.

While Heidegger and other reactionaries may have conceived of the "authentic" dimension of National Socialism in terms of artistic categories, as a self-producing national work of art, and while such a conception certainly sheds light on important aspects of National Socialism, we may go too far in describing National Socialism as "national aestheticism." In the rise and fall of National Socialism, there were clearly other ideological, economic, social, and political factors at work which are not explicable in terms of the hypothesis of the "aestheticization of politics." Nevertheless, for an adequate understanding of *Heidegger's* vision of National Socialism, Lacoue-Labarthe's hypothesis remains important. For Heidegger himself seems to have conceived of the "inner truth" of National Socialism in terms of artistic categories found in Nietzsche and Hölderlin's works. Hence, as we shall see, Heidegger envisioned himself as providing the understanding of the work of art (Hölderlin's poetry) necessary to lead the "revolution" in the direction of an appropriate relationship between German *Geist* and modern technology.

While Heidegger sought to revitalize German *Geist* so that Germans could live with technical devices without becoming their slaves, and while for a time he used Jünger's masculinist language, he resisted all fascist talk of wedding modern technology to the "hot impulses" of "lived experience" (*Erlebnis*) as a saving alternative to Cartesian intellectualism and mechanism. Instinctual *Erlebnis* could hardly save Germany from Cartesianism, he noted dryly, since Descartes himself had invented the idea of *Erlebnis*. [GA, 45: 149] The idea of *Erlebnis* was a metaphysical precursor to naturalism embodied in the doctrine of the "blond beast," which represented everything Heidegger regarded as wrong-headed about the usual Nazi interpretation of Nietzsche. Heidegger proclaimed that neither intellect nor instinct would save modern man, who has been so "hexed by machinations [*Machenshaften*]." [GA, 45: 183] In his confrontation with Nietzsche, Heidegger sought to reawaken Germany's sense of need for a saving work of art to restore weight, measure, and meaning to things—meaning which had been dissolved by the corrosive disclosure of being as nothing but raw material for enhancing technological power. Jünger had used Nietzsche's writings to justify the advent of such power; hence, Heidegger read Nietzsche in order to show that there was an alternative reading, one that could support the National Socialist fight against modernity and rampant industrialism.

C. NIETZSCHE'S VIEW OF ART AS DELIMITING THE OVERPOWERING CHAOS

Heidegger sought both to preserve Nietzsche's authentic philosophical insight from its naturalistic overlay, and to show the extent to which he could derive—in a confrontation with vulgar Nazi interpretations of Nietzsche's thought—a post-metaphysical conception of art that would be consistent with the "inner truth" of National Socialism. As is often the case in Heidegger's "confrontations" with other thinkers, the reader is never sure whether Heidegger was speaking for the other thinker or for himself. In what follows, I assume that Heidegger was trying to bring forth from Nietzsche's writings those insights compatible with Heidegger's own attempt to make a new beginning in Western thought.

Despite preparing the way for such a new beginning by transforming our understanding of art, Heidegger argued, Nietzsche's aesthetic doctrine—like all doctrines of "aesthetics"—remained within the metaphysical tradition. Because "aesthetics" originated with thinkers such as Kant, it was tied up with metaphysical subjectivism and representationalism, both of which were inimical to Heidegger's conception of *Dasein* as being-in-the-world. Aesthetic theory inquired about the relation between the perceiving subject and the material art object, but for Heidegger what the authentic work of art enabled *Dasein* to apprehend was nothing perceivable with the eyes or ears. Rather, the work of art revealed the *being* of entities.

Such a revelation did not involve *mimesis*, imitation. Art, in Heidegger's view, did not "represent" anything, for representation was a subjective activity initiated by Plato when he spoke about the need for correct "envisioning" of the eternal forms. In defining art ontologically as the event of truth, not as *mimesis*, Heidegger was in harmony with the expressionist painter Paul Klee, who said, "Art does not reproduce what is visible, instead, it makes visible."[29] Decades earlier, Jakob Burckhardt, another man much admired by Heidegger, had commented, "Without art we should not know that truth exists, for truth is only made visible, apprehensible, and acceptable in the work of art."[30]

By maintaining that art involves ontological disclosure, Heidegger rejected not only the mimetic definition of art, but also the romantic view of it as the "expression" of the soul. For him, such degenerate conceptions of art arose in connection with the decay of metaphysics. [GA, 43: 105] In the nineteenth century, romanticism—a late version of Cartesian subjectivism—reduced art to a matter of private "taste." This notion led first to the ideal of the "aesthetic man," and later to the image of the decadent "dandy." Heidegger, like many reactionaries, regarded the emergence of this sort of romanticism as a sign of European degeneration.[31] He argued, however, that such degeneration resulted not from "mongrelization" of European "blood," but instead from the decline of the hierarchy, order, limits, and *Gestalten* necessary for an authentic and strong *Volk*. A signal manifestation of this leveling process was the rising call for democracy in Europe. In a passage omitted from the original version of his first lecture course on Nietzsche, *The Will to Power as Art* (1936–37), Heidegger remarked:

> Europe always wants to cling to "democracy" and does not want to see that this would be a fateful death for it. For, as Nietzsche clearly saw, democracy is only a variety of nihilism, i.e., the de-valuing of the highest values, in such a way that they are only "values" and no longer *Gestalt*-giving forces. . . . "God is dead" is thus not an atheistic doctrine, but instead the formula for the basic experience of an event of Western history. With full consciousness did I use this proposition in my Rektor's address in 1933. [GA, 43: 193]

Nietzsche's word "God is dead" announced the loss of the old God and proclaimed the need for a new one. Heidegger prefaced his course on Nietzsche with the aphorism from Nietzsche's book *The Antichrist*: "Almost two

thousand years and not a single new God!" Insofar as Nietzsche, like Hölderlin, looked for the return of gods to provide new meaning and rank-ordering
to human existence, "Nietzsche, along with Hölderlin, was the only believing
man who lived in the nineteenth century." [GA, 43: 192] Evidence for the
collapse of the world- and being-shaping power of the old God could be
discerned in the fact that in 1914–1918 all the "Christian" European countries called on the same God for victory in the Great War [GA, 43: 191], but
this old God could not impede the technological juggernaut which universal-
ized humanity and thus destroyed the historical uniqueness of European peoples. A new goal was needed to preserve humanity from the nihilism of
modernity; only a god could provide such a goal.

    All too often, Heidegger admitted, Nietzsche had spoken about the "de-
generation" of Western humanity in terms of biological metaphors that were
easily misappropriated by Nazi racists. Moreover, by writing about how art
could enhance the instinctual feeling for power, Nietzsche focused too much
on the subjective dimension of art as the act of creating, as the damming up
and expression of instinctual feelings. At first, Nietzsche favored Wagner's
music for its capacity for such expression. According to Heidegger, Wagner
believed that the artistic Dionysian spirit was a drunken frenzy (*Rausch*);
hence, his music knew no proper limits, constantly overflowed appropriate
boundaries, and was hopelessly intoxicated. [GA, 43: 90ff.] Nietzsche eventually came to see that Wagner's music was un-Greek, for the Greeks loathed the
unlimited, the disordered, the overflowing. Nevertheless, Heidegger continued, Wagner's "unchaining of 'affects' could be valid as a rescuing of 'life' with
respect to the growing disillusioning and devastating of *Dasein* through industry, technology, and economics in the context of a weakening and undermining of the formative force of knowledge and tradition." [GA, 43: 103/88]
Heidegger believed that Wagner, like Spengler, understood the symptoms of
Western decadence. Neither man, however, was gifted with the philosophical
acumen required to understand the metaphysical—not biological—sources
for that decay.

    If Wagner floated upward on the Dionysian surge, Nietzsche, in contrast,
sought to give the Dionysian form and limit. In his earlier writings, Nietzsche
tended to separate the Dionysian and the Apollinian, but later he united them.
In fact, after Hölderlin, Heidegger remarked, Nietzsche was the first to rescue
the "classic" from the "classicists" and "humanists." [GA, 43: 149] He, too,
realized that great artistic style was classical. Such style yoked together the
"primordial opposition" of chaos and law, overpowering presencing and limitation. [GA, 43: 150] Hölderlin, Heidegger maintained, knew this opposition
of Dionysius and Apollo more intimately than Nietzsche, and unless Germany
learned its lesson from Hölderlin about the proper relation of this strife between dream and thinking, chaos and law, history would take its revenge on
Germany. [GA, 39: 123]

    Many National Socialists, in Heidegger's view, oscillated between two
views of art. First, art was that which stimulated the feeling of intoxication and
the obliteration of ego-boundaries. Wagnerian operas were a good example of

this view of art. Second, art was regarded as that which imposed eternal forms either upon the materials present in public works of art (such as Speer's architectural wonders) or upon the souls and bodies of the masses in public spectacles. Heidegger believed that the first alternative was the romantic version of the Dionysian as expression of soul, while the second alternative was a degenerate version of the Apollinian view that art was a "representation" or imitation of the eternal forms of truth. In contrast to such views, Heidegger described the genuine character of art as the yoking together of chaos and law:

> Where the free disposing [*Verfügung*] over this yoke is the self-forming law of the happening [*Geschehens*], there it is great style; where there is great style, there art is authentically actual in the purity of its essential fullness. And only according to what art is in its essential actual essence should it be appraised, should it be allowed to conceive itself, as a *Gestalt* of the entity, i.e., of the Will to Power. [GA, 43: 150]

Heidegger here defined *Gestalt* in terms of its Greek meaning as *eidos*, the appearance that defined an entity as what is was. *Eidos* was form, *morphe*, limit, which "constrains and places an entity into that which it is, so that it stands in itself: *Gestalt*." [GA, 43: 138] The great artist, in Heidegger's view, did not impose his subjective will upon entities, but instead was claimed by being as the site through which works of art could be produced in order to let entities show themselves anew. The artist, in other words, sought to curb and limit the presencing of entities in such a way that the entities could manifest themselves in their own determinate ways. Heidegger rejected the romantic notion of the artist as a genius, but he spoke as if being itself were now the genius—although without the sense of "agency" attributed to the human genius. While the human genius "produces" works, being "lets things be." Yet, while Heidegger spurned the romantic notion of the artist as overcome by feelings, he did say that the great artist, such as Hölderlin or Nietzsche, often went mad because he was overwhelmed by the ontological power at work through him: "*Da-sein* of historical man means: to be posited as the breach into which the preponderant power [*Übergewalt*] of being bursts in its appearing, in order that this breach itself should shatter against being." [GA, 40: 172/137]

Seeking to define art in non-subjectivistic, non-representational terms, Heidegger conceived of the artist as a kind of instrument in a disclosive event that transcended human aims and interests. To grasp the proper relation among being, humanity, and art, Heidegger argued that he and his countrymen needed to move beyond the aesthetic conception of art which still impeded Nietzsche's thinking. To think of art anew, however, required "a still more primordial transformation of our whole *Dasein*. . . ." [GA, 43: 160] Such a transformation would go far beyond what had been achieved by the political activities of National Socialism. Only after that transformation would the Germans learn that the "creativity" of art is not mimetic or representational, not the subjective achievement of an artistic genius. Instead, the "creativity" of art must be understood in ontological terms as that which *lets entities be*, in the sense of letting them be manifest: "Art itself is one, indeed the essential, way in which the entity gets made into an entity [*wie das Seiende zum Seienden geschaffen wird*]. And because this creating, limit-bearing, and *Gestalt*-grounding befalls art, the determination of the essence of art can be achieved only when it comes to the question as to what is the creative [dimension] in art." [GA, 43: 161]

As we shall see later, in our consideration of the history of productionist metaphysics, Heidegger attempted to rethink the idea that technology was an instrument for human ends. To accomplish this, so he believed, he had to call into question the whole metaphysical tradition which conceived that "to be" meant "to be produced." This tradition started with the Greeks, received impetus in the medieval notion of the supreme being as the Eternal Creator or Producer, and culminated in the technological era, which claimed that "to be" meant to be produced by and for the human subject. In 1954, at the conclusion of his famous essay on technology, Heidegger remarked:

> Because the essence of technology is nothing technological, essential reflection upon technology and essential confrontation with it must happen in a realm that is, on the one hand, akin to the essence of technology and, on the other hand, essentially different from it.
>
> Such a realm is art. But certainly only if reflection on art, for its part, does not shut its eyes to the constellation of truth after which we are *asking*. [VA I: 35/35]

In 1936, Heidegger wondered whether Hegel was right in saying that art is something past, without power for the modern spirit: "is art still an essential and necessary way in which that truth happens which is so decisive for our historical existence, or is art no longer of this character?" [HW: 67/81] Heidegger believed that technology and art were related in that both were truth-events: both were ways of letting entities be. Ordinarily, the Greek word *techne* is translated as a skilled making of the sort which anticipated the amazing production process of industrial technology. Heidegger argued, however, that *techne* had a twofold meaning. On the one hand, *techne* could be interpreted as involving *melete*, the authentic carefulness required to preserve something and to let it be. On the other hand, *techne* could be interpreted as involving *poiesis*, a producing or bringing-forth of something. Both aspects of *techne* involved intimate knowing of what things are. [GA, 43: 202] Skilled making was possible only because the artisan was gifted with the *techne* to disclose and to care for the thing to be made.

During the mid-1930s, Heidegger often spoke of the "letting-be" and the "producing" of *techne* as a violent knowing which involved an "incursion" against the overpowering presencing of the entity. Great art is such *techne*, which discloses what the entity as a whole is. [GA, 43: 98] Nietzsche's Overman was the artist who established the *limits* necessary to bring to a stand the overpowering presencing of entities, and thus restored rank, order, and meaning to things. Great art was a preserving, a measuring, a shaping (*Gestaltung*) of entities as a whole. [GA, 43: 105] For a time, Heidegger maintained that the act of founding a state was complementary to the act of producing a work of art: both contributed to setting up the boundaries, limits, and goals necessary for the self-manifesting of entities and for the enduring existence of a historical people. Although supposedly speaking of Nietzsche, Heidegger must have had his own political decision in mind when he spoke about the importance of establishing primordial goals for humanity:

> Goal-grounding is grounding in the sense of the awakening and freeing of those powers which lend to the posited goal the all-surmounting and all-mastering force of bindingness. Thus, only in the sphere opened up and erected through the goal can the historical *Dasein* become primordially developed [*wachsen*]. Finally, and this means incipiently, *to this belongs the growing of the forces which support and inflame and take the risk of preparing the new sphere, of advancing into it, and of structuring what is enfolded within it.* [GA, 43: 194; my emphasis]

But this shaping must not be conceived as the "stamping" of a subjective form upon a chaotic matter, although certainly this was how Nietzsche himself seemed to speak at times. Heidegger, too, sometimes spoke as if the artist gave shape or outline to previously indeterminate things. For the most part, however, he maintained that the delineation of the *Gestalt* of things by the work of art was achieved primarily by the things themselves, not by the artist. Rather, the artist existed as the "clearing" in which the self-limiting, self-defining disclosure of entities could take place. [GA, 43: 138] Hence, the power of *Gestalt*-giving was no merely human faculty. The violent activity of the poet, thinker, builder, and statesman is "a binding and articulating of the powers by virtue of which the entity discloses itself as such, as man moves into it." [GA, 40: 166] Man was compelled to take upon himself the power already at work in him: to disclose entities by forming a world in which they could stand forth.

Heidegger viewed the Overman as the one claimed by overpowering being to limit its own overpowering character, so that being could come to a stand, could appear, within a form, a limit, a shape. But all too often Nietzsche spoke of the Overman as impressing a form upon a chaotic matter, instead of acting as the midwife for the birth of the world-ordering *Gestalt* in which entities could display themselves according to their own appropriate shapes. Hence, for his idea that the presencing or being of entities appropriated humanity as the entity for their self-manifestation, Heidegger drew more upon Hölderlin than upon Nietzsche. For Hölderlin, art and nature (*physis*) were harmoniously linked. He spoke of artistic man as the "flower of nature," who brought nature into articulated manifestness. For Hölderlin, "the formative and artistic need is a veritable service that men render to nature." In Hölderlin, Heidegger heard "the song of the earth," the disclosure—through the poetic song—of the relatively autonomous and spontaneous forms of entities.

Nietzsche had stressed that art involved the act of limiting and bounding the overwhelming chaos. Great art was the creative tension of the Apollinian and the Dionysian. The fact that Nietzsche often spoke as if the artist himself "produced" the boundary-setting forms, and as if these forms virtually "created" the entities encountered by humankind, was no accident. For Nietzsche stood at the end of the history of productionist metaphysics; hence, he spoke of creating in subjectivistic and anthropocentric terms, as if humanity had taken over the role of the deceased Creator God. In his "confrontation" with Nietzsche, Heidegger argued that the work of art is not a human invention but instead an ontological event which occurs for the sake of the disclosure of entities. The disclosure of entities, Heidegger argued, must not be interpreted as occurring for some instrumental purpose, e.g., for the sake of enhancing human power, but instead must be regarded as ultimately purposeless. Things appear because they appear. Heidegger rejected, then, not only the Biblical creation story but also the idea that humans were given "dominion" over the earth. Instead, he believed, humanity should view itself as serving the presencing of entities, not as dominating them. Hölderlin, so we are told, understood more profoundly than did Nietzsche the ontologically disclosive nature of art. Hence, Hölderlin's poetry might open the way beyond the era of technological nihilism by defining *techne* not as nonstop industrial production but instead as appropriate disclosure of the being of entities.

Chapter 8. Hölderlin and the Saving Power of Art

Toward the end of his life, Heidegger maintained that his thought was essentially related to Hölderlin's poetry. [Sp: 214/281] This relationship is strikingly visible in Heidegger's 1934–35 lectures, *Hölderlins Hymnen "Germanien" und "Der Rhein."*[1] These lectures are crucial for understanding Heidegger's vision of National Socialism as originating a new world of work. Before resigning as Rektor, he had intended to offer a course on politics and the state. His decision to replace this course with the lectures on Hölderlin signaled at least two things. First, it marked his conviction that subsequent political decisions in Germany would go awry unless inspired by Hölderlin (whom he regarded as Germany's Homer) and given the necessary spiritual direction by Heidegger himself. Second, it indicated that he now regarded the work of art, not the method of philosophical *Wissenschaft*, as the prerequisite for an authentic German world. Beginning in the mid-1930s, then, Heidegger concluded that the work of art could help to make possible the non-representational, non-calculative, meditative thinking which would usher in the post-metaphysical age.

A. THE POET'S ROLE IN TRANSFORMING GERMAN *GEIST*

Heidegger believed that the generation of people who were university students between 1909 and 1914 did not fully understand Nietzsche's exhortation to see science through the optic of the artist, and to see art through the optic of life. At the time, this exhortation was interpreted to mean that science should not be a dry, abstract enterprise, but that it should become both more artistic and more useful for life. To Heidegger, this view of Nietzsche's exhortation, while helpful in some ways, involved an understandable misinterpretation on the part of young people hungry for cultural renewal. Unfortunately, Heidegger recalled, "no one was there who could have given us the right interpretation because for this is necessary the re-asking of the basic question of Western philosophy, the question about being. . . . " [GA, 43: 272] Beginning in 1934–35, however, Heidegger began to proclaim that the "right interpretation" could emerge from his articulation of Hölderlin's poetic vision.

We encounter the same complexity and violence in Heidegger's commentaries on Hölderlin as we do in Heidegger's commentaries on Nietzsche. One is never sure whether Heidegger or Hölderlin is speaking, although Heidegger would have asserted that "the matter itself" (*die Sache selbst*) was speaking. In Heidegger's view, this "matter" concerned Germany's present struggle to initiate a new beginning in the face of modern technology. Hence, he could systematically ignore the fact that Hölderlin's poetry was often inspired by and in dialogue with the French Revolution.[2] Thus, while Heidegger's reading of Hölderlin displays a formidable textual knowledge and insight into a difficult poet, that reading is also at times selective and idiosyncratic, refracted through the optic of Heidegger's specific philosophical and political concerns.

For Heidegger, poetry (*Dichtung*) was a "saying" which gathered entities and let them manifest themselves. And Hölderlin, at least for Heidegger, was "the *poet of poets*." [GA, 39: 30] The *Dasein* of a *Volk* sprang from poetry; as its poetry, its "saying," declined, the *Volk* fell into technological nihilism. National Socialist ideologues wrongly described poetry as the "expression of experiences" (*Ausdruck von Erlebnissen*), in Heidegger's view, for such talk was consistent with the view of humanity as a clever animal capable of "expressing" its experiences. Even dogs "express" themselves, Heidegger sardonically remarked. [GA, 39: 28] In a slap at ideologues who believed that their vision of National Socialism would save Germany from English liberalism, Heidegger instead asserted that the idea of "expressing" experiences was an aspect of the "determinate mode of being of 'liberal' man." [Ibid.]

The ideologues of poetic "expression" mistakenly believed that the revolution had occurred in 1933, blind to the fact that "we do not know our authentically historical time. The world-hour of our *Volk* is hidden from us." [GA, 39: 50] For Heidegger, the *real* revolution, in other words, had not yet begun. In 1933–35, he was to argue that the poet, the philosopher, and the state-founder were the "authentic creators" necessary for bringing forth the new historical world at the proper time. [GA, 39: 51] By the late 1930s, however, increasingly skeptical about Hitler, he was no longer including statesmen in the ranks of authentic creators. Genuine "creating" is linguistic: only poets and thinkers speak a new world into being. Such world-historical creativity was not produced by individual geniuses, however; creativity was the fruit of the linguistic-ontological power at work *through* creative individuals. In Heidegger's view, therefore, humanity does not possess language or "saying" as a tool; rather, language possesses humanity. [GA, 39: 67] The gift of language poses at once the greatest gift and the greatest danger for humanity, for the loss of a proper relation to language leads to spiritual and social catastrophe.

What, then, did Hölderlin, the poet's poet, have to say to the German people in its hour of need? To answer this question, we must briefly turn to Heidegger's interpretation of moods. He maintained, as we saw earlier, that moods make it possible for entities to show themselves. [GA, 39: 82][3] Moods do not represent things; they open up a realm in which things can first be represented. [GA, 39: 140] Hölderlin's poetry opened up the mood which enabled Germans to experience (*erfahren*) the departure of the old gods. But it also could elicit a mood in which Germans could be carried away (*entrücken*) to new gods and carried back (*einrücken*) to their earthly homeland. Responding to what the mood of the poet discloses, the thinker in turn *names* what the poet has allowed to be revealed.

Primal moods, including those elicited by the poet, disclose not only the presencing of entities but also the finitude of thought, its inability to capture either that presencing or the totality of entities thus present. Such moods disclose human finitude by showing that *no reason can be given for why things are*. Long before it occurs to us to ask about the origins of things, we have been thrown into a world in which we are affected and influenced by the overwhelming presence of entities. A primal mood places *Dasein* before the abyss of mortality and finitude, because it reminds *Dasein* of the absencing or the nothingness that hides itself in the very event of the presencing of entities.

In *Being and Time*, Heidegger maintained that *Angst* revealed both the finitude of *Dasein* and the inexplicability of entities. For early Heidegger, and in some cases for later Heidegger as well, *Angst* seems to have been an essential or transhistorical mood. Beginning around 1929, however, he began speaking of primal moods as historical in character. Later, he was to name at least four moods which governed how different peoples could encounter entities.[4] Astonishment or awe was the basic mood of the Greeks. [GA, 45: 162ff.] Certainty was the basic mood of the Cartesian age. The two basic moods which characterized technological humanity were horror (*Erschreckung*) and boredom (*Langweilige*). Whereas at first glance these moods may seem antithetical, boredom and horror are really two sides of the same coin: both moods reveal the meaninglessness of things in a world from which the gods have fled. Hölderlin's poetry encouraged the German *Volk* to discover the extent to which their lives were pervaded by these moods. Only by confronting their terrible situation, so Heidegger believed, could the Germans be opened up for a new beginning.

B. GERMANY'S ANOINTED TASK

In much of his poetry, Hölderlin turned to the ancient Greeks for inspiration. Heidegger looked to the Greeks for insight into the possibility of preparing for the advent of new gods. While recognizing that the Greek world could never be revived, Heidegger called for the *Volk* to initiate a beginning that was as radical and enduring as that carried out by the ancient Greeks. This new beginning was to be not an empty repetition of Greek culture in German dress but an authentic repetition of the world-founding *act* of the Greeks. Heidegger explained the essence of this creative act in terms of Nietzsche's Dionysian and Apollinian (*Apollinische*) categories, but qualified Nietzsche's use of these terms by reading them through Hölderlin, who he believed had meditated upon their relationship more deeply.

Every historical *Volk*, Heidegger argued, has both an endowment (*Mitgegebene*) and an anointed task (*Aufgegebene*).[5] The Greeks, existing within the mood of astonishment or wonder, were endowed with the Dionysian gift, "the exciting nearness to the fire of heaven, the being-touched by the might [*Gewalt*] of being [*Seyns*]. For them, the anointed task is the subduing of the unsubdued in the struggle of the work; grasping; bringing-to-stand." [GA, 39: 292] Struck by the Dionysian presencing of entities, the Greeks were com-

pelled to shape that presencing in terms of Apollinian metaphysical categories, forms, and structures. It was these categories which initiated the history
of productionist metaphysics—the history which culminated in the view that
"to be" means "to be produced" by and for the self-certain human subject.
This extreme Apollinian world-structuring became, in turn, the endowment of
the Germans: the "capacity for grasping; preparing and planning of domains;
and the calculating, the ordering up to the organization. For them [the
Germans], the anointed task is being-touched by being." [Ibid.] Whereas the
Greeks had to allow the Apollinian (their "task") to shape the Dionysian (their
"endowment"), the Germans had to let the Dionysian (their "task") free up
the Apollinian (their "endowment"): "This is the 'paradox.' Since we fight the
cause of the Greeks, but on the opposite front, *we become not Greek but German.*" [GA, 39: 293][6]

The basic mood for the Greeks was astonishment or awe; the basic
moods for the Germans were horror and boredom: "In *astonishment*, the basic
mood of the first beginning [in Greece], the entity comes to stand in its form
[*Gestalt*] for the first time. In *horror*, the basic mood of the other beginning [in
Germany], there is revealed behind all progressiveness and mastery of the
entity, the dark emptiness and goallessness and avoidance of the first and last
decisions." [GA, 45: 197] Just as the Greeks initiated Western history by
profoundly experiencing and responding to what was disclosed in their primal
mood (awe), so, too, the Germans could initiate a new historical beginning if
they experienced and responded appropriately to what was disclosed in their
primal moods (horror and boredom). Hölderlin's poetry could enable the
Germans to become accessible to the *pain* involved in the horror of modernity. Pain, so Heidegger believed, is "the basic form of the knowledge of
Geist." [GA, 39: 135] The "saying" of the poet could help to transform the
mood of horror into one of "holy affliction, mourning but prepared [*heilig
trauernden, aber bereiten Bedrängnis*]." [GA, 39: 141]

Clearly, this holy affliction could be neither explained in political terms
nor resolved by political actions. Rather, it called for a primal encounter with
being. For Heidegger, being named the self-concealing presencing in light of
which entities revealed themselves in their various ways. His image for this
self-concealing presencing was the mask: "This [the mask] is an exemplary
symbol of Dionysius, i.e., understood metaphysically and in Greek terms: the
primordial relatedness of being and not-being (presence and absence) to one
another." [GA, 39: 190] The Germans, like the Greeks, were in principle
capable of experiencing the self-concealing presencing of being in a primordial way. Lack of such experience, however, had deadened the Germans and
had led them into technological nihilism.

Paradoxically, Heidegger argued, the mourning involved in experiencing
the "holy affliction" (departure of the gods, concealment of being) would also
be the event of experiencing being anew—the event which would prepare the
way for the arrival of new gods. As Hölderlin's close friend Hegel explained it:
the hand that wounds is also the hand that heals. Hölderlin shared this vision
poetically, prophesying, "Where danger is, also grows the saving power."

Only by becoming vulnerable to the horror and grief involved in technological nihilism could the *Volk* discover that such nihilism was historical in character, so that an alternative disclosure of entities could be possible. The mood of holy affliction would presumably dissolve the Apollinian mask which had frozen Germans in their rigid technological stance. The "saving power" could arise only when, casting off their technological armor and ceasing to fall into the distractions of mass culture and the myth of "progress," Germans surrendered to the pain involved in the death of God.

Like Nietzsche and Hölderlin, Heidegger experienced the departure of the old God as an event of the greatest possible significance. Yet, in Heidegger's view, Hölderlin not only suffered the pain of this loss, he also "experienced poetically a creative decline of the truth of Being up until now." This decline was "creative" in that it mysteriously enabled Hölderlin to bring forth a creative rejoinder to it, a rejoinder which illuminated "new powers." [GA, 39: 150] Such a rejoinder, for Heidegger, was not merely an "aesthetic" act, but " 'politics' in the highest and authentic sense." [GA, 39: 214] Hölderlin's poetry was thus for those people who were called as "builders of the new building of the world." [GA, 39: 221] In creating such a world, Germans would not be forced to disown their technological skills; they would develop new social forms that would make possible a new relationship to working and producing in the post-modern age. Unfortunately, Heidegger did not successfully explain in concrete terms how his insights about producing could be translated into the political world.

His Hölderlin lectures clearly aimed at influencing political events. Now, however, somewhat chastened by the events of 1933–34, he distinguished between his understanding of "politics" and the understanding which governed current events. Thus, for example, he distinguished between the fatherland spoken of by "noisy patriotism" and Hölderlin's idea of the fatherland as the *Volk* of this earth in its historical being: "This being [*Seyn*], however, becomes poetically founded, thinkingly articulated and placed in knowledge, and in the deed of the one who grounds the state rooted in the earth and in historical space." [GA, 39: 120] Heidegger went so far as to claim that "the *'fatherland' is being itself*, which from the ground up bears and articulates the history of a *Volk* as an existing one. . . . " [GA, 39: 121] Here, he interpreted "being" as "innerness" (*Innerlichkeit*), which he regarded as Hölderlin's basic metaphysical word. [GA, 39: 249] Innerness named the "mystery" of being, "the primordial unity of the animosity [*Feindseligkeit*] of the powers of pure originating." [Ibid.] Through this self-unifying Heraclitean struggle, decisions are made about gods and earth, men and their artifacts. Everything "is" only through the striving of the self-concealing "powers" of being. Without the saying of the poet, everything would remain confused in the darkness of "holy wildness." [GA, 39: 257]

Only a poet could make the encounter with Dionysian wildness that the Germans so desperately needed a fruitful one. Hölderlin spoke of this "wildness" as "nature," a term very close to what Heidegger meant by "being." Heidegger read Hölderlin as believing that entities can show themselves only as a result of a Heraclitean struggle that both sets apart and unifies nature and humanity. As that which "founds" nature, poetry is the "clang" of the weapons of nature itself, the struggle which brings itself forth in words. [Ibid.] Poetry, then, must be understood as creation without an object: poetry is the "saying" that holds together the agonistic relationship between nature and humanity in such a way that a space is opened wherein entities may show themselves. By "saying" this struggle, Hölderlin also helped to elicit the mood of holy affliction necessary if the *Volk* was to encounter the depth of its despair resulting from its lack of contact with the nourishing "holy wildness." Only a poet could make it possible for the *Volk* to encounter holy wildness without either destroying it or being destroyed by it.

C. HEIDEGGER'S CHANGING VISION OF GERMANY'S TRAGIC DESTINY

Being, the self-concealing origin (*Ursprung*) and hidden presencing of entities, is what Heidegger understood as the primal "mystery." Profoundly disturbed by the loss of this mystery in modern times, his favorite poet Hölderlin often sang of "homecoming," the return to Germany's mysterious saving origin. But Germany could once again enter into the precinct of this holy origin only if there arose a poet who was destined to "found" (*stiften*) that origin in the sense of limiting and defining it. In a manner similar to his reading of Nietzsche on artistic creation, Heidegger read Hölderlin's poems "*Germania*" and "*Der Rhein*" as saying that poetic founding gives boundary and limit to the overpowering stream of the origin. Expanding on this interpretation, Heidegger argued that this very stream appropriates the poet as the unselfish counterwill necessary to bring it "to a stand," and thereby to open up a stable, historical world in which entities may manifest themselves. The ontological stream establishes a way for itself, in other words, by appropriating the poetic visionary. In so doing, the stream becomes

> a fate [*Schicksal*], an enduring [*Leiden*] in the sense of suffering [*Erleiden*]. But this does not mean the merely passive [event of] being-struck by another, but instead suffering as *first to struggle with and to create being in the enduring [im Leiden erst das Seyn erstreiten und schaffen]*. Being as fate has its origin not in the pushing of something decreed, assigned, as purely unalterable "lot," . . . but instead surmounting the breach and willing-back from this into the origin characterize the being of destiny. [GA, 39: 235; emphasis in original]

For Heidegger, the poet's struggle with the originating stream was akin to the hero's violent struggle with destiny, which is the essence of Greek tragedy. Like Hegel, he regarded Sophocles' *Antigone* as the "foundation of the whole of Greek *Dasein*." [GA, 39: 216] Antigone resisted the command of her uncle, who—representing civil authority—refused permission for her to bury her brother, who had been killed in civil war against him. Torn between her loyalty to the state and her obligation to the law of gods and family, Antigone—in courageous struggle with her destiny—illuminated and held open the enduring rift between what Heidegger was to call "world" (the open historical-political realm) and "earth" (the self-concealing dimension of "nature"). Antigone's death was the unavoidable accompaniment of a greatness that involved distinguishing world and earth in such a way that each could be itself. Only as thus set apart could world and earth enter into a dynamic relationship, a mysterious and hidden unity.

In Heidegger's reading of Greek tragedy, great tragedians such as Sophocles saw the necessity of preserving this dynamic strife, but decadent ones such as Aeschylus could not. At the end of his Oresteian trilogy, Aeschylus portrays Athena as balancing the vengeful claims of the Erinyes (representing fate), on the one hand, against the pleas of suffering Orestes, on the other. Athena persuades the Erinyes to accept a new dispensation in which reason and mercy provide a counterweight for fear and vengeance. From a Heideggerean viewpoint, such a "reconciliation" was a false one, because it broke the dynamic tension which both set apart and united fate and humanity in the "truer" tragedy of Sophocles.

In his first reading of Sophocles' *Antigone* in 1935, Heidegger focused on the choral strophe which speaks of man as the strangest and most violent of all things. He is strange not only because he is vulnerable like no other entity to the overpowering power of being as fate (*dike*), but also because he strikes out against that power and compels it to come to a stand. In attempting to master being, man must ultimately "shatter" against it. *Techne*, the violent shaping which limits being, must ultimately give way in the face of *dike*, the fate which man can never control: "Therefore, the violent one knows no kindness and conciliation (in the usual sense); he cannot be mollified by success or prestige. . . . In willing the unprecedented, he casts aside all help. To him disaster is the deepest and broadest affirmation of the overpowering." [GA, 40: 172/137]

It would seem, then, that either suicide or violent death, prerequisites for tragedy, is the unavoidable fate of the hero seeking to take a stand in the overpowering stream of destiny. Heidegger believed that Hölderlin (like Nietzsche) went mad because he stood in that stream and was overpowered by Dionysian presencing. In 1934–35, Heidegger, in effect, portrayed the German *Volk* as a tragic hero. The *Volk* was confronted with the technological fate initiated by the Greeks, but since this fate was not an ironclad decree, the *Volk* could make a creative rejoinder to it. Such a rejoinder was possible, however, only if the poet revealed to the *Volk* that it was their anointed task. The *Volk* could not simply cast aside the technological-Apollinian "endowment" and return to an earlier age. Instead, the *Volk* had to find a way of preserving the creative and dangerous tension between that endowment and the Dionysian "anointed task," a tension expressed so well in Sophoclean tragedy.

Heidegger drew another analogy between the Greek tragic hero and the German *Volk*. The Greek hero had to experience great suffering that was in some measure self-imposed in order to become open for new knowledge. Hence, Oedipus puts out his own eyes, but gains new insight as a result. Just so,

the German *Volk* must "go under" by suffering the pain of modern technology
before it can "rise again": "The going-under is thus a historically decisive moment which can extend itself to a century, because there the uncreated [*Uner-
schöpfte*], the uncreatable [*Unerschöpfliche*] of the new beginning, the *possible*,
can bring itself to power. . . . " [GA, 39: 122] Those Germans who believed that
the "revolution" had been achieved with Hitler's accession to power were,
then, wrong: 1933 was only the *beginning* of Germany's genuine struggle with
its destiny. All talk of a "thousand-year Reich" was foolish insofar as it underes-
timated the tenuous nature of all attempts to found and to preserve new historical worlds. Each new generation of the *Volk* would have the "task" of remaining
open for the Dionysian origin; each would also newly and perpetually confront
the technological-Apollinian "endowment."

In the 1940s, as Heidegger moved toward the language of "releasement"
(*Gelassenheit*), he lectured again on the strophe from *Antigone*. This time, however, he focused not on the "violent one" who compels fateful being to come
to a stand, but instead on the speech of the chorus at the end of the strophe:
"May such a man never frequent my hearth; may my mind never share the
presumption of him who does this." But in 1935, Heidegger ignored this line.
Only years later did he conclude that it was central to the meaning of the
tragedy. Kathleen Wright has observed how, in Antigone's last words with
Ismene, Heidegger saw that

> Antigone understands that that against which nothing can be done is the utter
> strangeness of human being in the world. . . . Unlike those others who turn
> against the overpowering power of Being and who take the necessity of disaster to
> be only a necessity that follows from particular human actions, Antigone accepts
> the necessity of disaster as a necessity of her very own being. . . . Antigone is,
> Heidegger says, uniquely strange . . . because she is self-consciously *to drama*, the
> tragic drama of being human. [Cf. GA, 53: 144ff.][7]

There is another way of being human, the older Heidegger realized, than
sailing away from the familiar to the turbulent sea and committing acts of
violence against the overwhelming power of being. This other way is the way
of the hearth. "The hearth (*hestia*)," Wright tells us, "is another word for
Being, one which counters the first word for Being, *to deinon*, Being as the
overpowering power of Being which is related to man's strange, and inevitably
catastrophic way of being."[8] Whereas in 1935 Heidegger had portrayed violent man as the "strangest," several years later he came to depict Antigone as
the "strangest" not because she is cast out of her home like a man compelled
to adventure, but instead because "she preserves and celebrates in the tragedy
of her life a *memory* of the familiar (*das Heimische*)."[9] Remembering what had
been forgotten, remembering the home-ish, remembering the sheltering origin—that became for Heidegger the most important thing a person could
"do" in the technological age. Just as Antigone's struggle and death reminded
her people of the dangerous but inevitable struggle between family and state,
so too Hölderlin's struggle and breakdown ought to remind the *Volk* of the

central struggle between endowment and anointed task, as well as between "earth" and "world." The authentic "struggle," so the later Heidegger concluded, was not the violent action of the man but rather the receptive attitude of the woman staying by the hearth—remembering what needed to be remembered. Such remembering or "homecoming," which Heidegger felt was the essence of Hölderlin's poetry, would prepare a place for the arrival of the new gods. Their arrival would mean the end of the history of productionist metaphysics and the beginning of a new history. For Heidegger, the history of metaphysics took the shape of a spiral in which the final stage was the working out of the possibilities "given" at the beginning to the Greeks. In arriving at the "end," then, the West returned to the origin from which a new and "fateful gift" (*Geschick*) might emerge.

D. EARTH, WORLD, AND AUTHENTIC CREATING

In "The Origin of the Work of Art" (1935–36) Heidegger elaborated upon the conception of artistic creation which he initiated in his lectures on Hölderlin. Once again, he portrayed the work of art as playing a crucial role in a basic conflict. The work of art, he explained, opens up and sustains the ontological strife between "world" and "earth." "World and earth are always instrinsically and essentially in conflict, belligerent by nature." [HW: 44/55] But, just as in his Hölderlin lectures, Heidegger had difficulty in thinking through to a definition of both "world" and "earth," as well as the relationship between them.

In the first version of the art-work essay (1935),[10] Heidegger conceived of world and earth as corresponding to the two aspects of "truth," understood as unconcealment. In every instance of such unconcealment, concealment also obtains. There can be no total disclosure of things. If "world" names the unconcealment dimension of truth, then "earth" names the concealment dimension. Heidegger concluded, however, that this distinction was inaccurate: "the world is not simply the open that corresponds to clearing, and the earth is not simply the closed that corresponds to concealment." [HW: 43/55] Rather, the duality of unconcealment-concealment characterizes world *and* earth, each in its own way. Let us examine each in turn.

The world is a historical, hence, partial and perspectival, unconcealment of entities. In a given historical epoch, entities may manifest themselves in a way that did not occur in another epoch. People living in any particular world, however, and especially in the technological world, presuppose that nothing is concealed in their world: everything is present. But, Heidegger argued, such a total revelation of things is impossible, not only because of the finitude of each historical world, but also because things themselves—which belong to the earth—are intrinsically self-concealing and resist every attempt to make them wholly present.

By "earth," Heidegger came to mean, in part at least, this self-concealing dimension of all entities. Earth resists the world's disclosive assault and thus can never be brought completely into the realm of history. Earthly things are not simply self-concealing, however, they do also emerge into presence. A blossoming flower, for example, unconceals itself by putting forth its colors. Animals and plants involve a spontaneous surging forth. Their modes of being include giving birth, growing, blooming, and perishing. Because earthly things always turn in upon and conceal aspects of themselves even while coming into presence, Heidegger described earth as "abysmal" (*Abgrundig*). Earth, then, is not merely self-concealment but a spontaneous self-emergence which always involves a self-concealing dimension.

Heidegger's meditations on the meaning of "earth" reveal a tension in his notion of "creating." At times, he tended to conceive of creating as *schaffen*; later on, he focused on creating as *schöpfen*. Both *schaffen* and *schöpfen* may be translated as "to create," but the former term emphasizes that the role of the artist is to shape or fashion a thing, while the latter term emphasizes that the role of the artist is to draw out or to disclose what is somehow already there. *Schaffen* characterizes the art of the violent one who carries *Gestalt* into being and brings it to a stand; *schöpfen* characterizes the art of one who remembers the origin and can listen to the being of the entity to be disclosed through the work of art. In portraying earth as the self-generating dimension of things, Heidegger suggested that it constitutes the inexhaustible plentitude of "natural" forms and processes which humans may be able to "draw out" (*schöpfen*), but never to "create."[11]

We must keep in mind that the self-revelation of things which occurs through the world opened up by the artist is not identical with the self-emergence of things, even though Heidegger tried to conflate these two senses of coming-into-presence. By coming to define "earth" as the non-historical, spontaneous, self-generating aspect of things, Heidegger corrected his tendency in *Being and Time* to reduce "nature" to a primarily passive, intraworldly totality of entities. Earth cannot be reduced to a dimension of world. Yet earth also cannot be understood naively as being radically separate from world. They maintain each other in their primal conflict. Moreover, both earth and world arise from what Heidegger called the primal "open," which seems to be akin to what Hölderlin meant both by "holy wildness" and by "innerness": that generative, agnostic source which precedes and makes possible earth and world. Heidegger wrote: "Innerness is that primordial unity of the animosity [*Feindseligkeit*] of the powers of pure springing-up." [GA, 39: 250]

Yet even as Heidegger sought to give to earth a dimension of spontaneity that was not assigned to "nature" in *Being and Time*, in "The Origin of the Work of Art," he sometimes suggested that entities could show themselves only in contrast with a work of art, such as the Greek temple. Here, natural things seem once again to *belong* to the world opened up by that art work, just as natural things *belonged* to the world as described in *Being and Time*. Heidegger did not sufficiently clarify the extent of the earth's "autonomy" and "self-definition" over against the disclosive power of world.[12] Resolving the question of the relation between earth and world is crucial for understanding what Heidegger meant by authentic "producing." We shall explore these issues in more detail in chapter fourteen.

In describing earth as the self-concealing, self-emergent dimension of things, we have considered only *one* of the meanings Heidegger assigned to earth. As Michel Haar points out, there are at least three other meanings to Heidegger's new ontological category of "earth."[13] The second meaning of earth, the "native ground" of a historical *Volk*, also pertains to the issue of whether the work of art, specifically poetry, "uncovers" the shapes of things, or "fashions" those shapes. As "native ground," earth is always specific, and never merely the territory depicted by the cartographer. This native ground, in Heidegger's view, contains the "destiny" of a people. Earth involves what a *Volk* can become. The poet opens up that "into which *Dasein* as historical is already thrown. This is the earth and, for a historical people, its earth, the self-closing ground, on which it rests with all that it already is, although still hidden from it." [HW: 62/75] In envisioning the future of his people, the poet, in effect, draws upon these hidden possibilities. Creation as drawing-forth (*Schöpfung*) of such slumbering possibilities is different from creation as the imposition of arbitrary linguistic categories upon the earth. The discovery of pre-existing possibilities of the *Volk* is not the same as the discovery of the pre-existing shapes of living things. Both *Volk* and things are "of the earth," but the *Volk's* historical mode of being is different from the mode of being characteristic of plants and animals. Nevertheless, with the notion that earth contains pre-figured historical possibilities, Heidegger hints that the earth may also contain the possible outline of "natural" things.

The third meaning of "earth" suggests that there is an autonomous, extrahistorical dimension even to merely "material" things, such as granite. Earth is not a passive, formless "stuff" or "matter" to be "formed" at will by the artist and technician, but rather possesses intrinsic qualities of its own which can manifest themselves appropriately only within the struggle constituted in and by a work of art. In the work of art, the conflict between earth and world is revealed. As world dawns, it discloses that which is measureless, and thus discloses the need for measuring. But earth rises up within that world, as that which bears everything, which shelters itself in its own law, and which is self-contained. [HW: 51/63] The conflict between the disclosing which wants to measure and the self-arising which contains its own law involves an essential conflict, a rift (*Riss*) which makes up the "innerness" of the event of being, the event of truth. The art work can never *represent* the primal rift, the struggle between earth and world, but it does somehow *embody* it. *Gestalt* or figure names the way in which this rift is "set in place" in the work of art.[14] As the particular embodiment of this enduring rift, the work of art constitutes the primordial sketch or outline which measures and gives meaning to entities: "Figure is the structure in whose shape the rift composes and submits itself. . . ." [HW: 52/64] The figure, as embodiment of the primal rift, outlines and measures entities which, as arising from the earth, already possess their own boundary and limit, although because this boundary remains shrouded in the earth's "holy wildness," it can never be made wholly present. Earth manifests itself from within the figure in which the art work is "fixed," but the measure of this figure never overwhelms the self-concealing boundary of the metal, stone, wood, or words which make up the earthly element of the work.

Curiously, in 1935–36 Heidegger used the term *Gestell* to describe the *Gestalt* through which the art work—as the embodiment of the conflict between earth and world—was set into place. Here, *Gestell* did not yet have its later meaning of the violent setting-upon that is essential to the technological disclosure of entities. This later usage of *Gestell* is, in fact, Heidegger's term for Jünger's *Gestalt* of the worker. In 1935–36, however, Heidegger was still working through his understanding of the *Gestalt*-giving nature of the work of art. He wanted to conceive of art or *techne* not as the "stamping" of a form upon passive and malleable chaos, but instead as the measured disclosure of the forms of entities within the opening provided by primal conflict. In Heidegger's view, Jünger knew nothing of the unifying conflict at work between earth and world, a conflict which was also "working" in the work of art. It was this conflict which constituted the *Riss*, the outline, the *Gestalt* in terms of which entities could appear. The fixing of truth in the art work, i.e., the embodiment of the primal rift in a particular outline or form, occurs in such a way that the art work itself withdraws to reveal itself as a truth-event through which other entities can show themselves and thus "be." The art work is so self-sufficient because it embodies agonistically the polemical event of truth upon which other entities are dependent for their unconcealment.

It is by no means clear whether Heidegger's notion of artistic *Gestalt* as ontological "outline," which is quite suggestive in terms of poetry, can provide a satisfactory replacement for the old "matter-form" distinction so useful for the plastic arts. The issue becomes even more complex when we consider poetry. What exactly is the "earth" dimension of the poem? And what is the "world" dimension? In poetry, Heidegger said, words withdraw so that things may be. But is the self-withdrawing or self-concealing of the words analogous to the self-concealing of the marble in a statue, or the self-concealing of the spontaneous presencing of an animal? These puzzling questions still await answers.

The fourth and final dimension of Heidegger's concept of "earth" involves its relationship to Greek *physis*, on the one hand, and Hölderlin's "nature," on the other. Like Hölderlin, Heidegger looked to the Greeks for insight, but like Hölderlin he also sought to go beyond them. Hence, Heidegger commented, "We might think of the *physis* of the Greeks. We also ought to do so, granted that we understand and surmise this *physis* adequately. And yet—even this does not suffice. Hölderlin is not Greek [*Greichentum*], but instead the future of the Germans." [GA, 39: 255]

The task for Germany was to find in Hölderlin's poetry a new understanding of *physis*, or nature, an understanding that would make possible an alternative to the technological view of nature. What the Greeks had meant by nature had, over the centuries, been

> *denatured [denaturiert]* through two strange powers. First, through Christendom, through which nature first gets degraded to "something made" and simultaneously gets brought into a relation to a super-nature (realm of grace). Then, through modern science, which reduces nature into the power sphere of the

mathematical ordering of world commerce and of machine technology in the special sense. . . . [Such technologized science] has nothing to do with the inner truth of natural science. [GA, 39: 195][15]

Heidegger emphasized that the technological and scientific view of nature was the late outcome of the early Greek struggle to bring Apollinian limits to the overwhelming Dionysian surge of being. This struggle was poetic in character; hence, its technological outcome is also "poetic." The Greeks may have spoken more poetically of nature as the "divine fire," but

> the "astronomical" sun and the "meteorological" wind, which we people today suppose to know more progressively and better, are not less, only more awkwardly and more unpoetically, poetized than the "fire" in the poem. The poetizing of astronomy and meteorology, the "poetizing" of nature-explaining is of the sort of calculating and planning. *Planning is still a poetizing,* namely, the counter-presence and absence of poetry. [GA, 52: 40]

Although deeply influenced by Heraclitus, Hölderlin realized that he could not inaugurate a new beginning for the Germans simply by "reviving" the ways of the Greeks, for those ways had culminated in the very industrialism which so threatened the "homeland." Hence, so Heidegger argued, while recognizing that Hölderlin owed much to the Greeks, we must nevertheless keep in mind that Hölderlin was seeking to *transform* their vision into a specifically German one. [GA, 39: 128ff.] Taking his starting point from Heraclitus, Hölderlin defined nature or *physis* as the "All-One" (*en panta*). Heidegger maintained that this definition was neither a pantheistic blending of nature, history, and the gods, nor a division of them into radically separate domains. "Neither is [Hölderlin's definition] a mere renewing of the ancient worldview nor a mixing of the same with an indeterminate enlightened Christianity." [GA, 39: 150] To some extent, *physis* corresponds to the overwhelming stream of origin, holy wildness. Heraclitus realized that this originating stream required boundary and limit. *Logos,* in Heidegger's view, was Heraclitus's name for the "fire" which illuminates and gathers all entities. *Logos* is, in effect, "the same" as primal poetic "saying." The illuminating and gathering *logos* brings *physis,* nature, holy wildness "to a stand." Holy wildness or "nature," then, calls upon the poet to delimit that wildness in a way that enables all things to appear in their proper outline and measure. [GA, 4: 60] The poet, standing in the chaotic cleft of this self-concealing holy Open, brings forth the work of art needed to "gather" together the elements of a historical world.

Bringing forth such a world is not accomplished merely for the sake of humanity, but rather for holy wildness itself. In a letter of June 4, 1799, Hölderlin wrote: "the formative and artistic need is a true service that men render to nature."[16] Nature, in Heidegger's interpretation of Hölderlin, "needs" humanity. Yet it is nature that first grants the "open" in which the mortal poet can bring forth the "saying" to ground the world needed for the historical encounter between gods and mortals, and for the self-disclosure of the earth.[17]

By arguing that nature "needs" and "appropriates" humanity as the agency by virtue of which nature brings itself to a stand and thus reveals itself, Heidegger stood in close proximity to the tradition of the German idealists Hegel and Schelling, both of whom were friends of Hölderlin. Hölderlin's Holy, that which "grants the open," is close to what Heidegger came to mean by the appropriating event (*Ereignis*) which "gives" the modes of "beingness" and time which constitute the history of being: "The Holy, 'older than times' and 'above the gods,' grounds in its coming another beginning of another history. The Holy decides in the beginning over gods and men, whether they are and who and how and when they are. . . . The Holy grants the Word and comes itself into this Word. The Word is the *Ereignis* of the Holy." [GA, 4: 76]

One can scarcely overlook the extraordinary parallel between Heidegger's remarks about the relation of Holy to the Word, on the one hand, and St. John's comments about the relation of God to the Word: "In the beginning was the Word: the Word was with God and the Word was God. He was with God in the beginning, through him all things came to be, not one thing had its being but through him. All that came to be had life in him and that life was the light of men, a light that shines in the dark, a light that darkness could not overpower."

Clearly, it was not by chance that Hölderlin's poetry, especially his account of the All-One, the Holy chaos which brought forth all things, bears such a striking resemblance to such Christian motifs, and yet was also so amenable to Heidegger's lectures on Heraclitus. Hölderlin was torn between the Christian and the Greek understanding of the world. And while Heidegger became increasingly contemptuous of "Christendom" after leaving his seminary training, he retained and in many cases transformed the Biblical language of fallenness and salvation. His "confrontation" with Hölderlin's poetry and with Greek tragedy sought to provide a guiding religious, artistic, and ontological myth for Germany in its time of need. Attracted to the violent vision of Nietzsche and Jünger, he was also pulled toward the less violent themes in Greek discourse and in Hölderlin's poetry which found final expression in his own doctrine of "releasement." For the postwar Heidegger, resolute, impassioned decisions would not renew the German spirit; only a release from the craving for control associated with the culminating stage in the history of productionist metaphysics could redeem the German *Geist*.

E. THE DEATH OF GOD AT AUSCHWITZ?

Reading Heidegger's reflections of the 1930s on Hölderlin's poetry, one is often struck by the discrepancy between his lofty abstractions and the brutal reality of National Socialism. How could he have possibly imagined, even in the early stages of Hitler's regime, that there existed some inner relation between Hölderlin's "holy wildness" and the Nazi's blood lust? How could he have believed that Hölderlin's poetry would provide any guidance whatsoever for such a violent movement? There are no easy answers to these questions.

Heidegger continued to interpret Germany's history in the light of the tragic categories of Nietzsche and Hölderlin even after—and perhaps even *especially* after—Germany's destruction in World War II. In 1945, for example, Heidegger argued that the war had arisen because of the technological frenzy provoked by the "death of God." This "death," he assured his reader, had nothing to do with "ordinary atheism," but resulted from the fact that "the world of the Christian God has lost its effective force in history. (Cf. my 1943 lecture on Nietzsche's word 'God is dead.') If things had been different, would the First World War have been possible? And even more, had things been different, would the Second World War have become possible?" [DR: 25/ 485] In Heidegger's reading of the period from 1933 to 1945, the Germans, blinded by the death of God, had made a valiant effort to prepare for a new advent of the Divine, but were led astray when the technological Will to Power perverted the National Socialist "revolution."

In *La fiction du politique*, Lacoue-Labarthe has used Hölderlin's conception of tragedy not to support Heidegger's reading but to provide an alternative to Heidegger's interpretation of what happened both to himself and to Germany from 1933 to 1945. First, Lacoue-Labarthe argues, the postwar Heidegger erred by not applying the theme of the death of God to Auschwitz:

> In effect, God died at Auschwitz, in any case the God of the Greco-Christian West. And it is not by any sort of chance that those whom one wanted to annihilate were the witnesses, in this West, to another origin of God that had been venerated and thought there—if this is even, perhaps, [witnessing] to another God, remaining free from his Hellenistic and Roman captivity and impeding thereby the program of the accomplishment [of nihilism].
>
> This is why this event, the extermination, is with regard to the West the terrible revelation of its essence.[18]

Lacoue-Labarthe maintains that Auschwitz represents what the poet called a *caesura*, a radical break in history which occurs when God and humanity become radically separated. According to Hölderlin, such a *caesura* is a basic law of the tragic character of finite existence.[19] Auschwitz reveals the "essence" of the West, because at Auschwitz and other extermination camps the Western drive toward total domination achieved its apogee: humans disposed of other humans like industrial waste, burning them in crematoria built according to the latest technology. The radical break between God and humanity which occurred at Auschwitz resulted from a hubristic humanity which tried to raise itself to the level of the all-powerful God. Germans also exhibited *hubris* in their attempt to become their own origin by "creating" themselves as an all-embracing, national work of art. In this attempt, they defied their own finitude and sought an underived, unmediated kind of existence that belongs only to God. According to Lacoue-Labarthe, in the tragic *caesura* provoked by such *hubris*, God reveals Itself as an abyss, a total absence which signals either the possibility for a new beginning or else the end of all historical possibilities. Heidegger saw this abyss opening up in 1933. He regarded subsequent events, including the death camps, not as instances of the *caesura* of which Hölderlin spoke, but instead as the symptoms or consequences of real *caesura*: the death of God which Hölderlin and Nietzsche had both surmised in the nineteenth century.

Once again calling on Hölderlin, Lacoue-Labarthe refers to a second law of tragedy: the hero, faced with the absence of God, must finally submit to his own finitude and mortality. The hero's *hubris* invites a cleansing punishment. Heidegger agreed that such submission is vital to the possibility of an authentic *Volksgeist*. Lacoue-Labarthe argues, however, that Heidegger offered a partial or confused interpretation of the "tragic" situation in Germany. Heidegger finally concluded that National Socialism was an episode in the technological Will to Will; but was Germany guilty of *hubris*? If so, Germany should have suffered a cleansing punishment, which might be said to have come from the victorious allies. But Heidegger never clearly assigned blame to Germany for its behavior between 1933 and 1945. In Lacoue-Labarthe's eyes, the people who really suffered as a consequence of German *hubris* were, finally, not the Germans at all but the Jews, whom Heidegger conveniently left out of the tragic "equation." In almost a literal sense, according to Lacoue-Labarthe, the Jews were the tragic "purgation" or "cleansing" (*katharsis*) of Germany and the West. Had postwar Heidegger been more courageous, we are told, he not only would have connected Auschwitz explicitly with German *hubris*, but would have also regarded the Holocaust itself—not Germany's pre-1933 confusion—as the real *caesura* in Western history.[20]

Jean-François Lyotard has recently questioned Lacoue-Labarthe's use of Hölderlin's theory of tragedy to interpret the meaning of Auschwitz.[21] Lyotard argues that Auschwitz cannot have represented a tragic *caesura* because nothing has really changed since World War II: the West has not been released from the technological impulse. In this respect, Lyotard agrees with Heidegger's assessment of World War II. Moreover, Lyotard asks, if the Germans were guilty of *hubris*, why did the Jews die in the extermination camps? If the Germans were playing the role of the hubristic hero, why were they not the ones forced to acknowledge their finitude and accept the tragic law of destruction and death? Clearly, Germany's defeat was not sufficient to transform the impulse toward mastery and control.

Lyotard offers a different interpretation of the meaning of Auschwitz. Drawing on the Western image of the Jews, he argues that the Jews died at Auschwitz because the Greco-Roman West has long resented them as reminders of the impotence of those projects (commercial, legal, scientific, technological, military, cultural) whose aim is to deny human dependence on the nameless Divine.[22] The Jews, who experienced the trauma of being uprooted and made hostage to the unnameable Divinity, represent the memory of the unconquerable obstacle to the Western quest to make everything representable, present, and thus controllable. Hence, according to Lyotard, the technological impulse to make everything "present" for use is a result of this attempt by Western man to make himself God. Lyotard contrasts this view with Heidegger's claim that the technological impulse is a consequence of the history of foundationalist-productionist metaphysics, i.e., the Western attempt to discover the absolute ground or foundation which explains how things have come to be.

As bearers of the memory of human finitude and dependence, Lyotard argues, the Jews had to be excluded. Their extermination, the most extreme attempt to erase this unwanted memory, was carried out in secret so that the very memory of the obliteration of memory would also be forgotten. Seen from this point of view, World War II was a massive *distraction* designed to conceal what was really going on.[23]

Even if Auschwitz cannot be conceived in terms of the artistic laws of tragedy, however, Lyotard still agrees up to a point with Heidegger that Auschwitz exemplifies the dreadful "art work" typical of the union of art and politics in the technological era. While an authentic work of art—such as a Greek temple—founded a world by letting entities present themselves appropriately and distinctly, the perverse technological work of art—such as a hydroelectric station—differentiates nothing, but rather reveals things indiscriminately as sources of power. Both temple and power station are instances of *techne*; both reveal entities; but the former constitutes a world, while the latter contributes to an un-world.[24] Consider, Heidegger remarked about a decade after the end of the war, "the monstrousness that speaks out of the two titles, 'The Rhine,' as dammed up into the *power* works, and 'the Rhine' as uttered out of the *art work*, in Hölderlin's hymn by that name." [VA I: 15/16]

However, Lyotard rejects Heidegger's view that the Nazi crematoria were metaphysically "the same" as today's hydroelectric dams. Such a view conveniently ignores that the crematoria were for the efficient destruction of corpses (mostly Jewish ones), while dams are for efficient production of electricity. Ignoring this distinction is justified, in Heidegger's eyes, by the fact that both crematoria and dams are allegedly manifestations of the same Western impulse toward total domination. But may we so easily glide over the difference between crematoria and hydroelectric dams? While the former destroys human life, the latter makes some contribution to human life. Moreover, by equating crematoria with dams, by saying that they are both manifestations of the "Western Will to Will," Heidegger refused to entertain the possibility that there was something specifically *German* about the political decisions which led to the crematoria. Like many Germans, Heidegger apparently could not comprehend the difference between German war deaths at the hands of Russian soldiers, on the one hand, and the deaths of Jews at the hands of Nazi functionaries, on the other. While both events were slaughters, the German soldiers died as combatants, but the Jews died as innocent victims of anti-Semitic violence.

Heidegger's insensitivity to this issue may be less surprising when we consider how little he commented on the concrete aspects of human social interaction. Even his political speeches about the need for a new workers' society are remarkably devoid of practical insight and specific recommendations. Edith Wyschogrod has argued that Heidegger could speak of crematoria and hydroelectric dams in the same breath because of his focus on the history of *productionist* metaphysics. His concern with the transformation of *things*—

how they are produced, analyzed, and used—in Western history led him to neglect the transformation of *social institutions and practices*. "The priorities established remain nothing short of astounding. For Heidegger, such facts as the damming of the Rhine are important data attesting devastation, but the unnumbered dead of the two world wars and the creation of death and slave labor camps as institutional forms are never so much as mentioned in his major essays."[25]

Heidegger's equation of death camps and power plants confirms, in Lyotard's eyes, the hypothesis that Heidegger was like other Germans in wanting to forget the historical significance of the Jews. Yet Lyotard allows that Heidegger, by claiming that Western humanity had forgotten the originating event and had thus ended in technological nihilism, may have appropriated the role of the Jewish prophet for his own secular purposes: just as Jewish prophets reminded their people to remember the unnameable Origin, so too Heidegger reminded his people to remember the nameless Origin. The crucial difference here, of course, is the difference between Jerusalem and Athens—the difference between captivation by the Divine and enthrallment to Being. According to Lyotard, Heidegger's scandalous postwar silence about the Holocaust revealed his refusal to deconstruct his most basic assumption: that the history of the West is identical with the history of being initiated by the Greeks. The history of the Jews testifies that there is *another* dimension to that history: faithfulness to the Divine Law.

Lyotard's critique of Heidegger is reminiscent of Emmanual Levinas's famous charge that Heidegger went astray, despite his brilliance, by elevating being above the moral law, ontology over ethics. To some extent, Heidegger was following his predecessors Hegel and Nietzsche in claiming that the world-historical individual is "beyond good and evil." By portraying ethical matters as secondary considerations which arise within and which are limited to a particular historical world, however, Heidegger ran the risk of justifying whatever ethical form of life happened to emerge in a world "founded" by a new work of art. The demented "world" of National Socialism reveals what may be "justified" when artistic considerations are allowed to triumph over supposedly outmoded ethical ones. Heidegger's refusal to describe his behavior between 1933 and 1945 in terms of moral guilt stemmed from his belief that his "ontological calling" to found a new world removed him from the moral censure that pertained only to ordinary people. Of course, since the German people themselves were "extraordinary" in being called to the dangerous mission of founding a new world, since they risked so much in this noble venture, they too—in Heidegger's eyes—were not morally culpable. Heidegger's assessment of German behavior has, not surprisingly, provoked consternation even on the part of those sympathetic to his thinking.

In his "Letter on Humanism" (1946), Heidegger defended his views on ethics and, indirectly, his self-understanding regarding his role in National Socialism. He explained that he did not intend to develop a systematic "ethics," since he regarded ethics as a branch of metaphysics.[26] Modern ethical systems were concerned with "values," which he regarded as positions posited by the human subject (collectively or individually) enhancing the Will to Power. The idea of "ethics," then, was part and parcel of the nihilistic modern age. Instead of ethics, what Western humanity needed was a new *ethos*, a new way of dwelling on the earth that would establish proper limits for human behavior. Such an *ethos* could come only as a gift; it could never be founded by merely human actions. For Heidegger, Hölderlin's poetry sought to hold open the possibility of the arrival of a gift that would restore meaning and order to a technologically powerful but nihilistic world. Heidegger, then, did acknowledge the importance of establishing limits and law for human conduct. To this extent, his view can be said to agree with that of Lyotard. He also believed, however, unlike Lyotard, that the "law" (not the Law) governing a world is merely historical, not eternal. As one can see in his Schelling lectures (1936) [SA], he also agreed that "evil" arose from human self-assertion over against the divine law. But, for Heidegger if not for Schelling, this "law" was bound up not with the Eternal Divine but with historically limited gods—manifestations of "holy wildness."

Selfishness in the form of radical anthropocentrism and racism, so Heidegger concluded at some point in the 1940s, was the root of the evil at work in the historical reality of National Socialism. In Heidegger's eyes, then, Nazi Germany had betrayed its highest possibility by becoming a virulent manifestation of the anthropocentrism defining the final epoch in the history of metaphysics. Selfishness, egoism, greed: these were, for Heidegger, the inclinations which led people to choose evil over good. In calling for Germans to surrender their selfishness and to decide to become part of a new historical world order, Heidegger believed that he was fighting against the dark tendencies that were rampant in the materialistic world. National Socialism had been animated by an authentic insight, so he believed, but had fallen by the wayside. Unfortunately, however, Heidegger could never straightforwardly admit that by "deconstructing" what he regarded as the Enlightenment's insidious principles of universal economic and political rights, by declaring that traditional Judeo-Christian moral beliefs had been vacated by the death of God, by claiming that new historical worlds arise from a primal source that is "beyond good and evil," and by working to found such a world based on artistic-ontological, not moral, considerations, he helped to make possible the triumph of a truly radical evil.

At this point, in the face of Heidegger's postwar refusal to abandon what seems such a reactionary understanding of Western history and his equal failure to renounce unequivocally a political movement that wrought such unparalleled misery, some readers may well ask how students of Heidegger have been able to continue to assert the value of his thought. Why did he continue to harbor such an ambiguous attitude toward National Socialism? Why did he express a reluctance, bordering on defiance, to condemn the Holocaust, or to grapple seriously with the historical and cultural anti-Semitic issues (not merely the metaphysical ones) that were obviously so central to National Socialism? More broadly, how are we—given such self-damning silence—to evaluate his extraordinary yet idiosyncratic reading of Western history as the unpredictable "play" of being? May we not question the historical judgment of a man who engaged in what Hugo Ott has described as extravagant "self-mythification" regarding his own place in the scheme of Western history?

The self-mythifying Heidegger believed that he had been destined to proclaim the saving vision of his hero, Hölderlin, and that he himself was thus the world-historical figure who would transform the fate of the West. Consider Heidegger's astonishing account, presented in 1942, of the "destiny" linking him with Hölderlin: "Perhaps Hölderlin, the poet, must become the determining fate [Geschick] of confrontation for one who thinks [Heidegger], whose grandfather, according to the records, was born in a manger [in ovili] (in the sheep-fold of a farmer), which lay in the upper Donautal near the edge of a stream under the cliff. The hidden history of saying knows no accidents. All is destiny [Schickung]."[27]

The grandson of the man born in a manger in Hölderlin's beloved Swabian countryside knew that he was destined to change the course of history! Even as Germany was being destroyed in 1944, Heidegger did not hesitate to say, "In truth, however, my works belong not to my person, but instead they serve the German future and belong to it. Their safe-keeping should demand a corresponding carefulness."[28] And in July, 1945, after Germany had been almost annihilated by the victorious Allies, Heidegger could dare to write: "Now everyone thinks of the decline. We Germans cannot go under, however, because we have not at all yet arisen and must only persevere through the night."[29] Heidegger's postwar utterances never deviated from the conviction, described by Ott as showing "monstrous hubris," that his vision of history remained the key to the past and future of the West.[30]

Given Heidegger's apparent hubris and his refusal, even at the urging of his old friends Karl Jaspers and Rudolf Bultmann,[31] to acknowledge his guilt for the role he and his thinking played in promoting Germany's catastrophe, we may readily see why his critics have been so suspicious both of his understanding of Western history and of the interpretation of modern technology he derived from it. Nevertheless, despite the undeniable problems connected with Heidegger's life and thought, his conceptions of history and technology have exercised a major influence on many respected thinkers. Thus, before we draw any conclusions about the value of his analysis of Western history and technology, we must also examine his own philosophical account of those related phenomena.

To this point in our study, we have been concerned primarily with the political dimension of Heidegger's concept of modern technology. We have seen that that concept was by no means the product of pure "thinking," but instead arose in connection with his concern about Germany's fate in the age of modernity and industrial technology. Motivated by such concern, he developed a political philosophy that shared much in common with, but was also far more sophisticated than, the reactionary political movement associated with the rise of Hitler. In providing the narrative of Heidegger's application of his concept of modern technology to politics, we have necessarily examined important aspects of his ontology, particularly regarding the role of the work of art in "founding" a new world. Nevertheless, we have postponed the more detailed examination of his ontology necessary for both a full understanding and an evaluation of his concept of modern technology.

The history of Heidegger's attempt to apply his thought to fascist politics is a dreadful story, but it is, in fact, only part of the narrative. Concurrently with his political actions in the 1930s, Heidegger was also formulating a theoretical account of Western history and technology that would serve as justification for his political decisions. This account concerned the following issues: (1) the primacy of work in human *Dasein*'s understanding of what it means for things "to be"; (2) how productionist metaphysics was linked with a certain understanding of "work" by the ancient Greeks; (3) how that metaphysics culminated in the technological era which has transformed humanity into a power-craving animal and which has altered the structure of everyday life. Despite his descriptions of how the old world was being obliterated by the advance of the technological one, Heidegger did not finally despair. Rather, he held out hope that a saving power could grow from out of the dangerous depths of technological nihilism. His meditations on the nature of art as authentic "producing" pointed to an alternative to the inauthentic production characteristic of the technological epoch.

Having analyzed Heidegger's account of the history of productionist metaphysics and of the nature and consequences of and the alternatives to the technological outcome of that metaphysics, we will be in a better position to examine the question of the validity of his conception of modern technology. My examination of these issues in Division Two will be informed by knowledge of Heidegger's own political interpretation and application of his thinking. In line with basic hermeneutic principles, I would also suggest that readers might profit by rethinking Division One in light of what is said in Division Two. Those who may be initially unfriendly toward Heidegger's reactionary political views may come to have a more sympathetic understanding of how he viewed the situation in 1933. Such an understanding of Heidegger's theory of Western history is necessary for us to make some sense of, if not finally to agree with, his political attitude.

DIVISION TWO

Heidegger's Critique of Productionist Metaphysics

Chapter 9. Equipment, Work, World, and Being

In Division One of this work, we saw how Heidegger's concept of modern technology arose within the context of his negative evaluation of and political resistance to the influx of modernist ideas and industrial technology. We also saw that he conceived of modern technology as the outcome of the history of productionist metaphysics. Because in Division One we focused on the political dimension and application of Heidegger's concept of modern technology, however, we postponed the more comprehensive treatment of his ontology until now. Informed by our findings in Division One, we shall examine in Division Two Heidegger's ontological conception of modern technology, including his understanding of the origin, nature, and influence of productionist metaphysics, and his notion that art opens the way to what he regarded as authentic producing: letting things be. This examination will enable us to comprehend fully and thereby to evaluate fairly his conception of technology.

The shift of focus here may seem abrupt. We leave behind Heidegger's meditations on Hölderlin's mysterious poems and silence about the Holocaust in order to return to his analysis of the nature of human existence—an analysis that he completed during the somewhat simpler and more innocent days before 1933. This analysis of the productive-disclosive character of the human understanding of being, however, anticipated much of what was to come later.

A. THE WORLD OF THE WORKSHOP

Heidegger had a lifelong concern with the nature of working and producing. He manifested this concern in what may be the most famous portion of *Being and Time*: the analysis of the workshop. Knowing what we do about Heidegger's affiliation with National Socialism and its pre-1936 critique of industrialism, we may not be surprised at his focus in the mid-1920s on the nature of handicraft. In addition to suggesting his dissatisfaction with industrial modes of production, Heidegger's analysis of equipment and the workshop was threefold: (1) to provide a starting place for understanding human existence as being-in-the-world; (2) to reveal the primacy of the equipmental-instrumental understanding of the being of entities as opposed to the objective-scientific understanding; (3) to lay the groundwork necessary for revealing that the unifying "meaning" (*Sinn*) of the being of entities is grounded in temporality. He prefaced his workshop analysis with the following considerations. To understand being as such, and its relation to time, Heidegger argued, we must choose a particular entity in order to investigate its mode of being. We must choose wisely, so that our analysis of the mode of being of the chosen entity will provide insight applicable to the modes of being of other entities as well. Given that the human entity, *Dasein*, is the entity characterized by its understanding of its own being and that of other entities, Heidegger concluded, the human entity would be an excellent choice to be the subject of the initial ontological inquiry.

Heidegger used his version of the phenomenological method for his "existential analytic" of human *Dasein*. He defined phenomenology as ontology: that kind of interpretation which allows an entity to show itself in the way appropriate to that entity itself. Every interpretation involves a fore-conception, a fore-having, and a fore-sight. This "fore-structure" is at work not only in theoretical interpretation but in the interpretation involved in my using a tool. How is it that I can so readily use tools? Because "any interpretation which is to contribute understanding must already have understood what is to be interpreted." [SZ: 152/194] My prior understanding of the being of tools enables me to use them appropriately. Heidegger argued that even if I am in a workshop about whose products or procedures I am wholly ignorant, I nevertheless recognize that it is a workshop. And in every activity in everyday life, I am always engaged in interpreting things. Heidegger expressed this ongoing interpretation as the "as-structure." I can make use of a doorknob *as* a doorknob because in advance of my encounter with the particular doorknob I am operating in accordance with the fore-structure that reveals entities as instruments or tools for my purpose. To take a tool *as* a tool requires that I know that it *is* a tool. But my everyday understanding or interpretation of entities is non-theoretical, or "pre-ontological." Heidegger sought to make explicit the ontological structure of this pre-ontological understanding of being.

Adhering to the phenomenological call to return "to the things themselves" (*zu den Sachen selbst*), he chose to ground his analysis upon a description of what people actually do in everyday life. Thus, he eschewed traditional definitions of humanity, e.g., "rational animal," which he believed were loaded with unexamined presuppositions, including the idea that theoretical cognition was basic to human life. He did presuppose, however, that practical activity was more basic than theory. In light of this presupposition, he hoped to disclose the basic feature of human existence: "being-in-the-world." In everyday being-in-the-world, things show themselves as instruments or tools conducive to human needs and purposes.

Consider the cobbler in his shop. He uses tools to make shoes. How is this possible? Because the cobbler understands in advance the equipment and supplies with which he works in terms of the network of relationships and possibilities that constitutes his world. He understands what his tools and his products "are for." In reaching for his hammer, he has already interpreted it. He understands his hammer *as* a tool, not just as an object lying around. It is possible for the hammer to be interpreted as a mere thing, to be denuded of its equipmental character so that it appears as an object present-at-hand.

Philosophers have traditionally presumed that entities are really first present-at-hand and can become tools under certain circumstances. Heidegger insisted, however, that this reverses the true situation. The fundamental way in which entities "are" for us is as ready-to-hand. Only by an act of abstraction can *Dasein* remove itself from its involvement with the activities of everyday life and adopt instead the attitude of a passive spectator or observer, for whom what was once a useful device now becomes a mere "object" with certain properties analyzable by specific scientific procedures, and so on.

Ordinarily, equipment is so ready-to-hand, so "handy," that we are not explicitly aware of it as such. For example, in hammering away at the sole of a shoe, the cobbler *does not notice the hammer*. Instead, the tool is in effect transparent as an extension of his hand. And for Heidegger the human hand plays a crucial role in the disclosure of entities. For tools to work right, they must be "invisible," in the sense that they disappear in favor of the work being done. It is the peculiarity of what is chiefly ready-to-hand that, in order to be authentically ready-to-hand, it must, as it were, withdraw in its readiness-to-hand. That with which the everyday dealings proximally dwell is not the work tools themselves; instead, what is primarily [the object of] concern and thus also ready-to-hand is the work, what is to be produced at the time. [SZ: 69–70/99]

Tools are useful because they are reliable. In their reliability they disappear in favor of the work to be done with them. Tools become visible, paradoxically, precisely when their reliability vanishes: when they are *missing*, when they *don't work*, or when they *get in the way*. When the cobbler reaches for a tool and cannot find it, when a breakdown of some sort occurs in the work activity, the work world suddenly becomes illuminated in a way that it is not when he is engaged in working. The set of reference relationships that constitute his world becomes revealed precisely because the smooth functioning of the work that is a constitutive element in those relationships is now upset by the missing tool. The cobbler realizes that he cannot finish the shoe on which he is working without the tool; that fortunately the tool itself is sold by a local supplier who purchases it from a manufacturer in a distant city; that the tool is used to make the sole of a shoe; and that the shoe itself is one of a pair on order by a customer, who will use them for walking about.

All the elements in the work world are internally related. There is no such thing as an isolated tool; tools occur within an equipmental-referential context *in terms of which* a particular thing can reveal itself as a tool. Without this meaningful referential context, this familiar domain in which we have lived from the start, this "world," tools could not *be*. As Heidegger notes,

> Even the workshop of a handworker whose work is wholly unfamiliar to us is in no way encountered chiefly as a mere conglomeration of things all thrown together, but instead in the closest environmental orientation are shown handworking tools, materials, produced and ready pieces, unready pieces, those found still in the works. Primarily, we experience the world in which the man lives, however foreign, yet still as world, as the disclosed totality of references. [GA, 20: 255/188]

Tools are always assigned to some purpose or another, and hence are experienced as "in order to" (*Um-zu*). This relationship of assignments has the character of signifying, and the totality of relationships is called "significance" (*Bedeutsamkeit*). [SZ: 87/120] Significance, the meaningful totality of reference relationships constituting the world, is grounded in *Dasein*. Whereas other beings involved in the world are functional and instrumental, *Dasein* itself is that *for the sake of which* (*Worumwillen*) the referential totality operates. Without the world opened up by human existence, beings would not "mean" anything. Hence, the phenomenon of "world" is not to be understood as the totality of natural entities or as the domain of creatures made by God, but instead as the structure of reference relationships constituted by and for human existence, a structure that enables entities to manifest themselves or "to be" in various ways.

Dasein exists in the world in such a way that its own being is an issue for itself. Heidegger said that because *Dasein* is concerned about itself, other people, and things, the very being of *Dasein* is "care" (*Sorge*). In other words, the human way of manifesting itself is to be engaged with things, in making and doing and using, and with others, in speaking and acting and sharing, and with oneself, in deliberating and thinking and choosing. Since an overriding concern in everyday life is maintenance of the social and economic structures necessary for survival, most things manifest themselves to *Dasein* in instrumental terms, as ready-to-hand for human ends. Some things show up as artifacts which have been produced, but non-artifactual natural things also reveal themselves instrumentally: "Accordingly, in the environment [an] entity becomes accessible which is not in need of producing, but is always already ready-to-hand. Hammer, tongs, and needle in themselves refer to—they consist of—steel, iron, metal, mineral, wood. In equipment that is used, 'nature' is discovered along with it by that use, 'nature' in the light of products of nature." [SZ: 70/100]

By emphasizing the instrumental understanding of entities, Heidegger reversed the traditional priority of theory over practice, of thinking over doing. Although he interpreted human *Dasein* as the entity who understands entities *as* entities, then, Heidegger made clear that for him the practical understanding of a person engaged in work is prior to the theoretical understanding of a person looking on in a detached manner. He argued that the detached kind of "seeing" involved in philosophical reflection and scientific theorizing is derivative from the engaged kind of seeing—circumspection (*Umsicht*)—that is involved with everyday practices. Pragmatic behavior, then, cannot be regarded merely as an inferior grade of theoretical knowing. Ever since Plato's time, the goal of theoretical knowing has been to formalize and to make explicit all aspects of itself. Total self-transparency, however, turns out to be impossible, because human beings are *thrown* into a world of deeply imbedded, historical practices that can never be made fully explicit. A person begins theoretical inquiry only long after he or she has been involved with non-theoretical everyday practices. In principle, such inquiry comes "too late" to lay bare all the relational factors that constitute *Dasein*'s being-in-the-world. These relations, moreover, "resist any sort of mathematical functional-ization. . . . " [SZ: 88/122]

The obstacles involved here are not only that human beings have an intrinsic tendency toward concealment, and that their everyday practices have what Polanyi has described as a "tacit" character, but also that humans exist in a way that is always an issue for themselves. This existence discloses enti-ties not only instrumentally and cognitively, but also in terms of non-rational moods and feelings. Furthermore, because humans are embodied, they are open to and involved with the whole of entities in a way that cannot be properly represented by the model of theoretical knowing.

Heidegger argued that we disclose entities not only in terms of our under-standing, but also in terms of our moods (*Stimmungen*). As we saw earlier, moods are not merely psychological "colorations" projected onto things; in-stead, moods articulate humanity's openness for the being of entities. "A mood is a style [*Weile*], not merely a form or a mode, but a style in the sense of a melody, which does not float over the so-called authentic being-present-at-hand of man, but instead provides the tone for this being, i.e., attunes and determines the kind and how of this being." [GA, 29/30: 101]

A mood then is not an entity, such as a psychological "state" or "condi-tion," but instead is the "*basic style in which the Dasein is as Dasein*." [Ibid.] Moods characterize the way in which we "find" ourselves in the world; they define how we are "doing" or how things are "going." Usually we presume that what we're dealing with is primary, while the mood in which those deal-ings take place is secondary or merely "subjective." But Heidegger maintained that the mood is the "medium" which discloses things. [GA, 29/30: 101–102] Moods let things "matter" to us in a way that could not happen with under-standing alone, and indeed provide motivation for such understanding in the first place. "All knowing is only an appropriation and a form of realization of something which is already discovered by other primary comportments. Knowing is rather more likely to cover up something which was originally uncovered in non-cognitive comportment." [GA, 20: 222/101] Heidegger noted approvingly that for Augustine and Pascal, love and hate are more fundamental than knowledge. [Ibid.] He also claimed that moods disclose our facticity, our "thrownness" into a particular historical situation. Heidegger's insistence upon the ontological function of moods was at least as controversial and illuminating as his claim that everyday instrumental practices are more fundamental than theoretical knowledge.

In fact, at least during the era of *Being and Time*, Heidegger so favored the importance of everyday practices that he even claimed that assertions are grounded in them. Prior to making an assertion about something, *Dasein* has always already interpreted it as being something or other—a tool, an object, another person. In making and doing, *Dasein* exhibits its understanding of what things are. This pre-ontological understanding of being was Heidegger's clue to his initial answer to the question of the unity of being, i.e., to the question of how the copula ("is") could be used in so many different ways. Early Heidegger explained the "indifference" of the copula, its incapacity for discriminating the various ways in which entities "are," by saying that assertions are a secondary mode of disclosing entities. [GA, 24: 301/211–212] More basic than assertions (e.g., "This hammer is too heavy") are instrumental interpretations which make appropriate distinctions among entities (e.g., noticing the hammer is too heavy, tossing it aside, and picking up another hammer). In his own copy of *Being and Time*, Heidegger later wrote: "Untrue. Language is not stored up [*aufgestockt*], but instead *is* the primordial presencing of the truth as there." [GA, 2: 87/117) Hence, he came to emphasize something which he had already said during the era of *Being and Time*: "*Language makes manifest* [*Sprache macht offenbar*]." [GA, 20: 361/262] The apparently extralinguistic capacity for using tools is in fact imbedded within a pre-articulated level of intelligibility which makes language—including "body language" and gesture—possible in the first place.

Gerold Prauss has argued that such remarks by later Heidegger indicate that he withdrew his earlier support for the distinction between the pragmatic attitude involved in the ready-to-hand way of disclosing things, and the theoretical attitude involved in the present-at-hand way of disclosing things.[1] Prauss maintains that such a distinction cannot be plausibly maintained. Science is not merely theoretical but practical, at least insofar as it employs experimental devices. Likewise, practical behavior always involves an element of understanding, even if it is not explicitly theoretical in character. Instrumental activity, then, does not *precede* language but instead is made possible by linguistic understanding. Later Heidegger concluded, moreover, that language may not be conceived as a human instrument, but instead as that disclosive-ontological power at work *through* human existence. The "turn" in Heidegger's thought was linked with his decision that he was misguided in trying to define "being" in terms of the transcendental structure of the human entity, for such an approach to being ended in fruitless considerations of the "subjectivity of the subject."

Already in *Being and Time*, however, we may discern a tendency to undermine the subject as the source of intention and purpose, despite Heidegger's claim that human *Dasein* is that "for the sake of which" all practical activity is undertaken. As Prauss points out, Heidegger defined the very being of tools as ready-to-hand. [SZ: 69ff./98ff.][2] But by describing tools "in themselves" as ready-to-hand, did not Heidegger tend to conceal the fact that tools become tools only when someone *uses* them? Our everyday conviction may be that the means-character belongs to the tool itself, but this conviction is misguided. The tool itself is made for a purpose that is achieved only when someone uses it. The user is informed in advance by knowledge about his or her situation, about the capacity of the tool, about the task to be accomplished. The practical activity of using the tool, then, is not productive of but instead derivative from the knowledge which leads one to pick up the tool in the first place. Critics such as Prauss want to restore to the subject some of the autonomy which Heidegger seems to have removed in his account of the worker as a relatively un-self-conscious role-player within a pre-existing referential framework.

Later Heidegger de-emphasized pragmatic-instrumentalist activity, which he at one time made the basic feature of human behavior, because he realized that this instrumentalist account of human existence was too reminiscent of the instrumental attitude present in the technological understanding of things. Moreover, while early Heidegger had described science as a detached way of seeing things, later Heidegger concluded that a pragmatic, power-oriented motive was at work in modern science from the very beginning. Could it be, Heidegger wondered, that the instrumentalist attitude discernible in the early Greeks and in the craftwork of the contemporary artisan was not an *a priori* but a historically determined attitude linked to the technological Will to Power? We shall examine this topic in more detail in the next chapter.

B. HEIDEGGER'S DEBT TO KANT AND ARISTOTLE

In his reflections on instrumental activity, language, understanding, mood, and projection, early Heidegger was indebted to but was also trying to move beyond the transcendental tradition founded by Kant. Transcendental philosophy seeks to disclose the conditions necessary for the human experience and understanding of entities. In the style of Kant, Heidegger posed the following question: How is it possible for people to understand entities as equipment in the first place? When I reach to open a door, I have already understood the doorknob as an instrument for the purpose of opening the door. To encounter the doorknob *as* a tool for use, the conditions necessary for this encounter must themselves *not* be encountered: they must recede into the background so that the doorknob itself can be present. For Kant, the transcendental conditions necessary for experience of objects are (a) the pure intuitions of space and time, and (b) the pure categories of the understanding. One can never experience these conditions; instead, they must be deduced as the conditions which make it possible for the "thing in itself" (*Ding an sich*) to be represented as an object (*Gegenstand*) over against the knowing subject. In his radical reinterpretation of Kant, Heidegger argued that there is no "thing in itself." What we encounter is not a "representation" of the thing in itself, but instead the very thing itself from a limited perspective. As finite and receptive, humans can encounter only aspects of a thing, but nevertheless they experience the thing itself. For Heidegger, the transcendental conditions necessary for the experience of entities *as* entities were (a) the self-manifesting (being) of the entities themselves, and (b) the temporal-linguistic clearing of *Dasein* in which that self-manifesting could occur.

Heidegger was influenced not only by the Kantian but also by the Aristotelean conception of the "transcendental." Aristotelean realists claimed that the transcendental refers not to the conditions necessary for experience, but instead to the ontological features which are not themselves entities but which nevertheless characterize all entities. Aquinas said that there are five such "transcendentals": being, unity, beauty, truth, and goodness. Every thing is, is unified, is beautiful, is true, and is good. Early Heidegger tried to combine the best features of both of these transcendental schemes. He wanted to avoid

Kant's notion that our experience involves only "representations" of things, but he also wanted to avoid naive Aristotelean realism which failed to account for the role played by the "subject" in the being of entities. In his *Habilitationsschrift*, he argued that transcendental categories constitute the "logic" of scientific knowledge. [FS] He asked: Could these transcendental categories have the ontological function of structuring the ways in which entities could show themselves and thus "be"? In his attempt to reconcile Kantian (epistemological-scientific) transcendentalism with Aristotelean (ontological) transcendentalism, Heidegger defined the categories of being as "transcendental meaning." Such "meaning" constituted the logical "space" or even the "world" that enabled entities to be encountered intelligibly. Without such a "world," entities could not present themselves and thus could not be said "to be."[3]

Heidegger's work on scientific logic was the source for the transcendental conception of meaning in *Being and Time*: "*Meaning is that wherein the intelligibility of something maintains itself*" (*Sinn ist das, worin sich Verständlichkeit von etwas hält*). [SZ: 151/193] Here, "meaning" names a transcendental structure: the temporal-linguistic "clearing" which constitutes human existence. Within the limits established by this clearing, entities can manifest themselves and thus "be" in determinate ways. The transcendental meaning-structure of human existence makes possible all human encounters with entities *as* entities. The bearer of this disclosive transcendental structure is not a disembodied cognitive ego, but instead a concrete historical entity concerned about itself and dependent upon the entities that surround and support it. While humans exist as the meaning-structure in terms of which entities can be disclosed as entities, that meaning-structure is empty without the presencing of the entities disclosed within it. Heidegger's concern about being, then, was directed to being "insofar as being enters into the intelligibility of *Dasein*." [SZ: 152/193] Critics such as Karl Löwith argued that by making "being" so dependent on humanity, and by making humanity so radically different from all other entities, Heidegger followed the anthropocentric trail blazed by his Christian and Cartesian predecessors.[4] In fact, Heidegger did state that the gap between humans and other entities is so great that of the former one should say "they exist," while of the latter one should say "they are."

Although Heidegger radically distinguished *Dasein* from other entities, nevertheless he seems to have portrayed the structure of *Dasein*'s being as analogous to the structure of living things, such as plants and animals. In doing so, he drew upon one of the most important aspects of Aristotle's *Physics*, the doctrine of movement (*kinesis*). For Aristotle, *kinesis* refers to the very being of living things. For example, in a plant's growth and life, some aspects of the plant show themselves, while others are hidden. The seed gives way to the hidden sprout, which in turn gives way to the emerging leaves, and so on. The plant's very being, as Thomas J. Sheehan has argued, is always characterized by absencing or hiddenness, as well as presencing or appearing. "But such relative absentiality is precisely what lets the entity *be* a moving entity. Therefore, to know a moving entity as what it truly is means to keep present to mind not only the *present entity* but also the *presence of the absential-*

ity that makes it a moving entity. The presence-of-its-absentiality is the moving entity's Being-structure. We may call it 'pres-ab-sentiality.' "⁵

Heidegger used the Greek terms *dynamis* and *physis* to describe the fact
that living things conceal themselves even as they are revealing themselves. In
the pro-ducing or appearing of a living entity, there is also a putting away, "as
the blossom is put *away* by the fruit. But in this putting *away*, the placing into
the appearance—*physis*—does not give itself up. On the contrary, the plant in
the form of the fruit goes back into the seed, whose being is nothing else but a
going-forth into the appearance. . . . " [WGM: 367/267]

In 1928 Heidegger translated *dynamis* as *Ereignis*, "the event of an entity's self-disclosure." Similarly, he translated the Greek term *aletheia*
("truth") as the event of unconcealment in which an entity manifests itself.
Entities can manifest themselves only through the clearing opened up by and
through human existence. There is an analogy between the self-disclosive
motion of living things, on the one hand, and the disclosive motion of human
existence, on the other. Human ek-sistence involves the peculiar movement of
transcendence (ek-sistence, standing-out, *ekstasis*) and of return to entities. As
Sheehan explains,

> Man's transcendence (his relative absentiality) is correlative to the privative di-
> mension of the autodisclosure of entities (*their* relative absentiality); and man's
> "return" from transcendence to worldly entities (his presencing to them) is cor-
> relative to the positive dimension of the autodisclosure of entities (their presence).
> If man has access to entities because he is in excess of them, that excess is
> correlative to the recess-dimension of entities. The interplay between access, re-
> cess, and excess (in other terms: *aletheia*, *lethe*, and transcendences) is the heart of
> Heidegger's thought, and there, as he says, *Alles ist Weg*, everything is a matter of
> movement.⁶

Heidegger's conception of truth as *aletheia* (unconcealment) was influenced by Husserl's doctrine of "categorial intuition." Categories were the
transcendental structure of meaning that enables entities to manifest themselves. Later Heidegger said that "what occurs for the phenomenology of the
acts of consciousness as the self-manifestation of phenomena is thought more
originally by Aristotle and in all Greek thinking and existence as *aletheia*, as
the unconcealedness of what-is-present, its being revealed, its showing itself."
[ZSD: 87/79] What makes study of being so difficult is that neither the
presencing (*Anwesen*, being) of entities nor the temporal-historical absencing
(*Abwesen*, *-lethe*) in which that presencing occurs is itself an entity. Rather,
presencing-absencing are the transcendental conditions for encountering entities. Presencing-absencing conceal themselves, withdraw from view, in favor
of the entities that show up. For this reason, Heidegger concluded that the
problem of "truth" was much more profound than it was ordinarily thought to
be. He regarded the traditional definition of truth ("correspondence between
assertion and state-of-affairs") to be derivative from truth understood as unconcealment, *aletheia*. One can assert X about Y ("The cat is on the mat")
only because the cat has manifested itself in the first place. As the entity

through whom this event of unconcealment takes place, human *Dasein* constitutes the disclosedness (*Erschlossenheit*) or the temporal-historical "clearing"
(*Lichtung*) or the transcendental realm of meaning.

C. THE ROLE OF TEMPORALITY IN THE UNDERSTANDING OF BEING

The role of temporality, emphasized in the title *Being and Time*, can be
broached by referring once again to the human capacity for understanding and
using tools. I can use a hammer because I have already understood it in terms
of its being as readiness-to-hand. Hence, I understand an entity in light of its
being. But in light of what do I understand being itself? Heidegger replied,
"The understanding of being already moves in a *horizon* that is everywhere
illuminated, giving luminous brightness." [GA, 24: 402/284] *Time* constitutes
the transcendental horizon illuminating being itself. Hence, time composes
the existential-transcendental structure of human meaning, interpreting, and
understanding. I can make use of a tool because the tool shows itself within
the temporal horizon of the present. In an important sense, we can say that
even more primordial than the presencing or being of entities is the temporal-
linguistic clearing in which that presencing can occur: being as unconcealment (*aletheia, physis*) is indebted to that which conceals itself (-*lethe*).

Influenced by Dilthey, Heidegger argued that being was not eternal but
instead essentially finite: temporal and historical. Even Plato's metaphysics of
the eternal forms, so central to productionist metaphysics, contained an element of temporality. Plato conceived of the *eidos*, the eternal form, as permanently *present*. And permanent presence is a static conception of the temporal
dimension of presencing, in terms of which entities can be ready-to-hand.
Plato was the first great metaphysician, because he transformed the primordial
event of presencing (being) into an eternal foundation (beingness): the permanently present form. This transformation initiated the history of productionist
metaphysics. When Nietzsche proclaimed that "God is dead," he meant that
the old Platonic-Christian metaphysical ideals of eternity had collapsed.

Correlatively to his claims about the finitude of being, Heidegger also
argued that humans are radically finite, limited, dependent, mortal. Humans
need entities to survive. Moreover, humans can experience entities as such
only insofar as the entities show or present themselves. Finally, even though
the presencing of entities cannot be accomplished without temporality, human *Dasein* does not own that temporality, but rather is owned by it. Humans
are "thrown" into their finite existence. Anxiety can reveal the truth about
mortal finitude, but people flee from anxiety in order to conceal that unwel-
come truth. Heidegger called this movement toward concealment "falling."
Human *Dasein*'s temporal "movement"—existence—always involves such
falling. Temporality is more or less constricted, limited, perspectival in character. Hence, there is no complete or perfect disclosure of entities. In everyday
life, temporality is constricted in such a way that entities can show themselves
primarily only as instruments.

When everyday falling becomes aggravated, for example, when a person tries to disown his (or her) mortal existence and pretends to be immortal, a person exists in a way that Heidegger called unowned or inauthentic (*uneigentlich*). Inauthentic existence reveals things as distractions which will conceal the truth about *Dasein*'s mortality. By way of contrast, *Dasein* can be transformed by authentic temporality which calls on *Dasein* to own its mortal openness, to become authentic (*eigentlich*).[7] Authentic temporality discloses one's concrete situation in terms of possibilities that are ordinarily hidden. These possibilities are to be understood as one's genuine heritage manifesting itself as the authentic future. Early Heidegger usually spoke of inauthenticity and authenticity in personal, decisionistic terms, but he later concluded that they must be understood primarily as features of entire historical cultures.[8] Hence, it is no failure of character which compels contemporary humanity to lay waste to the earth in search of infinite power. Rather, humanity is destined to reveal all entities as standing-reserve for technological domination.

Charles Guignon uses the term "foundational historicism" to describe Heidegger's paradoxical and ultimately unsuccessful attempt to unite the search for transcendental, essential structures of human existence with his belief that human existence and understanding are essentially historical in character.[9] Early Heidegger assumed that his analysis of contemporary human existence, e.g., everyday life in the workshop, could provide insight into the enduring character of human existence. The existential analytic aimed to show that temporality constituted the transcendental structure of *Dasein*'s being. Had his philosophical aim been transcendental in Kant's sense, Heidegger would have stopped at this point, for he would have discovered the essential transcendental structure of human existence.

Heidegger, however, questioned whether one could arrive at "eternal" truths about anything, especially about an entity—human *Dasein*—whose very being involved a radical kind of temporality. Always an issue for itself, *Dasein* is constantly in the process of posing and answering questions about its identity. To exist means to be a never-ending process of interpretation, with never a hope of arriving at an "essential" identity. As later Heidegger was subsequently to conclude, human existence is interpretation "all the way down." There is no "foundation" for it. Yet even while searching for the *a priori* structure of human *Dasein*'s transcendental existence, Heidegger also undertook the second aspect of foundational historicism, the deconstruction of the sedimented layers of historical understanding and practice. This deconstruction sought to disclose and to liberate *Dasein* from unexamined presuppositions about its own existence, presuppositions which were—in Heidegger's view—responsible for the crisis of the Western world.

According to Guignon, the interplay between the historicist and the transcendental aspects of foundational historicism took the following form. On the one hand, Heidegger's deconstruction of the Western philosophical tradition enabled him to authenticate his discovery that human existence constituted the temporal clearing for the being of entities. Plato caught a sideways glimpse of this when he spoke of the "good" (*agathon*) as that which makes possible the presence of entities. On the other hand, Heidegger's idea that time provides the transcendental structure for the human encounter with the presencing of entities provided him with the "clue" needed for his deconstructive reinterpretation of the history of Western metaphysics. Heidegger's attempt to "provide an alternative to both ahistorical transcendental philosophy and unbridled historicism" had to fail, in Guignon's view, for two reasons.

First, human existence is tangled in distortions and concealments which make it impossible to arrive at a "transparent" explanation either of its own existence or of the origins and course of its history. Second, if all understanding is shaped by its historical situation, then any "clues" which a person might use to disclose transcendental features of human existence will also be relative to the situation of that person. In other words, the being of *Dasein* might have changed through history; hence, what may be true of twentieth-century *Dasein* might not be true of *Dasein* in other historical epochs. There is no way to arrive at the "timeless" structures or "existentialia" of human existence. Moreover, "if our understanding is always caught up in the flow of history, there is no access to a privileged 'clue' which will enable us to arrive at ultimate conclusions about the meaning of Being."[10] Heidegger's recognition of the incompatibility of a historically oriented search for *a priori* features of human existence was another factor in his "turning."

In this turn, Heidegger emphasized that what he had been calling "the understanding of being" was misleading in that such understanding could be interpreted in the traditional way as a human faculty or capacity. "The understanding of being," he explained, is not to be construed as an achievement of the subject, as when we say, for example, that "he finally understood the problem." Instead, "the understanding of being" is in effect identical with the event of being itself: the event of disclosedness or presencing by virtue of which entities show themselves. To speak of "the understanding of being" indicates that this disclosive event requires a site—human existence—through which to occur. Later Heidegger insisted, however, that the basic ontological feature of humanity was to be appropriated by and for being as the site for its self-disclosure. Hence, while early Heidegger often spoke as if human *Dasein* were an end in itself and autonomous regarding the authenticity of its understanding of being, later Heidegger emphasized that humanity was in the service of being and was not in control of the changes that occur in how things manifest themselves historically. Rather, these changes arise from the historical play of being itself, a play that began in the days of the ancient Greeks and culminated in the technological era.

The shift of Heidegger's attention from the structure of human *Dasein* to the play of being itself influenced his understanding of the nature of modern technology. Early Heidegger argued that the production and use of tools is "not only a basic mode of comportment of man, but [also] a decisive determination of the existence of ancient *Dasein*." [GA, 33: 140–141] Later Heidegger concluded that he may have overemphasized the continuity between Greek and contemporary attitudes toward the production and use of things. Clearly, Greek instrumentalism was by no means the same as the instrumentalism of modern technological culture. Nevertheless, the Greek productionist orientation toward being set Western history moving down the path which ended in modern technology. But was this productionist orientation the result of "a basic mode of comportment of man," or was it the result of a change in the way entities showed themselves to the Greeks? In other words, was the productionist orientation of the Greeks not "basic" or *a priori* at all, but instead a historical transformation from the non-productionist, non-metaphysical orientation of the pre-Platonic to the productionist, metaphysical orientation of post-Platonic Greeks? Heidegger's affirmative answer to this latter question helped to shape his mature concept of modern technology.

As we shall see in the following chapter, early Heidegger's tendency to portray a productionist attitude as *a priori* and "basic" to human existence led to unintended consequences. For example, as we shall see in the next chapter, some commentators have concluded that because *Being and Time* portrays everyday life as being primarily pragmatic and productionist in orientation, that book constitutes one of the final stages in the history of productionist metaphysics.

Chapter 10. *Being and Time*: Penultimate Stage of Productionist Metaphysics?

Hubert L. Dreyfus has argued that the instrumentalist orientation of *Being and Time* unwittingly promoted the technological disclosure of entities.[1] In this chapter, we shall see that while Dreyfus's argument wins support from many passages in *Being and Time*, there are other passages from early Heidegger's writings which help to explain and thus to temper the apparent instrumentalism of *Being and Time*.

A. HEIDEGGER'S EARLY INSTRUMENTALIST ONTOLOGY

The instrumentalist orientation of *Being and Time* can readily be seen in the following statement: "*Readiness-to-hand is the way in which entities as they are 'in themselves' are defined ontologico-categorially.*" [SZ: 71/100] We have already noted some problems inherent in this claim that tools really *are* instruments for human use. Expanding on this instrumentalist attitude, Heidegger remarked—apparently approvingly—that for the peasant and worker in everyday life, "the wood is a forest of timber, the mountain a quarry of rock; the river is water-power; the wind is wind 'in the sails.' " [SZ: 70/100] More than two decades later, having arrived at his concept of modern technology, he said that in the technological era "the earth now reveals itself as a coal-mining district, the soil as a mineral deposit. . . . Air is now set upon to yield nitrogen, the soil to [yield] ore, ore for example to [yield] uranium. . . . " [VA I: 14–15/14–15]

At first glance, these statements seem to say the same thing. According to Dreyfus, however, *Being and Time* occupies an ambiguous middle ground between the full-scale technological understanding of being, on the one hand, and the early Greek instrumentalist understanding of being, on the other. The instrumentalism of *Being and Time* is not yet fully technological, because it portrays equipment not in terms of mathematically structured technology but instead in terms of the craftsman's shop. Moreover, *Being and Time* does not speak of nature as undifferentiated standing-reserve. Nevertheless, despite whatever similarities the workshop described in *Being and Time* might have to a pre-industrial workshop, Heidegger located the workshop in a "world" that seems increasingly technological.

At first describing the workshop as a relatively autonomous "region,"

Heidegger went on to explain that this region was an element in the *totality* of regions, the "worldhood of the world." [SZ: 86/119] Dreyfus comments that Heidegger thereby "expands the local [workshop] context to a single over-arching totality. He recognizes that this tendency to totalize is a specifically modern phenomenon whose full meaning he realizes has not yet been re-vealed."[2] Early Heidegger discerned this totalizing movement in modern hu-manity's way of being "close" to things. Only *Dasein* exists in a world; hence, only *Dasein* can be close to or remote from entities that show themselves within that world. A rock may be touching another rock, but it is still not "close" to the other rock in the way that a person is to a family member, even if he (or she) is far away. Heidegger used the term *ent-fernen*, roughly trans-lated as to "un-distance" something, to describe the ontological process of bringing things close. In modern times, this process has become a drive to vanquish remoteness altogether. The attempt to make everything equally close and available arises from the increasingly one-dimensional ontology of moder-nity: everything appears to be nothing but various kinds of matter which can be used and switched about at will. "All the ways in which we speed things up, as we are more or less compelled to today, push us on toward the con-quest of remoteness. With the 'radio,' for example, *Dasein* has so expanded its everyday environment that it has accomplished an un-distancing of the 'world'—an un-distancing which, in its meaning for *Dasein*, cannot yet be visualized." [SZ: 105/140]

Early Heidegger argued that uprooted modern humanity no longer "dwelt" authentically upon the earth. Later, in his lectures on Hölderlin, he said that dwelling occurs only when entities are "gathered" (*versammelt*) into a world in which the integrity of things is preserved. Such a world would be intrinsically "local," bound up with place in a way wholly foreign to the planetary reach of modern technology. According to Dreyfus, *Being and Time*—despite later Heidegger's dislike of planetary technology—anticipated "total mobilization" by conceiving of the local workshop-world as *a* region within *the* all-encompassing region: the *referential totality*.[3]

Early Heidegger emphasized the primacy of an instrumentalist ontology to counter what he regarded as two basic problems of modernity: objectifica-tion and subject-object dualism. While scientists and philosophers asserted the primacy of the detached, cognitive ego-subject, Heidegger maintained that it was an abstract derivation from the practically oriented, socially defined "anyone" self. While liberal politicians insisted on the radical autonomy of the subject, Heidegger countered that authentic individuation was in part a collec-tive decision of a generational cohort. Later Heidegger would concede that despite its atomism, subjectivism at least distinguished between the subject and the objects of its cognition and action. Self as ego-subject was characteris-tic of the *modern* era, but in the subsequent *technological* era the subject-object distinction vanished as subjectivism vanquished all "otherness." Even humans came to be regarded as the most important raw material. Early Hei-degger's description of everyday life as role-playing within the all-embracing referential totality anticipated his later view that humanity had become raw material. While early Heidegger spoke of *Dasein* as the "for-the-sake-of-which" of worldly activity, he also seemed to discern the extent to which *Dasein* was submitting to the kind of life demanded by the character of modern production. *Being and Time* makes clear that instrumental activity is the basic way of being-in-the-world: "When concern holds back from any kind of producing, manipulating, and the like, it puts itself into what is now *the sole remaining mode of being-in* [my emphasis], the mode of just tarrying at. . . . In this kind of *'dwelling'* as a holding oneself back from any manipulation or utilization, the *perception* of the present-at-hand is consummated." [SZ: 61–62/88–89] Apart from the activity of manipulating and producing things, we are told, the *sole* alternative is to treat them abstractly: either as objects for ordinary curiosity or as objects for scientific scrutiny. In emphasizing the primacy of productive activity in everyday life, Heidegger was challenging the predominance of the constricted Cartesian disclosure of everything as present-at-hand objects for the theorizing intellect. Descartes's metaphysics, Heidegger explained, was determined not by his preference for mathematics, but instead by his orientation "toward being as constant presence-at-hand, which mathematical knowledge is exceptionally well suited to grasp." [SZ: 96/129] Cartesianism was a variant of Plato's thesis that the "really real" is the permanently present *eidos*. Descartes claimed that for a thing "to be" meant for it to be re-presented by the self-certain subject. Descartes thus helped to define modern science as the quest to formalize everything, to make everything totally present for knowledge. Heidegger maintained that this drive to make everything wholly present for knowledge was an ingredient in the technological drive to make all things wholly present as standing-reserve.

More than one critic has remarked upon the utilitarian vision of human existence offered by *Being and Time*. Manfred S. Frings, for example, argues that *Being and Time* is "an expression of ardent desires for, and flamboyant *glorifications of work* and the work-a-day-world spawned by the political isolation of Germany after 1918."[4] Early Heidegger's phenomenological description of everyday life disclosed it as ceaseless work broken only by occasional spells of distraction—not so different from what Ernst Jünger was to say about the totally mobilized technological world. To be sure, the technological disclosure of things involved a still more radical instrumentalism than that found in *Being and Time*.

Nevertheless, early Heidegger spoke about the compulsive, anonymous, everyday character of work in a way which seems to have been at odds with his later conviction that authentic working and producing were somehow possible. It is not clear, then, whether for early Heidegger the craftworker in the shop was inextricably drawn into self-forgetfulness of the falling, inauthentic kind, or whether the craftworker could be "authentic."[5] Certainly in Division Two of *Being and Time*, Heidegger spoke of authenticity as a way of being that did not float above everydayness, but rather transformed everyday practices. Everyday practices, then, are not intrinsically inauthentic; rather, they are relatively "undifferentiated" and may thus be capable of becoming either authentic or inauthentic. Perhaps this belief that everyday practices could be transformed by authentic existence was what led him to call for an authentic retrieval of Germany's possibilities, a retrieval that he believed, for a time, was being carried out by Hitler.

Early Heidegger's "either/or" of readiness-to-hand and presence-at-hand suggests that he had not yet clearly articulated what he came to regard as the predominant mode of disclosure in the contemporary era: modern technology. When he began examining this mode of disclosure in detail around 1930, he seems to have attributed to it certain aspects of the utilitarianism of readiness-to-hand, on the one hand, and the objectification of presence-at-hand, on the other. The pragmatic attitude of the artisan, supposedly basic to human existence, turned out to be the work mania associated with the technological era. As Prauss has remarked, we can only imagine "how disturbing it must have been for Heidegger to have seen that precisely technology, which he again and again sees drawing near as the greatest danger for man, is already laid out in the relation of man to being, which in *Sein und Zeit* he regarded as the primordial [relation], in circumspective dealings with the ready-to-hand."[6]

One consequence of Heidegger's insight was that he changed his attitude toward science. Early Heidegger, as we shall see toward the end of this chapter, had a positive understanding of the nature of science. Later Heidegger, however, concluded that the objectifying tendencies of modern science contributed to the instrumentalism of modern technology. Far from being a disinterested. unpragmatic way of disclosing things, modern science is motivated by the urge to dominate things. The "objectification" at work in the scientific attitude, then, makes possible ever greater means of controlling and utilizing the thing being investigated.

B. EVIDENCE FOR A NON-INSTRUMENTALIST ONTOLOGY IN HEIDEGGER'S EARLY THOUGHT

The very idea that early Heidegger's thought was somehow consistent with the instrumental, technological view of things seems quite inconsistent with his political critique of industrial modes of production. Early Heidegger was certainly not supportive of the radically utilitarian attitudes of industrial technology; nevertheless, he did suggest in *Being and Time* that the only two alternatives in everyday life were either compulsive working or detached observing. Later Heidegger's remarks about work and equipment do not conceive of work as a compulsive or all-consuming activity. In his 1941–42 lectures on Hölderlin's poem "As When on a Holiday . . . " ["*Wie wenn am Feiertage . . .* "], for example, he argued that we must avoid defining a holiday (*Feiertag*, literally "rest-day") as the mere cessation of work, or as the relaxation needed to make us ready for work once again. Rightly understood, "rest" is the entrance into wonder—into wonder "that a world worlds around us, that the entity is and rather than nothing, that we ourselves are and yet scarcely know who we are, and scarcely know that we do not know all this." [GA, 52: 64] The rest-day is the moment in which we become free from and unencumbered by ordinary things, so that we may encounter the *extraordinary in the ordinary.* Be the "ordinary," Heidegger meant the things and people closest to us, which we cannot encounter appropriately because we are driven about by utilitarian concerns. All too easily, the everyday becomes the inauthentic (*un-eigentlich*), that which is not properly appropriated (*eignet*) by us and which does not rightly appropriate us. In everyday life, things and people "do not have to have this trait of inauthenticity . . . , but they have it for the most part. . . . " This is especially true when "care [*Sorge*] is understood only as affliction and grief, as commercial efforts and as the unrest of machination, instead of recognizing that care is another essence [*Wesen*], namely, the belonging to the preserving of a relationship to the essential of all entities—i.e., to the authentic, which is always the unordinary." [GA, 52: 65]

The unordinary, we are told, is not the sensational or extravagant, but instead the simple and constant presencing (*Anwesen*, being) of the entity, thanks to which it retains its measure and can demand of humanity that it be the measure-holder. Later Heidegger developed the idea that a thing such as a jug is not merely an instrument for human ends, but instead helps to constitute the very world in which human existence unfolds. Humanity, then, does not "constitute" the jug as a subject constitutes its object; instead, human existence and thing join in a mutual dance or play in which a world can maintain itself. All of this, however, still seems quite a way from the apparent instrumentalism of the early Heidegger.

Such instrumentalism, however, was to some extent *merely* an appearance. Heidegger emphasized the primacy of everyday instrumentalism and handicraft production precisely in order to suggest that authentic producing somehow involves handiwork. While small workshops were being eliminated by factories, and while the skills of many artisans had been degraded by modernist influences, nevertheless handiwork had to be understood and appreciated in its ontological dimension if there was to be any hope of discovering an alternative to modern technology.

There are several passages in his early works which show that the young Heidegger did not regard human existence solely in utilitarian terms. For example, even in *Being and Time* we read (though admittedly in only one place) of the possibility of a disclosure of entities that is neither instrumental (ready-to-hand) nor objectifying (present-at-hand):

> As the "environment" is discovered, the "nature" thus discovered is encountered, too. Disregarding its mode of being as ready-to-hand, this "nature" itself can be discovered and defined simply in its pure presence-at-hand. From this uncovering of nature, however, there also remains hidden nature as that which "stirs and strives," surprises us, captivates [us] as landscape. The plants of the botanist are not the flowers of the hedgerow; the geographically fixed "origin" of a river is not the "spring in the ground." [SZ: 70/100]

In the 1930s, Heidegger pointed out that Hölderlin's hymns to the rivers Rhine and Ister spoke of their "source" in a radically different way than did the cartographer. It may be argued that under the influence of Husserl's definition of philosophy as rigorous science (*Wissenschaft*), early Heidegger did not focus on the ontologically disclosive power of great art and poetry. Already in 1919, however, he stated that Enlightenment rationalism failed to evaluate the poet as "a *Gestalter* within a genuine experiential world." [GA, 56/57: 133] Because of his appreciation of the world-formative importance of Greek and German literature, his religious training, and his contempt for bourgeois commercialism, he rejected the charge of critics that he viewed life in utilitarian terms. In lectures given in 1929–30, for example, he remarked that he had never meant to suggest that "the essence of man consists in handling spoon and fork, and in riding on the streetcar." [GA, 29–30: 263] In a passage that expresses even more directly his opposition to the utilitarian view of life expressed by modernists such as Le Corbusier, as well as by those committed to *Die Neue Sachlichkeit* and the *Bauhaus* movement, Heidegger stated:

> We speak of machine-building [*Maschinenbau*]. But not everything which can and must be built is a machine. Thus it is only another sign of the groundlessness that dominates today's thinking and understanding, when one offers us the house as a machine for living [*Wohnmaschine*] and the chair as a sitting machine. There are many who see in such absurdity a great discovery and the harbinger of a new culture. [GA, 29–30: 316]

Thus, while there is an undeniable utilitarian cast to *Being and Time*, we would be wrong in concluding that this utilitarianism was identical with what Heidegger regarded as the commercialism and materialism of the modern world. Instead, the utilitarianism he favored was that of the peasant or craftsman at home in the world. They did not see things as "nothing but" raw material, despite their ruggedly pragmatic ways. Heidegger's decision to begin *Being and Time* with an analysis of the craftsman's workshop may be regarded as indicative of hostility toward industrial technology. Certainly most Marxist critics have read Heidegger as a political reactionary who hated industrial reality, because of its alienating forms of production as well as its class warfare, which undermined the possibility for a genuine *Gemeinschaft*. Hence, according to such critics, his analysis of the workshop and handicraft proclaimed his preference for a pre-industrial, romantic world of handicraft activity. While there may be some truth to this assessment of Heidegger's preferences, the noted neo-Marxist Karel Kosik says that we must nevertheless be wary of it: "The critique that sees in *Being and Time* the patriarchal world of backward Germany has fallen for the mystification of Heidegger's *examples*. Heidegger, however, is describing problems of the twentieth century capitalist world which he exemplifies—quite in the spirit of romantic disguising and concealing—by the blacksmith and forging."[7]

Kosik argues that *Being and Time* met with such "extraordinary acclaim" because it described the ready-made world which is "the universal surface level of twentieth century reality."[8] Heidegger's analysis of the workshop, then, must be read in at least two different ways: first, as an account of the importance of the role played by the human hand in producing things, i.e., in "letting them be"; second, as an account of the extent to which industrial technology was already transforming handicraft producing and the way of life associated with it. Read in this second way, *Being and Time*'s account of the routinized and anonymous character of everyday life anticipates in certain respects his subsequent description of humanity as stamped by the *Gestalt* of the worker. Read in terms of the transformations brought by modern technology, the anyone self exists in "a ready-made world of devices, implements, and relations, a state for the individual's social movements, for his initiative, ubiquity, sweat. . . . The individual . . . has long ago 'lost' any awareness of this world as a product of man. Procuring permeates his *entire* life."[9] According to Kosik, then, *Being and Time* recognizes that in the technological era, everyday life is governed by a completely utilitarian way of dealing with things. The oft-mentioned gap between *Being and Time*'s description of the down-to-earth workshop, on the one hand, and its account of the inauthenticity of everyday urban existence, on the other, may be bridged if we recognize that Heidegger was already cognizant of the extent to which the work-world was being transformed by modern technology.

C. GREEK HANDICRAFT AND THE ORIGINS OF PRODUCTIONIST METAPHYSICS

Yet, even if Heidegger's analysis of the work world and everyday life did reflect current industrial and social conditions, that analysis still attempted to discover something *essential* about the process of producing and about the nature of things produced. Understanding producing and products was, in his view, so important because "the basic concepts of philosophy" arose from Plato and Aristotle's productionist metaphysics. [GA, 33: 137] If Heidegger was to be able to point the way to a new mode of working and producing, he would have to deconstruct the productionist metaphysics which made possible and justified the industrial mode of producing. In providing a critique of Western industrialism, Heidegger knew that he was in direct competition with Marx's thought. The problem with Marxist critiques of industrial production was that they lacked sufficient insight into the metaphysical origins of such work: "One must become clear about what it means, that man has a relation to the works which he produces. Hence, in a certain book, *Being and Time*, the talk is of dealings with equipment; not in order to correct Marx or to set up a new national economy, especially from a primitive understanding of the world." [Ibid.]

Heidegger usually regarded the "primitive" as the most powerful, but in this case he used the term pejoratively as a reply to those critics who maintained that he was calling for a return to pre-industrial craft conditions. Despite appearances to the contrary, he usually conceded that Germany could not return to an earlier mode of production. Germany's task was to return to its primal origins in order to initiate an authentic future in which a more satisfactory relation could obtain between productive activity, on the one hand, and the German *Geist*, on the other. For a time, at least, he hoped that it would be possible to transform (*aufheben*, to cancel, to preserve, *and* to super-sede) industrial producing in a way that would halt alienation and class war-fare, and that would promote an authentic *Gemeinschaft*. Heidegger believed that such a transformation could occur only if light were shed on the meta-physical origins of industrial modes of production. Hence his quest to under-stand the characteristics of "productionist metaphysics."

Despite his conviction that he could discover in the ancient Greek world an alternative to productionist metaphysics, Heidegger also maintained that such metaphysics arose *within* the Greek world. The early Greeks conceived of producing as "releasing" or "freeing" something so that it could manifest itself, but primarily as an instrument or means for some human end. [GA, 24: 162–163/115] Hence, the productionist orientation which came to a climax in the technological era was already at work in everyday life in ancient Greece.

The metaphysical schemes of Plato and Aristotle, Heidegger argued, were based on the view that the structure of all things is akin to the structure of products or artifacts. Aristotle's metaphysics, for example, is "productionist" insofar as he conceived of all things, including animals, as "formed matter." The most obvious example of such "formed matter" is the work produced by an artisan who gives form to material. Plato and Aristotle seemingly projected onto all entities the structure of artifacts.

Heidegger entertained the following objection to this thesis. Since the Greeks conceived of themselves as part of an eternal and uncreated *kosmos*, how could they have extended their productionist metaphysics to those things which were *not* produced? Heidegger explained that the instrumentalist orien-tation shaped in advance the Greek encounter even with "natural" things lying around in their world:

> In other words, it is first of all in the understanding of being that belongs to productive comportment and thus in the understanding of what does not need to be produced that there can grow the understanding of an entity which is pres-ent-at-hand in itself *before* all production and *for* all further production. . . . In production, therefore, we come up against just what does not need to be pro-duced. . . . The concepts of matter and material have their origin in an under-standing of being that is oriented to production. [GA, 24: 163–164/116]

Productionist metaphysics, including Plato's doctrine of the ideal forms and Aristotle's form-matter distinction, had a wide-ranging capacity for ex-plaining things. This explanatory capacity is "the basis for the universal signif-icance assignable to the fundamental concepts of ancient ontology." [Ibid.] Despite their productionist metaphysics, however, the Greeks did not "objec-tify" things because the Greeks were not yet "subjects." Nevertheless, they helped pave the way for Cartesian dualism, the "fateful constriction" of which

> dominates the entire previous tradition of philosophy. It became prepared in a certain way through Greek philosophy insofar as here the world became experi-enced not in the extreme sense of mathematization, but yet according to a natural pull [*naturliche Zuge*] of knowledge, as *pragmata*, as the *wherewith of having-to-do*.

But [the world] did not become ontologically understood in this sense, but rather in the widest sense as a thing of nature. [GA, 20: 250/185]

Let me make two remarks about this passage. First, notice that early Heidegger said that the instrumental-pragmatic orientation arose from a "natural pull." The history of productionist metaphysics began, in order words, because of the essentially pragmatic orientation of the Greeks. Later Heidegger, as we have seen, was to say that productionist metaphysics began not because of any natural "pull," but instead because of the self-concealment of being itself. Nevertheless, later Heidegger also acknowledged the enduring importance of handiwork for the authentic disclosure of things. The "pull" toward handiwork, then, must be understood not as "natural" (biological, adaptive) in character, but instead as "ontological." Hence, the instrumentalist orientation of *Being and Time* was intended, in part, to emphasize that in unreflective activity people are drawn to use their hands in the activity of disclosing things. Such unreflective activity, so Heidegger believed, arises from an authentic possibility of human existence.

Second, Heidegger argued that while Greek instrumentalism had certainly not yet arrived at the fully mathematicized technology characteristic of modernity, nevertheless the instrumentalism of the early Greeks was linked to their metaphysical view that "nature" is made up of a totality of things which are like artifacts insofar as they are formed matter. Which came first? The instrumentalist-pragmatic attitude, or productionist metaphysics? In an important respect, pragmatic dealings with things would have to have preceded the emergence of a metaphysics conceived in terms of such dealings. In other words, Plato could not have developed his notion of the ideal forms as eternal blueprints or models for temporal things unless he had been part of a culture which prized the activity of artisans.

It is not easy, however, to reconcile this apparent regard for excellent artisanship with the notoriously contemptuous attitude displayed by Greek elites toward handiwork. Max Scheler has argued, for example, that the Greek institution of slavery was what enabled philosophers such as Aristotle to view the natural world in terms of teleological forms rather than merely as an instrument for human ends.[10] Hence, a hierarchical social structure made it possible for Plato and Aristotle to shift from the pragmatic, ready-to-hand orientation of the lower classes to the theoretical, present-at-hand orientation of the elites. The elite Greeks did not view the world either as an object of labor or as an object of a creative act that man had to continue, but instead as "the realm of living and noble, energetic forms to be seen and contemplated and loved."[11] Scheler concluded that this contemplative attitude, the origin of all theory, was radically divorced from instrumentalism; hence, there is no inner connection between the theoretical study of nature, on the one hand, and the technological domination of it, on the other.

Whereas Scheler argued that the non-instrumental attitude of theory had been made possible by Greek slavery, Heidegger made no reference to the hierarchical structure of Greek society in his analysis of the relation between everyday Greek instrumentalism and Greek metaphysical theory. Once again presuming that he had penetrated beneath the surface of such merely socio-economic analyses, he insisted that Greek philosophy was productionist in orientation because it derived its basic metaphysical insight from the everyday production of artifacts. It was this derivation of the theoretical attitude from instrumentalist activity which gave rise to the ambivalent attitude toward science discernible in Heidegger's early thought.

On the one hand, as we shall see a bit later, he spoke highly of science as a way of disclosing things. On the other hand, he spoke of science as leading to a constricted way of seeing things, a way that is epitomized in Descartes's thought. While early Heidegger regarded the theoretical attitude as relatively detached and uninvolved by comparison with the instrumentalism of everyday life, later Heidegger concluded that science was in fact governed by a profoundly instrumentalist attitude. Later Heidegger, then, realized more clearly that the everyday instrumentalism described in *Being and Time* was in important respects not a "neutral" dimension of human existence which could provide *a priori* insight about any historical possibility of human society. Instead, the all-embracing instrumentalism described in *Being and Time* was specific to society in the final stages of productionist metaphysics. Hence, while the theoretical or scientific attitude which derives from that instrumentalism may at first glance appear to be detached and non-instrumental, in fact it is very much in the service of the instrumentalism which defines everyday life in the technological era. It was insight into the dominating, all-embracing character of everyday modern instrumentalism which led later Heidegger to say that the technological disclosure of things arose prior to the development of modern science, even though the technological application of science to industry occurred only many years *after* the rise of modern science.

Heidegger argued that the ontological category of presence-at-hand was first derived from the production process involved in the fabrication of ready-to-hand artifacts. Then, Plato and Aristotle tended to define the *kosmos* as a totality of entities present-at-hand, of which humans were an important instance. Aristotle defined man as a certain kind of thing present-at-hand: a rational animal. Centuries later, Descartes helped to solidify the interpretation of humanity as a present-at-hand creature, but simultaneously he defined this creature's rationality as the metaphysical ground for all things. Finally, in the technological era, the present-at-hand human subject has become reduced to a means useful for enhancing the power of the technological system. If people in the technological era are treated instrumentally, this is because the Greeks defined humans in terms of categories which originally applied to artifacts such as equipment!

In the mid-1930s, in his lecture "The Origin of the Work of Art," Heidegger added a new dimension to his account of the relation between productionist metaphysics and handicraft production. This new dimension was the work of art, which is a work of the hand but cannot be understood as a "product" in the way equipment is. Equipment, that which is needed to produce things, "has long since occupied a peculiar pre-eminence [starting with the Greeks] in the interpretation of entities." [HW: 27/38] In light of their productionist metaphysics, Greeks defined even living things as "formed matter," for they interpreted everything as being analogous to artifacts formed from material. This form-matter distinction, Heidegger argued, was inadequate not only for describing natural things, but also for describing artifacts.

If not "formed matter," then, what is a thing? By way of answering this question, Heidegger first distinguished among three kinds of things: (1) mere things (such as rocks), (2) equipment, and (3) works of art. (He concluded that animals, plants, and humans cannot be categorized as "things.") He defined equipment in terms of its usefulness: "It [usefulness] is the basic trait from which this entity [equipment] regards us, that is, flashes at us and thereby is present and thus is this entity." [HW: 18/28] Here, once again, Heidegger maintained that the usefulness of the equipment is somehow intrinsic to the thing, as if usefulness were not ultimately definable in terms of the people who made and use the equipment in question. Equipment occupies a curious middle ground between mere things (such as stones) and works of art. On the one hand, equipment (such as a pair of shoes) is self-contained like a mere thing, but unlike the mere thing, shoe-things have to be manufactured. On the other hand, equipment is also like a work of art in that it is formed by hand, but is not self-sufficient in the way the work of art is. [HW: 18/29] Because equipmental things are midway between mere things and works of art, people have tended to define both mere things and works of art in terms drawn from their interpretation of equipment. Thus, on the one hand, one might define a mere thing as like equipment in being formed matter, but unlike equipment in being denuded of usefulness. On the other hand, one may define the work of art as being like equipment in being formed matter, but unlike equipment in being useless. Curiously, in being useless, mere things and works of art are closer to each other than to equipment. Still, there is no denying that both works of art and equipment are artifacts, while mere things are not.

According to Heidegger, the attempt to define mere things and works of art in terms of categories originally applied to equipment failed for at least two reasons. First, it failed to define usefulness; second, it presupposed the universal validity of the form-matter distinction. Instead of defining the work of art in equipmental terms (e.g., as a "useless artifact"), he argued that we should define equipment in terms of how it is disclosed through the work of art. Consider, Heidegger said, how van Gogh's painting of a pair of shoes (allegedly owned by a peasant woman) reveals the very *being* of the shoes as equipment. In addition, the painting discloses a dimension of usefulness that is usually overlooked, especially in the technological era: reliability (*Verlässlichkeit*).

> By virtue of this reliability the peasant woman is made privy to the silent call of the earth; by virtue of the reliability of the equipment she is sure of her world. World and earth exist for her, and for those who are with her in her mode of being, only thus—in the equipment. We say "only" and thereby fall into error; for the reliability of the equipment first gives to the world its security and assures to the earth the freedom of its steady thrust. [HW: 23/34]

If we overlook the reliability of useful things, we fail to see that equipment can arise and have application only within a stable, ordered world and grounded in the self-generating earth. Because of this oversight, we have an impoverished idea of equipment, as an isolated thing which has been fabricated by impressing form on matter. In fact, however, "matter and form and their distinction have a deeper origin" than that conceived of in the metaphysical tradition. [HW: 24/35] This "deeper origin" refers to the event which made philosophical reflection possible in the first place: the primal Greek encounter with being. Only by virtue of that encounter could Aristotle ask about what and how things *are*, and answer that they are "formed matter." In focusing on the metaphysical question of the foundation or structure of things, however, instead of on the ontological event of the presencing by virtue of which things appear at all, Aristotle contributed to the development of productionist metaphysics. So far removed is technological humanity from the Greek encounter with being, so Heidegger argued, that we treat not only equipment but everything else in purely instrumental terms. Lacking insight into the being of entities, we reduce all things to the undifferentiated "stock" needed to fuel the expansion of the technological system.

Despite the fact that the Greeks made such a contribution to productionist metaphysics, Heidegger also believed that in their highest moments they were gifted with a non-instrumentalist attitude of awe and wonder in the face of the fact that things *are*. Aristotle observed that all philosophy begins with such wonder. The metaphysics of Plato and Aristotle were great, awe-provoked responses to the primordial encounter with being, but they already began to speak of being as an extraordinary kind of thing, i.e., a permanently present form, or a metaphysical ground. Heidegger's aim was to "retrieve" the Greek tradition in order to "understand the Greeks better than they understood themselves." [GA, 24: 157/111] Such understanding, he hoped, would lead thinking into the vicinity of the pre-Platonic, pre-metaphysical encounter with being. Only in such an encounter could there arise a new mode of producing that could save Germany from modern technology.

Heidegger meditated on basic Greek words, with the aim of discovering in them glimmers of an originary understanding of what things are. For example, in examining the term *chre*, "to use," he explained that "using" a thing really means letting it be what and how it is. Letting something be means caring for the thing and responding to its own intrinsic nature.

> "Using" does not mean the mere utilizing, using up, exploiting. Utilization is only the degenerate and debauched form of use. When we handle a thing, for example, our hand must fit itself to the thing. Use implies fitting response. . . . [A]uthentic using first brings the thing used into its being [*Wesen*] and holds it therein. Thought in this way, use itself is the summons that something be admitted into its own being and that using not desist from it. Using is: admitting into being, preserving in being. [WHD: 114/187]

Yet, if even the Greeks, who existed so near to the original encounter with originary presencing, fell into metaphysical thinking, what chance do we moderns have to encounter being anew? How can we "let things be" or develop a "fitting" response to them today? Do not things inevitably show up as standing-reserve for the increase of power? Conceding that in today's world a fitting response to things is very difficult, Heidegger nevertheless argued that the "origin" is not far away but very near. Employing imagery drawn from his Christian upbringing, but properly secularized to fit his own ontological aims, Heidegger asserted that being "is" closer to us than any entity, only being conceals itself from us today. To prepare ourselves for the new advent of being, we must begin paying attention to those practices that have not yet been completely absorbed into the technological imperative. Perhaps recalling the craftsmen and peasants in his own family, he suggested that handicraft workers maintained a relationship with the being of things that was impossible for industrial workers:

> One can object that today every village cabinetmaker works with machines. One can point out that today gigantic industrial factories have risen alongside the craftsmen's workshops. . . . [Such objections fall] flat, because [they have] heard only half of what the discussion has to say about handicraft. The cabinetmaker's craft was chosen as an example, and it was presupposed thereby that it would not occur to anyone that through the choice of this example is the expectation announced that the condition of our planet could in the foreseeable future, or indeed ever, be changed back into a rustic idyll. . . . However, it was specifically noted that what maintains and sustains even this handicraft was not the mere manipulation of tools but the relatedness to wood. But where in the manipulations of the industrial worker is there any relatedness to such things as the shapes [Gestalten] slumbering within the wood? [WHD: 53–54/23]

In *Being and Time*, however, Heidegger did not speak of "slumbering shapes" to be freed from wood or rock, but instead of the ore to be taken from the mine. Given his instrumentalist depiction of nature, early Heidegger sometimes sounded like a proto-technological man: "In the same way we uncover nature in its violence and power not through reflection about it, but in struggle with it and fight against it, in *becoming master over it* [Herrwerden über sie], so too do nature-myths include a history of this struggle with nature, i.e., they are the interpretations of a primordial relation to it." [GA, 25: 21; my emphasis]

Even during the 1930s, however, when he had supposedly moved beyond such an instrumentalist viewpoint, Heidegger still affirmed that violence was needed to open up a world in the face of the overpowering force of *physis*, the presencing or being of entities. The violence exercised by the early Greeks in bringing presencing to a stand, i.e., in opening up a world in which entities could present themselves in a differentiated way, was not primarily instrumental but metaphysical in character. The ancient Greeks, in other words, were not merely "clever animals" striving to secure their existence, but instead were *Dasein* appropriated by being as the site through which entities could disclose themselves. The Greek violence against entities was tempered or at least justified by their higher aim of freeing entities so that they could manifest themselves appropriately within a particular historical world. During the cen-

turies following this originary Greek disclosure of the being of entities, however, a severe metaphysical decline set in. Instead of doing violence in order
to disclose entities in and for themselves, modern Western man does violence
in order to subjugate entities solely to his needs. Industrial modes of production compel entities to present themselves in a one-dimensional way, as
standing-reserve.

D. SCIENCE AND PHENOMENOLOGY AS AUTHENTIC DISCLOSING
   AND PRODUCING

Heidegger defined *authentic* producing precisely as a "freeing" disclosure
which allows the entity to come to stand on its own, independently of the one
who discovered or made it. For Heidegger, "producing" something meant
disclosing it, opening it up, letting it be. Making something, then, is only one
particular kind of "producing." Making is a mode of disclosure in which an
artisan "occasions" the process by which his or her material is gathered into a
thing. Early Heidegger often spoke of science as such an authentic mode of
disclosing or producing things. For example, mathematical physics opened up
a whole new way for entities to show themselves. By projecting nature in
terms of *mathesis universalis*, i.e., by opening up a horizon in which nature
could reveal itself in terms of mathematical quantities, mathematical physics
provided a new clearing in terms of which novel "facts" could be generated.
The decisive insight of modern science was that there are no such facts apart
from theory, understood here as the *a priori* disclosure of the being of entities.
The aim of scientific theory, so Heidegger argued in *Being and Time*, is to
"free" entities in such a way that they can be examined "objectively." [SZ:
363/415]

Philosophy itself, so argued early Heidegger, is a rigorous science (*Wissenschaft*). All such science "has its existentiell basis in a resoluteness by which
Dasein projects itself toward its potentiality-for-being in the 'truth.' " [Ibid.]
Scientific *Dasein* freely binds itself to the task of uncovering the entity for its
own sake:

> Thereby are discontinued all behavioral goals which aim at the [technical] applica-
> tion of the uncovered and known; and all those boundaries fall away that confine
> the investigation within planned technical purposes—the struggle is solely di-
> rected to the entity itself and solely in order to free it from its hiddenness and
> precisely thereby to help it into what is proper to it, i.e., *to let it be the entity which
> it is in itself.* [GA, 25: 26]

In this passage, early Heidegger mixed voluntaristic language about *Dasein*'s "free self-binding" with talk of letting the entity be. Increasingly, however, he was to say that instead of choosing to disclose things, *Dasein* was
chosen or destined to do so. Even the violence of the creative type about
which Heidegger spoke in the mid-1930s was not the possession of a human
"subject," but instead the artistic, political, or philosophical "creator" was in

fact seized and even violated by the disclosive power of being at work *through* him. For early and later Heidegger, great scientists are philosophers gifted with a new way of understanding what things are. Admiration of the ontological insights of Galileo, Descartes, and Newton balanced early Heidegger's critical attitude toward positivistic science. Positivism concluded not that mathematical physics was *a* valuable new way of understanding things, but instead that it was the *only* legitimate way. Heidegger did fight against the positivist doctrine that only the statements of empirical science were meaningful. In the 1920s he hoped to develop an alternative conception of science and scientific method.

Heidegger regarded Husserl as the founder of such a new method, phenomenology. Instead of simply "normalizing" that method, however, Heidegger sought both to purge it of residual traces of subjectivism and to expand its capacity for disclosing and "freeing" the being of entities. Later Heidegger became convinced that authentic science had become virtually impossible in the technological age. Hence, by 1935 he was no longer saying that Germany could be renewed only by a rebirth of authentic science, but instead by the advent of a saving work of art, such as the poetry of Hölderlin.

The foregoing considerations suggest that Dreyfus may be overstating his case by saying that early Heidegger's instrumentalism "removes every vestige of resistance . . . to the technological tendency to treat all beings (even man) as resources." The aim of early Heidegger's analysis of everyday instrumentalism and scientific objectification was to retrieve from them the possibility of an *authentic* way of producing or disclosing things. Moreover, two interrelated beliefs prevented early Heidegger from arriving at complete instrumentalism. First, because humanity is radically finite and dependent, it cannot achieve the longed-for goal of making entities totally present and available, despite the achievements of modern technology. Second, entities themselves resist human *Dasein*'s efforts to disclose them, a theme later Heidegger was to develop in more detail in his essay on the work of art. Faced with an entity's tendency to conceal itself, finite humanity had to struggle to disclose the truth about things. In 1928, Heidegger said that *Dasein*'s highest calling is to "free" entities, to "let them be" what they are. In freeing entities, *Dasein* achieves its own authentic freedom.

> Insofar . . . as freedom (taken transcendentally) constitutes the essence of *Dasein*, as existing, it is always essentially and necessarily "further" than any given factical entity. On the basis of this upswing, *Dasein* is in each case beyond entities, as we say, but it is beyond in such a way that it, first of all, experiences entities in their resistance against which transcending *Dasein* is powerless. The powerlessness is metaphysical, i.e., to be understood as essential; it cannot be removed by reference to the domination of nature, to technology, which rages about in the "world" today like an unchained beast; for this mastery of nature is the real proof of the powerlessness of Dasein, which can attain freedom only in its history. [GA, 26: 279/215]

The metaphor of the "unchained beast" was not accidental. Heidegger believed modern humanity's ontological understanding had declined to the point that humans could conceive themselves only as clever animals whose aim was security and power. Instead of existing as the mortal clearing in which entities could present themselves, humanity now conceived of itself as a beast of prey, fully justified in subjugating all things to human will. Just as fangs were deadly tools in the organic machine called the tiger, so too technical devices were weapons in what Jünger called the "organic construction": the modern worker-soldier. Heidegger's meditation on tools and technology aimed, in part, to refute the naturalistic interpretation of humanity, including the Marxist interpretation of man as the "laboring" animal. His retrieval of the meaning of authentic "producing" aimed to show that man is not truly "free" when he turns the earth into a gigantic factory for satisfying his boundless cravings for security, power, and pleasure. We shall examine Heidegger's critique of the naturalistic interpretation of human existence in chapter twelve.

Before doing so, however, we must first examine his account of the history of productionist metaphysics. He maintained that the decline of humanity to the level of a "clever animal" was occasioned by the self-concealment of being that occurred in the history of productionist metaphysics. Jünger's technological worker-soldier may seem to be a far cry from what Aristotle had in mind by his conception of man as the "rational animal." Heidegger argued, however, that Jünger's worker was the manifestation of what Nietzsche called the "blond beast of prey," and this technological beast was the final expression of Aristotle's metaphysical conception of man as the "rational animal."

Chapter 11. The History of Productionist Metaphysics

The limitless domination of modern technology in every
corner of this planet is only the late consequence of a very
old technical interpretation of the world, which interpretation is usually called metaphysics. [GA, 52: 91]

When Heidegger saw the paradox at work in his "foundational historicism,"
i.e., when he discovered that he could not reconcile a quest for *a priori* structures of human existence with a historicist conception of human existence, he
also began questioning whether the instrumentalist orientation to things was a
universal feature of human existence. If Greek instrumentalism did not result
from a "natural" or universal human tendency, then he would have to devise a
different account of the origin of productionist metaphysics. This new account
involved a major shift between Heidegger's early and later interpretation of
the history of being. Early Heidegger interpreted history in terms of a structural feature (instrumentalism) of the *human understanding* of being. Later
Heidegger interpreted history in terms of the unpredictable ontological play of
concealment and unconcealment. Early Heidegger argued that the production
of works was a "basic" mode of human behavior. Later Heidegger concluded
that he may have overemphasized the continuity between Greek and contemporary attitudes toward production of works. After all, the ancient Greek attitude toward production was by no means the same as that of modern
technological humanity. Nevertheless, despite the differences between Greek
and contemporary technology, post-Platonic Greeks were already stamped by
the productionist orientation that was to culminate in the technological era.
Already in Plato's time, the self-concealment of being had initiated the history
of productionist metaphysics.

Heidegger's claim that modern technology is the outcome of "the history
of being" has led many critics to ask: Does not technology result from scientific discoveries, new economic formations, nationalism, and other material
and cultural factors? In other words, isn't modern technology a development
within the history of *humanity*? In maintaining that technological systems and
devices result from the "self-concealment of being," did Heidegger not hypos-
tatize being, thereby turning it into a transcendental agency? Moreover, by
portraying humanity as the medium through which the play of being could
take place, did not Heidegger deprive humanity of the spontaneity and autonomy traditionally held to be characteristic of humanity? As we have seen,

Heidegger believed that such questions betray a humanistic bias which is itself the consequence of a lack of understanding of what it means to be human. While such humanism is held to liberate humanity, Heidegger argued, in fact humanism has helped to turn humanity into the most important raw material for enhancing the endless quest for power.

He conceded that his view of humanity, namely, that it was appropriated as the site for an ontological play beyond human control, was both controversial and difficult to grasp. He himself was always puzzled by the relation between being and human existence. In the following passage, he insisted that being was both independent of human existence and also somehow in need of it:

> The history of being is neither the history of man and of humanity, nor the history of the human relation to entities and to being. The history of being is being itself, and only being. However, since being claims human being [*Menschenwesen*] for grounding its truth in entities, man remains drawn into the history of being, but always only with regard to the manner in which he takes his essence [*Wesen*] from the relation of being to himself and, in accordance with this relation, loses his essence, neglects it, gives it up, grounds it, or squanders it. [N II: 489/82]

As we noted earlier, Heidegger's account of the history of being, and of humanity's relation to it, reads in places like a Greek tragedy, in which the protagonist—Western man—acts as if he were gifted with self-understanding and self-mastery, but in fact possesses neither. The hero turns out to be a victim of *hubris*, having wrongly assumed that he knew who he was and that his own intentions were the origins of his acts. Western man's blindness, however, is not self-inflicted but fated: the being of entities conceals itself to such an extent that Western man can no longer understand who he is or what other entities are. Happening "behind the backs" of historical humanity, the self-concealment of being tends to turn humanity into players in an ontological game. In tracing out the history of being, Heidegger sought to discover the character of this play. Such discovering was an act of *anamnesis*, i.e., the remembering of what had been concealed and thus forgotten.

The history of being involved a double concealment. First of all, being concealed itself insofar as being "is" not an entity but instead the event of the presencing or unconcealment of an entity. This event of presencing conceals itself, withdraws in favor of the entities revealed through it. The so-called pre-Socratic Greeks caught a glimpse of this presencing, but immediately began to turn it into an eternally present entity. The second level of concealment involved the covering-up of the self-concealing character of being. Heidegger's task was to open up a conversation in which people could realize that something important had been forgotten.

The increasing self-concealment of being constitutes the history of productionist metaphysics. This history involves three interrelated transformations: (1) the change of being from *energeia* to actuality or reality, (2) the change of truth from *aletheia* to certainty, and (3) the change of substance from the change of *hypokeimenon* to *subiectum*. Early Heidegger believed that the history of productionist metaphysics was initiated by Plato and Aristotle, but later Heidegger could already see the beginnings of it in Heraclitus and Parmenides. In chapter fourteen, we shall see how Heidegger's attempt to initiate an authentic, non-metaphysical mode of producing involved meditation on the pre-Socratic notion of *techne* as a knowing disclosure closely akin to art. In this chapter, we shall take up the story of productionist metaphysics with Plato and work our way down to Nietzsche. Heidegger himself told this story in a number of versions, not all of which are completely consistent. The following account is a synthesis of those versions.[1] It will restrict itself wherever possible to those aspects of the story which are most pertinent to the phenomenon of modern technology.

A. THE TRANSFORMATION OF BEING FROM *ENERGEIA* TO *ACTUALITAS* TO THE WILL TO POWER

Heidegger claimed that the distinction between essence and existence, possibility and actuality, was one of the most important in all of productionist metaphysics. Although Aristotle coined this distinction, Plato developed the famous doctrine that the eternal forms constitute the enduring *essence* of things. For Plato, *eidos* ("form") answered the question *ti estin*: what is [an entity]? [N II: 403/4] *Eidos* named a thing's "look" or "aspect"; defined and limited it; gave it form (*morphe*), shape, figure; and constituted its "nature" and "perfection" (*teleion*). *Eidos* was a thing's prototypical image that was prior to the appearance of the thing itself. *Eidos*, then, was the eternal blueprint or model of which sensible, changeable things were but imperfect copies. Earlier, we noted that the concept of *eidos* was based on a productionist model of reality. Consider a sculptor at work. For him or her, the shape (*morphe*) of the planned sculpture is grounded in the imagined or projected aspect (*eidos*). *Eidos*, then, named *ti en einai*: "What a thing already was [before it was produced]." Conceived metaphysically, *eidos* was the "really real," the permanently present "aspect" which provided the ontological foundation for finite things. [GA, 24: 149–151/106–108] As that which shines forth and radiates, Plato's *eidos* retained some aspect of the original character of being as *physis*: presencing/appearing. Nevertheless, *eidos* refers not so much to the event of presencing as it does to the permanently present form that provides the foundation for and thus gives coherence to entities in the world of the sense. Plato initiated productionist metaphysics by turning being into "beingness," presencing into something permanently present.

Let us now turn to the other half of the essence-existence distinction, existence, which was developed in detail by Aristotle. Many metaphysicians have long considered *existentia* to be both more basic than *essentia* and also more self-evident. Existence says of an entity: *hoti estin*, "that it is." [N II: 403/4] For Aristotle, existence was the major philosophical puzzle. As we have seen, Heidegger's own conception of the "movement" in human existence was indebted to Aristotle's account of the change and motion (*metabole, kinesis*) involved in the growth and appearance of things. Nevertheless, Aristotle never conceived of presencing as such, but instead he conceived of being as the unique "motion" through which an entity attained to stable presence. This stable presence, rest, was for Aristotle the end (*telos*) of motion. For example, a house stands there and exposes and unconceals itself, in that it stands forth in its outward appearance. Heidegger explained:

> Standing, it rests, rests in the "ex" of its exposure. The resting of what is produced is not nothing, but rather gathering. It has gathered into itself all the movements of the production of the house, terminated them in the sense of completed boundary—*peras, telos*—not in the sense of mere cessation. Rest preserves the completion of what is moved. The house there *is* as *ergon*. "Work" means what is completely at rest in the rest of outward appearance—standing, lying in it—what is completely at rest in presencing in unconcealment. [N II: 404/5]

For the Greeks, work did not involve accomplishing something through strenuous activity, nor was it conceivable in terms of cause and effect. The "workness" (*ergon*) of a thing meant its manner of presencing, *ousia, energeia*, the enduring unconcealment of its outward appearance. For Aristotle, *energeia* was virtually synonymous with *entelecheia*, derived from the word *telos*. *Telos* meant "the end in which the movement of producing and setting up gathers itself. . . . *Entelecheia* is having-(itself)-in-the-end, the containing of presencing [*Anwesen*] which leaves all production behind and is thus immediate, pure: being in presence [*Wesen in der Anwesenheit*]." [N II: 405/6] For Aristotle, according to Heidegger, *energeia* named the *ousia* of an entity.

To *ousia*, ordinarily translated as "substance," Aristotle gave at least two additional meanings. First, he defined it as the presence or the existence of something, expressed by the phrase "*that* something is." *Ousia* meant what *is* in the proper sense, what is available to the entity itself, what is present for itself, what lies forth: *hypokeimenon*, "substance." [N II: 430/27] Second, he defined *ousia* as the self-showing of the outward appearance, in terms of which the existing entity allows itself to come to presence as a certain kind of entity. In other words, secondary *ousia* was the *eidos* of an entity, its outward aspect by which was revealed the commonality of its aspect with other entities of the same species. While for Plato *ousia* was *eidos* or essence, the permanently present foundation or "beingness" of entities, for Aristotle *ousia* was *energeia* or existence, the event whereby an entity comes into and lingers in presence. For both Plato and Aristotle, the being (*ousia*) of an entity was that which made it possible, that which effected it and brought it into presence. Because Aristotle conceived of the entity as that which truly *is* in a way that Plato did not, Aristotle was "more truly Greek" than Plato. [N II: 409/9–10] However, because Aristotle could conceive of substance as *energeia* only in opposition to Plato's notion of substance as *eidos*, he was still unable to encounter being in its most primordial mode: as the interplay of presencing and absencing, *physis* and *aletheia*. Hence, the subsequent history of metaphysics was determined by the distinction between *eidos* (essence: enduring presence of the form) and *energeia* (existence: lasting presence of the entity).

Gradually, existence came to have precedence over essence. This happened because metaphysics increasingly conceived of being in terms of that which causes something or makes it possible. Heidegger maintained that *"the essential origin of being as making possible and as causing rules throughout the future history of being."* [N II: 420/19] In searching for that which effects or causes things, metaphysicians turned away from the sheer event of presencing that *precedes* the distinction between essence and existence. Increasingly, existence took precedence over essence, for that which exists is that which has been effected or caused, and which can likewise cause and effect. To speak about "being" came to mean to speak about a particular thing which had been created or produced. *"Both factors, the precedence of the entity and the assumed self-evidence of being, characterize metaphysics."* [N II: 411/11]

Roman translations of Greek terms gave enormous impetus to productionist metaphysics. The key here was that whereas for Aristotle *energeia* continued to mean in part the self-manifesting or "being" of an entity, for the Romans *energeia* came to be understood not in ontological terms, but instead in terms of cause and effect: the "action" involved in putting something together or manufacturing it:

> The *ergon* is no longer what is freed in the openness of presencing, but rather what is effected in working, what is accomplished in action. The essence of "work" is no longer "workness" [*"Werkheit"*] in the sense of distinctive presencing in the open, but rather the "reality" [*"Wirklichkeit"*] of a real thing [*Wirklichen*], which rules in working and is fitted into the procedure of working. Having progressed from the essence of *energeia*, being has become *actualitas*. [N II: 412/12]

This translation of *energeia* into *actualitas* was so decisive that "all Western history is in a manifold sense Roman, and never Greek." [N II: 413/13] *Actualitas* still contained the Platonic element of *eidos* or *idea*, that which makes a thing possible in its outward aspect. As the "whatness" of a thing, *idea* was related to *agathon*, that which makes possible. Hence, *idea* came to mean the cause, *aitia*, of a thing, the permanently present form which brings something into being. [N II: 413–414/13] While *aitia* originally meant "causing" only in the sense of being responsible for bringing something into presence, we have seen that the Romans interpreted *aitia* as "causing" in the sense of working upon something, effecting it, making it. When presencing was reduced to *actualitas* or reality, entities were considered to be what is most "real." Their presencing, the event of their self-manifesting, disappeared from view altogether. "To be" now meant to be the effect of some cause, to be made present by some action. To some extent, Aristotle had anticipated this view by defining being as "the making possible of presence, [as] effecting constancy or permanence." [N II: 414/14] Heidegger insisted, however, that

> the fundamental characteristic of working and work does not lie in *efficere* and *effectus*, but lies rather in this: that something comes to stand and to lie in unconcealment. Even when the Greeks—that is to say, Aristotle—speak of that which

the Romans call *causa efficiens*, they never mean the bringing about of an effect. That which consummates itself in *ergon* is a self-bringing-forth into full presencing: *ergon* is that which in the genuine and highest sense presences [*an-west*]. [VA I: 41–42/160]

Christian theologians, including St. Augustine and St. Thomas Aquinas, furthered this Latin interpretation of being as the productive basis for things. For such theologians, God became identified with the being of entities, i.e., the self-grounding Creator who produced all creatures. The "onto-theo-logy" which resulted from the wedding of Christian theology and Greco-Roman metaphysics proved disastrous for both, in Heidegger's view. God was reduced to the status of an all-powerful causal agent, hardly a Divine to Whom one could sing and dance [ID: 140/72], while being was reduced to a superior kind of thing, the creative ground.

In the metaphysical God, Heidegger discerned the form of technological humanity. God was depicted as the truly real, *actualitas* in the sense of the effecting, calculating, planning agency which produces the relatively stable and independent presence of creatures. With the "subjectivistic turn" initiated by Descartes, humanity began to arrogate to itself the role of God, the grounding ground, the producer of all things. Through Descartes's writings, humanity shed the bonds of the "truth" of revelation and entered into "the new freedom, as the self-law-giving which is secure for itself." [N II: 147/100] By giving the law to himself, in a way that anticipated Kant's notion of freedom, Descartes made it possible for humanity's projects to be grounded solely within the cognizing subject's own self-binding certainty. For Descartes, for something "to be" meant for it to be the object of the self-certain subject. From Descartes to Nietzsche there occurs the change of the determination of the being of entities from objectivity—the re-presentability of entities to the self-certain subject—to value—the capacity for entities to contribute to the subject's limitless Will to Power.[2] With Descartes's famous *cogito*,

> All consciousness of things and of being in the whole is referred back to the self-consciousness of human subjectivity as the unshakable ground of all certainty. The reality of the real is determined henceforth as objectivity, as that which is conceived *through* the subject and *for* it as something thrown and held over against it. The reality of the real is the representedness [*Vorgestelltheit*] *through* and *for* the presenting subject. Nietzsche's doctrine which makes everything which is and how it is a "property and product of man" merely carries out the furthest unfolding of Descartes's doctrine, according to which all truth is grounded on the self-certainty of the human subject. [N II: 129/86]

In effect, Descartes started the process whereby humanity came to conceive of itself as the God-like source of the reality, truth, and value of all things. Descartes enabled humanity to interpret the world as a picture (*Bild*), the reality of which was assessed according to how the image stood in relation to the standards of the productive-measuring subject. What is presented to the subject is set-there (*zugestellt*) not only as a pre-given object, but also as disposable (*verfügbar*) by the subject. As one of the founders of scientific methodology, Descartes was already concerned with how "objective" knowledge would make it possible for humans to gain control of objects.

By making the self-certainty of the cogitating ego the ontological basis for all things, Descartes made the indubitable presence of the reason-giving ego the standard against which to measure the presence (or reality) of all other things. Nothing really *is* unless it can be re-presented (*vorgestellt*) to the fully self-present subject and according to the exacting standards of that subject. The term *vorstellen* means literally "to place before" and is usually translated as "to present." As something placed before the subject, a *Vorstellung* is an object, *Gegenstand*, something which is projected by and which stands over against the subject. The term "object" itself is taken from the Latin word *obiacere*, meaning "to lie against." The English verb "to represent" can be construed to include many of these meanings. For Heidegger, "to represent" meant not to accept the thing as it presents itself, but rather to re-present the thing in the sense of portraying it in terms amenable to the standards and purposes of the re-presenting subject.

For Descartes, of course, the real was only that which could be represented in terms that were as certain as the self-certainty of the subject's own presence to itself. And what can be represented so precisely are only extended objects, which can be measured mathematically. Heidegger said that Descartes's mathematical representationalism was "the first resolute step through which modern machine technology [*Kraftmaschinentechnik*] and with it the new world and its people become metaphysically possible." [N II: 165/ 116] Missing from Descartes's thought, however, was the crucial element in the metaphysics of modernity: Will. It was Leibniz who carried out the "decisive beginning" of the metaphysics of subjectivity [N II: 237/179], by his doctrine of "monads" as subjects who strive to actualize all their experiences. By introducing an Aristotelean dynamism into Descartes's static metaphysics of the subject, Leibniz opened the way for Nietzsche's doctrine that the Will to Power constitutes the being of all entities. Heidegger noted that "only when we conceive beingness as actuality [*Wirklichkeit*] does there open a connection with doing and effecting [*Wirken und Erwirken*], i.e., with the empowering to power as the essence of the Will to Power. Accordingly, an inner relation obtains between beingness as subjectivity and beingness as Will to Power." [N II: 236–237/179] For Leibniz, all "actuality" was contained as representations or experiences within each individual monad. And monads, these spiritual points of experience, were self-willing agents which actualized and made present all the entities in their experience. Leibniz still maintained that monads were created by God, but it was not a great step from Leibniz to the doctrine which Heidegger attributed to Nietzsche: that everything we experience is merely the product of human will.

Kant contributed to the history of productionist metaphysics by his doctrine, borrowed from Plato *mutatis mutandis*, that the categories of human understanding make possible the objects of experience. But Kant also emphasized human finitude and receptivity in a way which pointed beyond subjectivism. Hegel, ignoring Kant's insistence on human finitude, paved the way for the culmination of subjectivism and, hence, of modern technology. Hegel deified reason and depicted Will as the agency through which reason would achieve its ultimate goal: absolute self-transparency and freedom. Self-willing reason was, for Hegel, the being of entities, the actuality of the actual. [N I: 299/222–223] Hegel described Will as the agency of God's goal of becoming completely present, completely self-manifest. Only Nietzsche, however, made willing an end in itself. For him, Will was the unconditioned subjectivity of life which strives to become ever stronger: the Will to Power was essentially the Will to Will, the aimless striving for ever more striving. This ever-expanding circle never opens up beyond the self-contained limits of blind striving. Humanity, stamped by this Will, is reduced to a clever animal striving for more power. Hence, "at the end of metaphysics stands the proposition: *Homo est brutum bestiale.* Nietzsche's word of the 'blond beast' is not an incidental exaggeration, but instead the characterization and motto for a context in which he knowingly stands, without penetrating its essential historical relations." [N II: 200–201/148] Nietzsche's thought, which is at work in the essence of modern technology, exhausted the possibilities of the history of productionist metaphysics. His Overman stood for the technological worker-soldier who would disclose all entities as standing-reserve necessary for enhancing the ultimately aimless quest for power for its own sake. For Nietzsche, then, for something "to be" meant for it to be "valuable" for enhancing the Will to Will. And human Will was the ontological basis for everything. "To be" meant to be generated by and for the Will to Will: human willing as that which causes, effects, and makes possible all things.

So far we have seen how Plato transformed being into beingness: the transcendent and eternally present form which makes things possible. And we have seen how Aristotle helped to initiate the idea that for something "to be" means for it to be effected, caused, or produced. This productionist model of being was gradually changed in Roman and medieval times, but received its decisive transformation in the subjectivism of Descartes, Leibniz, Hegel, and Nietzsche. Nietzsche represents the triumph of subjectivistic productionism: for him, to be meant to be produced by human Will and to be useful for enhancing that Will. Let us now consider briefly the parallel history of the transformation of truth from *aletheia* to certainty.

B. THE CHANGE OF TRUTH FROM *ALETHEIA* TO CERTAINTY

Without insight into how truth changed to certainty, we cannot fully understand the anthropocentric subjectivism of Nietzsche's doctrine of the technological Will to Will. According to Heidegger, Plato initiated the subjectivistic turn in the history of truth. In a controversial interpretation, Heidegger first conceded that Plato retained the notion of truth as unconcealment, *aletheia.* Indeed, the allegory of the cave depicted the *paideia* or "education" (*Bildung*) required if a person was to be transformed in such a way that he or she could encounter things in their unconcealment, in their truth. Plato, however, focused not on unconcealment as such but instead on the radiant appearing (*eidos*) of the ideas that were unconcealed and made available to the steady seeing of consciousness. Hence, with Plato *idea* and *eidos* became the "master" (*Herr*) of truth understood as unconcealment. [WGM: 136/187] In other words, instead of saying that truth (as unconcealment) made possible the very appearing of *eidos*, Plato said that *eidos* made truth (as correctness) possible. Here began the history of the "master-names" for being and truth which have dominated Western history. These master-names, such as *ousia*, *physis*, *energeia*, *eidos*, *idea*, *en*, and *logos*, have constricted and rigidified the ontological play of presencing-absencing, appearing-concealing, by turning that play into a permanently present structure which governs all historical discourse and behavior.

We recall that, according to Heidegger, the pre-Socratics were "intuited" or looked at by the entities themselves. With Plato, however, humanity took on the role of the viewer. What became important was the right way of catching sight of the *eidos*, the right mode of "seeing." The stages which marked one's journey from out of the darkness of the cave toward the sunlight, then, involved the forming (*Bildung*) of correctness of vision. Here, there arose the ideal of *homoeisis*, the correct correspondence between the outward appearance (*eidos*) and perception of that appearance. "In this change of the essence of truth a shift of the place of truth takes place at the same time. As unhiddenness truth is still a basic feature *of entities themselves*. But as correctness of 'looking' truth becomes *the label of the human attitude toward entities*." [WGM: 137/187] Like Plato, Aristotle wavered between conceiving of truth as unconcealment and as correctness of perception and assertion. Nevertheless, he promoted the shift toward the classical notion of truth as the right "correspondence" between assertion and state-of-affairs.

According to Heidegger, Plato's doctrine of truth initiated "humanism," the view according to which humanity establishes itself as the authoritative entity among entities. Humanism, then, began the long process whereby humanity understood itself not in terms of a subservient relationship to being, but instead in terms of a governing relationship to entities. Aristotle defined humanity as the "rational animal," that living entity which can understand itself and other entities. The rational animal, defining itself in terms of and finding itself in the midst of all other entities, became concerned with ensuring its survival: "This happens as molding of his 'moral' conduct, as salvation of his immortal soul, as unfolding of his creative powers, as development of his reason, as cultivation of his personality, as awakening of common sense, as training of his body, or as an appropriate coupling of some or all of these 'humanities.' " [WGM: 142/191]

Heidegger claimed that the "change of the essence of truth and of being is the authentic happening [*Ereignis*] of history." [GA, 54: 63] From its primordial meaning as unconcealment, truth became conceived as conformity in the service of promoting human well-being. This conformity became involved with the notion of truth as *rectitudo*, correctness, *adequatio*. Truth as *veritas* and *rectitudo* passes over into the *ratio* of man. The Romans, once again, were the vehicles for this degeneration of Greek insight. Heidegger summarized how the change in truth led to the emergence of the technological era:

> The Greek *aletheuein*, uncovering of the covered, which still for Aristotle governed the essence of *techne*, is changed into the calculating self-directing of the *ratio*. Henceforth, this determines, as a result of a new essential change of truth, the technicity of modern, i.e., machine, technology. This has its origin in the sphere of origin from which the imperial comes. The imperial arises from the essence of truth as correctness in the sense of the direction-giving, arranging security of the security of domination. The "holding-forth-true" of *ratio*, or *reor*, becomes self-erecting [*Sicherstellen*] which grasps out and before. *Ratio* becomes calculating, *calcul*. *Ratio* is self-arighting [*Secheinrichten*] on the right [*Richtige*, "correct"]. [GA, 54: 74]

Whereas *logos* was once conceived as the power which gathers entities and enables them to be present, the Romans reduced it to *ratio*, reason, the human faculty for making correct judgments, for giving the proper "reasons" for why something is the way it is. For modern humanity, something cannot "be" unless a reason can be "given" for it, e.g., by providing a satisfactory causal explanation for it. This process of anthropomorphizing led to the reduction of reason ("truth") to the status of instrumental intelligence, a human tool for human ends. In his lectures on Parmenides (1942–43), Heidegger offered a remarkable example of how the Romans transformed the meaning of "truth" and "falsehood."

He began by considering the opposition between true and false. The Greek word for "truth" was *aletheia*, unconcealment. The Greeks defined truth in terms of untruth, i.e., unconcealment, because they regarded hiddenness as more primordial than unhiddenness. If "truth" meant *aletheia* as unconcealment, we would expect "false" to mean concealment as *lethe*, but in fact the Greeks never used *lethe*, but instead used *pseudos* for "false." *Pseudos*, however, has nothing etymologically to do with hiddenness. Yet, thought in Greek terms, *pseudos* does involve a kind of concealment, a dissembling and feigning which lets something appear other than it does "in truth." Consider how the being of entities conceals itself so that entities can appear. Even in dissembling, then, something appears, as in the case of a Kierkegaardian pseudonym. [GA, 54: 49] *Pseudos* means a blocking or obstructing (*verstellen*), as in the phrase "the neighboring house blocks our view of the mountain." To remove this obstruction would enable the mountain to show itself. Truth, then, means clearing the obstruction which hides things. And *pseudos* turns out to mean something very close to *lethe*, that which blocks, conceals, obstructs, but also that which reveals.

At this point, Heidegger noted that the German word for "false" (*Falsch*) is "un-German," derived from the Latin *falsum*, which was linked to the verb *fallere*, "to fall." *Fallere* was, in turn, derived from the Greek *sphallo*, "to bring to fall, to fall, to make weak." The idea here seems to have been that a falsehood is a deception which brings a man down, makes him weak, causes him to stagger.

According to Heidegger, however, such making-weak or making-stagger was *not* what the Greeks meant by falsehood as *pseudos*. *Pseudos* named concealment, while *sphallo* named that which deceives and causes to fall. Such deception is possible only because entities can conceal themselves even while revealing themselves. The Romans, however, ignored altogether the idea that the opposite of truth (unconcealment) is concealment. Instead, they defined the opposite of truth as the deception which causes something to fall.

Why did the Romans conceive of the false in this way? Because, according to Heidegger, it was consistent with and made possible their imperial, domineering behavior. *Imperium* means "command," which originally meant to shelter and conceal, as in "the earth conceals the dead" or "commit your way to the lord (take protection in him)." [GA, 54:58] This dimension of commanding *(befehlen)* can be discerned in the word "commending" *(empfehlen)*. Gradually, however, the command came to mean *im-parare*, to order, to make arrangements, to possess in advance and thereby to take possession and to rule over. Even the Old Testament God has this command character, which is the basis for all lordship and domination (*Herrschaft*). The Greek gods, Heidegger observed, did not command but instead showed, referred, disclosed. When the Romans spoke of their gods in terms of *numen*, they meant that the gods were commanding in character. Commanding also belongs to the Roman term for "right" or "law," *ius*, which is related to *jubeo*, to order and command. Hence, the Roman idea of having "right," *iustitia*, was bound up with command and domination.

Heidegger then went on to describe Roman commanding in terms which intentionally paralleled the scientific and technological way of disclosing things. Science and technology open up in advance a realm in which entities can appear, but always only in accordance with the specifications laid down by the cognizing subject. To commanding there belongs a being-over or a surmounting of the underlings. Commanding is an overseeing, a constant vigilance, an active readiness for taking action: "This surmounting overseeing is that dominating 'seeing' comes to expression in an often-cited word of Caesar: *veni, vidi, vici*—I came, I oversaw, and I conquered. The conquest is already only the consequence of Caesarian overseeing and seeing, which has the character of acting. The essence of the *imperium* rests in the *actus* of constant 'action' ['*Aktion*']." [GA, 54:60] Just so, the constant action involved in the total mobilization of modern technology is induced by the dominating disclosure-in-advance of humanity as a clever animal craving power and all other entities as standing-reserve which are valuable for enhancing that animal's power. The scientific and technological eras, in other words, are the direct descendants of the Roman *imperium*.

To conquer, the Romans had to bring other peoples to a fall. Bringing to a fall can happen by direct assault, or by deception in the sense of going behind one's back. "Now, this 'bringing-to-fall' is the going-behind, the 'trick,' which word comes not accidentally from the 'English.' " [GA, 54:60] Through such trickery, those who are brought to fall are not annihilated, but instead are allowed to stand fastened in service of the commanding Romans. "To fasten" means in Roman *pango*, from which we get the word *pax*—peace. The Romans, in other words, conquered through clever deceit. For example, they were tolerant in allowing conquered people to retain their own gods and customs, but only insofar as those people became subjects of Roman imperial power. Bringing to fall by going behind the back, by trickery, then, was not merely a derivative, but instead the *essential*, feature of imperial action. Heidegger's listeners in 1941–42 could not help but hear in all this that the English were tricky imperialists aiming to make slaves of the Germans. But those same listeners were also told that Nietzsche, too, saw the Greeks through Roman, i.e., imperial, eyes: "Likewise, we think the Greek *polis* and the 'political' in a wholly un-Greek way. We think the 'political' in Roman, i.e., in imperial, terms." [GA, 54: 63]

The Roman word for "false," *falsum*, was taken from this imperial action of deceptive bringing to fall. The Romans translated the Greek *pseudos*, concealment, into their own imperial realm of bringing to fall. *Pseudos* now was reduced to the deceptively false, that which brings to a fall, while the primordial element of concealment involved in *pseudos* was completely lost. "That the West today still, and today most decisively, always thinks the Greek world [*Griechentum*] in Roman terms, i.e., in Latin terms, i.e., in Christian terms (as heathendom), i.e., romantically, i.e., in modern European terms, this is an event [*Ereignis*] which concerns the innermost midst of our historical *Dasein*." [GA, 54: 66]

The Roman misinterpretations of truth and falsity set the Western world on a power trip which affected all its institutions, not merely political ones. For example, the Roman Catholic sacerdotal "*imperium*" developed its own forms of domination in the *curia* and the Roman popes, and in church dogma. That dogma made the question of being superfluous by stating as "revealed truth" that creatures are the products of a personal Creator. Because Christendom refused to question its own dogmas, the idea of "Christian philosophy," Heidegger argued, is like the idea of a "square circle." [N II: 132/88] Medieval Christian theology didn't really offer knowledge about entities, but instead about the personal salvation of the immortal soul. Truth came to be associated with "right belief" and falsity with "heresy" and "unbelief." Hence, "the Spanish Inquisition is a form [*Gestalt*] of the Roman curial Imperium." [GA, 54: 68]

The Christian tradition paved the way for the subjectivistic turn by its emphasis on introspection, examination of conscience, and reflection as ways of guaranteeing security regarding salvation. This subjectivistic turn was directly related to the history of productionist metaphysics. Medieval humanity regarded the really real (*actus purus*) as the Divine efficient cause Who produced all creatures. But God was not only efficient cause but also final cause: the *summum bonum, actus purus* in the sense of the final end which anchors and gives meaning to all reality. A person, made in God's image, could become fully real only insofar as he or she could hold fast to this highest good through faith: "Through faith, man is certain of the reality [*Wirklichkeit*] of the highest real entity, and thus at the same time of his own real [*wirklichen*] continuance in eternal bliss." [N II: 425/23] The transition from the medieval

to the modern world involved humanity's decision no longer to seek its reality
or actuality in a divine causal agent and in a divine *summum bonum*, but
instead to become self-causing, self-producing, self-grounding as well as the
source of all goodness and value.

The transformation of truth from revelation to certainty was promoted by
late medieval anxiety about salvation. Heidegger's account of this transformation is reminiscent of Max Weber's largely discredited thesis about the relation
between Protestantism and the spirit of capitalism.[3] Weber argued that Protestant character promoted the ceaseless work and yearning for success that is so
important for capitalism. Protestant character was shaped by anxiety about
salvation. No longer guaranteed salvation by the Roman Catholic church, each
individual Protestant now had to face God alone. Calvin suggested that the
elect could be discerned in this life by their character: hard-working, diligent,
thrifty, ascetic, devoted to improving the human estate in order to fulfill God's
will. Some of Calvin's followers observed that such diligence was rewarded in
this life by monetary success, which could be read as a "sign" of one's elect
status. By internalizing the asceticism of this Calvinist work ethic, the Protestant hoped that his diligent character and worldly success would reassure him
of his salvation in the life to come.

This ascetic self-discipline transformed itself into a work mania which
gained ever greater control over European humanity, even as the religious
impulse behind this ascetic will-to-control was undermined by the processes
of secularization. According to Weber, industrialized humanity became
trapped within the "iron cage" of bureaucratic-technical rationality, the end
product of secularized asceticism.

> The Puritan wanted to work in a calling; we are forced to do so. For when
> asceticism was carried out of monastic cells into everyday life, and began to
> dominate worldly morality, it did its part in building the tremendous cosmos of
> the modern economic order. This order is now bound to the technical and eco-
> nomic conditions of machine production which today determine the lives of all
> the individuals who are born into this mechanism, not only those directly con-
> cerned with economic acquisition, with irresistible force. . . .
>
> Since asceticism undertook to remodel the world and to work out its ideals in
> the world, material goods have gained an increasing and finally an inexorable
> power over the lives of men as at no previous period in history. . . . [Today],
> victorious capitalism, since it rests on mechanical foundations, needs its [i.e.,
> religious asceticism's] support no longer. The rosy blush of its laughing heir, the
> Enlightenment, seems also to be irretrievably fading, and the idea of duty in one's
> calling prowls about in our lives like the ghost of dead religious beliefs. . . .[4]

Weber believed that the United States represented the "highest development" of the decline of the West into the controlled frenzy of producing
wealth as an end in itself. He concluded his study on the spirit of capitalism
with the following famous lines: "Specialists without spirit, sensualists without
heart; this nullity [for example, the United States] imagines that it has attained
a level of civilization never before achieved."[5]

Heidegger's account of the relation between Christendom, Protestantism, and the rise of capitalism (as one manifestation of the technological Will to Will) resembles Weber's account in several respects, although Weber would have been skeptical regarding Heidegger's apparently romantic attitudes toward work and regarding his preference for metaphysical schemes which discounted empirical data. Both writers seemed to argue that the world-conquering dynamic of modern capitalism resulted, in part at least, from the secularization of Christian beliefs and practices aimed at relieving anxiety about "justification" before God. For Heidegger, of course, insofar as developments within Christianity contributed to the shaping of Western history, those developments had to be understood as aspects of the history of being. [GA, 48: 186ff.] Both agreed that secularized asceticism so transformed the social world that the character of virtually everyone, even those who apparently lived outside the realm of the modern economy, was stamped by the impulse to produce and consume things at an ever faster rate. By speaking of how the instrumental rationality of capitalism had "disenchanted" the world, Weber seems to have agreed with Heidegger that something important had been lost during the course of Western history. Both authors, in other words, were critical of the notion that Western history was unambiguously "progressive," although Weber—unlike Heidegger—had a measure of faith in many of the progressive achievements of the Enlightenment. Moreover, Weber may have disagreed with Heidegger's astonishing application, in 1940, of Weber's thesis concerning the relation between capitalism and Calvinism: "We know that the empowering of power to its unconditioned dominance can conceal itself in the form of the strictest Christian morality and human happiness, since already for decades *Max Weber* has set into the light the connection of English imperialism with the Calvinist ethic." [GA, 48: 18]

Heidegger offered the following version of the connection between Protestantism and capitalism: In medieval theology, *iustitia* meant the correctness of reason and will.[6] Justification, then, was a product of intellectual voluntarism, the will to salvation. Various disciplinary techniques—meditation, prayer, fasting, self-denial, self-flagellation—were developed to promote the actualization of the soul's will to salvation. Self-denial, however, turned out to lead not to the allegedly desired goal of self-lessness, but instead to a heightened sense of individuality and self-assertion. Beneath the veneer of medieval piety, asceticism, and introspection, in the yearning for justification and salvation, then, there smoldered the self-affirming self which would later burst forth in Descartes's thinking and culminate in that of Nietzsche. "Put briefly: the *rectitudo appetitus rationalis*, the rightness [*Richtigkeit*] of will, the striving after rightness is the basic form of the Will to Will." [GA, 54: 75] Luther helped to shape the subjective consciousness necessary for modernity. The monk from Wittenberg asked whether and how a person could be certain of the eternally Holy, i.e., the truth, and could thus be a "true" Christian man. [Ibid.] As Luther discovered, one can never "do" enough to gain salvation. While he counseled that faith alone saves, his followers and those of Calvin were themselves governed by Luther's original impulse to "produce" his own salvation by "works." Moreover, by emphasizing the aloneness of the individual before God, Luther heightened the sense of egoism and self-centeredness which was to become the hallmark of modernity.

One such "work" took the form of scrupulous examination of conscience, a process which gave rise to the modern fascination with introspection and a "rich inner life." Concerning such religious introspection, Heidegger asked, what if

> through this kind of self-solicitousness and through this form of self-encountering the power of the self-reflection of subjectivity would first have become set free and thereby established in the modern world? Then Christendom, which as a result of the *techne*-like Creation doctrine believed and taught in it (seen metaphysically, it is also one of the most essential reasons for the coming of the dominion of modern technology), would have an essential part in the formation of the domination of the self-reflection of subjectivity, such that precisely Christendom can do nothing toward overcoming this reflection. Whence then the historical bankruptcy of Christendom and its Church in modern world history? Is yet a third world war needed to prove it? [GA, 55: 209]

Christendom could be of no avail in halting the technological Will to Power, because Christendom had helped to promote it. By defining a "true" man as one who was self-certain about his own righteousness and justification, Christendom had paved the way for the modern concept of security and salvation. Modern man appeals to his own rational capacities to become secure and certain about himself in the midst of the whole of entities. Instead of seeking salvation in an otherworldly heaven by the method of self-denial, salvation now "becomes sought exclusively in the free self-unfolding of all creative capacities of man." [N II: 133/89] This unlimited expanding of human power was, in Heidegger's view, "the hidden thorn" in the side of modern humanity. [N II: 145/99]

Influenced by the medieval quest for a method for gaining heavenly salvation, Descartes developed the first modern "method" for gaining the kind of truth and certainty necessary for guaranteeing earthly salvation for humanity. Concerning the meaning of Descartes's principle, *cogito sum*, Heidegger stated:

> The principle speaks of a connection between *cogito* and *sum*. It says that I am as the one representing, not only that *my* being is essentially determined through this representing, but that my representing—as the authoritative *repraesentatio*—decides about the presence [*Präsenz*] of everything represented, i.e., about the presence [*Anwesenheit*] of what is meant in it, i.e., about its being as an entity. The principle says: representing, which is essentially represented to itself, posits being as representedness and truth as certainty. [N II: 162/114]

Heidegger argued that Christian thinking paved the way for modern "psychotechnology," by which he apparently meant the disciplinary practices developed in industries, factories, barracks, educational institutions, corporations, and so on. [GA, 55: 285] Michel Foucault elaborated upon Heidegger's suggestion that the Western "power trip" manifested itself in the control not only of the "external" world but of the "internal" one as well. For Heidegger, the impulse toward self-control was bound up both with the transformation of *logos* to rationality, and with the subsequent and peculiarly modern tendency for rationality to bend back upon or to reflect upon itself. Ever since Descartes, and especially in German idealism from Kant to Nietzsche, human reason had become increasingly concerned about its own structure. "Logic" became the primary mode of philosophical discourse: thinking about the nature of thinking. One may object that in such reflection, humanity takes a wrong turn into an "unholy entanglement." [GA, 55: 208] Does not such pondering impede acting and joyous resoluteness? And does it not force one back into selfish "individualism"? Heidegger replied, however, that in the modern age, even a group of people can be reflective, can bend back upon itself as a collective or as a *Volk*, and can thereby become wholly self-preoccupied. Consider, he remarked, what he regarded as "the most monstrous reflection: that humanity on this planet still thinks only of itself." [Ibid.]

No entity can resist or stand on its own in the face of the self-reflective, anthropocentric power drive of human reason. Only being as such can never become objectified. Hence, Heidegger called for a new kind of thinking, a thinking which "let entities be" and which recalled being as such. Instead of existing in the "openness" of being so that entities could manifest themselves appropriately, technological humanity fills up that openness with methodological projects which compel entities to manifest themselves in a one-dimensional way: "What presences [the entity] does not hold sway, but rather assault rules. . . . Representing is making-stand-over-against, an objectifying that goes forward and masters." [HW: 100/149–150][7]

For Heidegger, modern science was the best example of this entrapping, enframing, ensnaring way of disclosing things. His early attitude toward science was somewhat ambiguous; he even described his own phenomenological method as scientific. Increasingly, however, he distinguished between "genuine" science, which was always philosophy, and "derivative" science, which was uncritical and largely in the service of technology. Far from being a dispassionate quest for truth, scientific methodology had become the modern version of the power-oriented salvific methodologies developed in the Middle Ages. Hence, Heidegger argued, even though modern science preceded the rise of modern technology by about two hundred years, modern science was already essentially "technological" in character, i.e., oriented toward power. [VA I: 23/22][8]

Science, we are told, works by "projecting" a ground plan on the basis of which entities may show themselves in a certain way, and only in that way: "Only within the perspective of this ground plan does an event in nature become visible as such an event." [HW: 72/119] Science is "rigorous" in restricting its investigative behavior within the limits imposed by its own stipulations. Methodology lays down the laws and rules which guide scientific investigation. This methodology becomes institutionalized in modern science, the ultimate goal of which is to ensure that its methodological procedures take precedence over whatever the object of research happens to be. [HW: 78/125] Stamped by scientific methodological requirements, modern scientists become engaged in ongoing activity (*Betrieb*) which has nothing to do with what used to be called the process of gaining "erudition" or "learning." [HW: 78–79/126] Science, then, or at least the degenerate science of modernity, seeks not to let the entity show itself in ways appropriate to the entity in question, but instead compels the entity to reveal those aspects of itself that are consistent with the power aims of scientific culture.

C. THE CHANGE OF SUBSTANCE FROM
HYPOKEIMENON TO SUBJECT

The rise of modern science was intrinsically linked to the transformation of substance from *hypokeimenon* to human subject. For the Greeks, *hypokeimenon* meant "substance" in the sense of that which lies-forth as the ground which gathers the various aspects of an entity. *Hypokeimenon* was the "basis" or "substance" of things. Living substance meant that which "actualized" itself, in the sense of bringing itself forth into presence. Substance used to be synonymous with "subject," a fact discernible today when we speak of a "subject of investigation." With the Cartesian "turn," however, which was fostered by Christendom's salvific, introspective power trip, humanity made itself into the "substance" of things: the self-certain subject, the grounding ground, the founding basis for all truth, reality, and value: "Man becomes the relational center of the entity as such." [HW: 81/128] In effect, modern man has absorbed into himself the "substance" of things, thereby reducing them to the status of mere representations. Things become "real" or "actual" insofar as they are "objectified" by the cognizing subject. Humanity becomes the efficient cause of everything objective. Here, then, was the culmination of humanity's drive to "make" itself in the image of God, i.e., God understood as the providential Creator Who "dominates and calculates the whole of the entity" as a "work" which He has produced. [GA, 54: 164} In effect, Western humanity reversed the meaning of the Biblical dictum that we are "made in the image of God." Instead, humanity projected its own self-image onto God. God became the name for an all-powerful man, the Heavenly Producer Who was the Basis for all things. The triumph of productionist metaphysics occurred when Marx, urged on by Feuerbach, pulled back this projection of human productive-creative power. "Man is the only god for man," Marx proclaimed. Man not only produces and re-produces himself, Marx continued, but he also creates the social world within which he lives.

It is worth emphasizing Heidegger's remarks about the "dominating" Creator God. During the 1930s and 1940s, Heidegger was a bitter opponent of Roman Catholicism in particular and Christendom in general, for he conceived of them as crucial factors in the concealment of being through the centuries. He usually distinguished between "Christendom," as a power-political movement wrapped in religious clothing, and "Christianity," the authentic religious impulse that was sustained for a few years before the "propaganda missions" of St. Paul. An evangelical minister named Buhr, who was present at Heidegger's Todtnauberg "science camp" in 1933, reports that Heidegger—in support of the National Socialist attack on Christendom—criticized the entire basis of the Judeo-Christian tradition:

> If one wanted to attack Christendom [Buhr reports], then it is not enough to limit oneself to the second article of this doctrine (of Jesus as the Christ). Already the first article, that a God has created and preserves the world,—*that the entity is merely something made as produced by a handiworker*—, that must be rejected first. There already lies the basis for a false devaluing of the world, world-despising and world-negating; this is further cause of that false feeling of safety, of security, grounded upon representations of the world which one has made onself, untrue against the great, noble knowing about the lack of safety of the *"Dasein."* Thus I understood it approximately at that time and have remembered it. This was to me, to the reader of Ernst Jünger (I think of *The Adventurous Heart*) not strange.[9]

In connection with Heidegger's assertion that the Creator God is like a craftsman who makes and preserves the world, the following question arises: Does this conception of God exhaust the Jewish tradition of the Creator? Or does the Jewish tradition have a non-productionist, non-metaphysical experience of God, one that was "corrupted" at the hands of St. Paul, St. John, and other early Christians influenced by Greek metaphysics, especially Platonism? If the Jewish God may be construed as non-metaphysical, then perhaps there is *another* possibility for renewing the West: an originary encounter with the God of the Old Testament. Furthermore, is the supposedly world-despising tendency of Christendom really consistent with the attitude of Jesus Christ, or is it the consequence once again of metaphysical dualists who attribute to the "soul" the same power over the "lowly" body that God has over Creation? Are there not resources within both Christianity and Judaism for a life-affirming, non-dualistic, non-anthropocentric conception of human existence?[10] In any event, as Paul Ricoeur has pointed out, even if the early Heidegger discerned authentic insights in early Christianity, he systematically avoided an encounter with the "massif" of the Hebrew tradition.[11]

Heidegger did suggest that the conception of God as Creator was a projection on the part of thinkers steeped in productionist metaphysics. Just as the Creator used calculating intelligence to construct a world over which He would have dominion, so too modern man—having arrogated to himself the attributes of this Creator—conceived of himself as the foundation and master of all things. Directly related to the rise of humanistic power economies was the rise of modern science, which portrays the world as an image or picture projected by human reason. Despite the fact that medieval humanity worshipped a Creator God, at least medieval people never reduced the being of entities to mere objectivity for the human subject. And for the ancient Greeks, far from being the subject who "represents" and "pictures" the world, "man is the one who is looked upon by that which is; he is the one who is—in company with itself—gathered toward presencing, by that which opens itself." [HW: 83/131] In the age of modern science, man asserts himself as the self-constituting, radically independent rational ego, whose aim is to gain control of everything. Heidegger used the term *subiectum* to describe humanity in the grip of this control-oriented anthropocentrism.

By *subiectum*, Heidegger meant the selfishness and self-seeking which were hallmarks of inauthentic human existence. [GA, 54: 203] This "egoity" (*Egoität*), the essence of subjectivism, was not identical with the subjective "egoism" or individualism of modern capitalist man. The essence of egoity lies in the representing thinking which transforms all entities into objects. "Only in the essential sphere of subjectivity does there become historically possible an age of cosmic uncovering and planetary conquest. . . . " [GA, 54: 204] Egoity may take the form of "individualism," but it can also take the form of the social collectivism of Marxism, or even the racist *Volksgemeinschaft* promoted by the misguided wing of National Socialism. In 1942–43, in a striking rebuke to Nazi ideologues who claimed that the *Volk*-revolution had overcome the threat posed by English egocentrism and individualism, Heidegger argued:

> Only after metaphysics, i.e., the truth of the entity in the whole, is grounded upon subjectivity and I-hood, does the national and the folkish have that metaphysical background from out of which it is in general capable of history. Without Descartes, i.e., without the metaphysical grounding of subjectivity, is Herder, i.e., the grounding of the folkness of the folks, unable to be thought. . . . In distinction to the "individualism" of the 19th century, . . . Nietzsche sees the arrival of a new stamping of humankind, which is characterized by the "typical" ["*Typische*"]. [Ibid.]

As we have seen, Jünger described the emerging technological man as the Nietzschean *Typus*. For Jünger, the age of world wars marked the beginning of the new age of total mobilization, in which nation-states would vie for planetary domination. For the sake of control, the nationalistic *subiectum* "brings into play [its] unlimited power for the calculating, planning, and molding of all things. Science as research is an absolutely necessary form of this establishing of self in the world; it is one of the pathways upon which the modern age rages toward fulfillment of its essence. . . . [HW: 87/135]

For Heidegger, modern science was one of the most important dimensions of representational thinking, i.e., thinking which seeks to dominate entities. While science portrays itself as an empirical discipline, Heidegger argued, it is in fact heavily theory-laden, so much so that "an original reference to the things is missing." Hence, while zoology and botany may tell us something about animals and plants, the question is: "Are they still animals and plants? Are they not machines duly prepared beforehand of which one afterward even admits that they are 'cleverer' than we?" [GA, 41: 40/41] Modern science, Heidegger proclaimed, fails to disclose things, but instead imposes its own framework on things and is thus implicated in the boundless Will to Will of modernity. [WGM: 99-100/350] Nevertheless, the objectifying scientific disclosure of nature is in fact one way in which things can manifest themselves [VA I: 54/174], but this way had become so arrogant and constricted that science had lost its original relation to nature: "Historically considered, modern science is certainly more progressive than that of the Greeks, if the technological mastering and thereby also the destroying of nature is progress—as opposed to the preserving of nature as a metaphysical power." [GA, 45: 53] The failure to preserve nature as such a metaphysical power was the real source of the "crisis" not only of modern science, Heidegger argued, but of modernity in general. Heidegger's lectures on Hölderlin were an attempt to disclose once again the power of nature, the ingathering *physis* which had been "disenchanted" both by Christendom and by modern science.

In 1936–37, Heidegger distinguished between philosophy and science by saying that the former is useless while the latter is useful. [GA, 45: 4] Science is useful for producing armament and foodstuff, airplanes and telephones. While "useless" in the short run, however, philosophy is nevertheless the most profound and important knowledge of the essence of things. Since the sciences now drive toward complete "technologization," they are separated in the most radical way from philosophy. In a particularly acrid denunciation, Heidegger intoned:

> Only then and in every case when one believes himself in the possession of the "truth" does he come to science and its constant activity. "Science" is the denial of all knowing of the truth. To suppose that today "science" is persecuted is a basic error: never has it gone better for "science" than today, and it will go better still for it than up till now. But no man of knowledge will envy the "scientist"— the most wretched slave of the most recent time. [Ibid.]

While Heidegger extended Hölderlin's phrase that humanity "dwells poetically on earth" to include people in the scientific age, more often Heidegger interpreted authentic "dwelling" as living in a world unified and gathered together by the "fourfold" of earth and sky, gods and mortals. Modern humanity lives in a world bereft of gods, old or new. But even in the impoverished modern age, poetry, the gods, and the Holy are only hidden; they have not gone forever. The poetic capacity for naming things and thus letting them appear in their own distinctive ways was still present to some extent in the great researchers who founded modern science: they were men of passion and strength. Unlike today's "normal" scientists, who had lost all touch with fundamental thinking and no longer understood the relation between science and poetry, Galileo, Kepler, Newton, and Descartes were at least capable of "poetizing" a new way for entities to reveal themselves, even if this way turned out to be constricting and finally contributory to the triumph of subjectivism. [GA, 45: 53] Such conciliatory remarks about genuine science contrast sharply with Heidegger's attacks on "scientism" and "positivism." These acknowledgments of authentic science were designed as a caution to those of his listeners who believed that his critique of science was one and the same with that of other conservative thinkers who maintained that the only way out of the rigid scientific-technological cage of modernity was by a revival of the "irrational" forces of blood and instinct. Heidegger distinguished between such irrationalism, on the one hand,

and his own quest to discover that which is prior to and more basic than the rational. For him, the irrational meant either that which was without proper limits, and thus insane; or else, as in the case of blood frenzy, the merely animalistic and bestial. From his viewpoint, this latter kind of irrationalism was simply a variety of Nietzsche's violent "blond beast" which was the degenerate culmination of the view that man is a "rational animal."

It was in Nietzsche's metaphysics, of course, that Heidegger believed that the history of productionist metaphysics came to an end, in the sense of working out its final possibilities. Nietzsche proclaimed that he was going to combat nihilism, i.e., the collapse of the old Platonic-Christian table of values, by bringing forth a new table of values that would enhance humanity's Will to Power. But Heidegger argued that instead of overcoming nihilism, Nietzsche simply reinforced it. Nietzsche was entirely cut off from being as such; for him, being was merely an empty "vapor," a mere nothing. For Heidegger, however, speaking of being as nothing, *nihil*, was the essence of nihilism. Nihilism was the "fundamental movement of the history of the West," since nihilism was the name for the process of the gradual self-concealment of being through Western history. [HW: 201/62] Nietzsche's word "God is dead" meant that the suprasensory values which had given meaning to life had vanished. While the onto-theo-logical God was made in the image of man stamped by productionist metaphysics, that God was nevertheless invested with a superhuman and extrahuman power and authority. With the collapse of God, nothing stood in the way for humanity to make itself into a God-like entity, the "clever animal."

Nietzsche's doctrine of the Overman, according to Heidegger, gave voice to the view of man as God-like producer and destroyer. Nietzsche, then, brought to a head what Heidegger regarded as the arrogance which character- ized the entire history of modernity. At first, human conscience and reason took over the role played by God, but these were replaced by "social instinct" (Marxism). Whereas medieval humanity had fled from the earth into a supra- sensory heaven, modern humanity fled from the earth into the ever-receding goal of "heaven on earth" in the form of "progress." Utilitarian prattle about "happiness for the greatest number" replaces the idea of everlasting bliss for the individual soul. Cultivation of religion is replaced by the creating of cul- ture, or by the spreading of civilization: "Creativity, previously the unique property of the Biblical God, becomes the distinctive mark of human activity. [Human] creativity finally passes over into business enterprise." [HW: 203/64]

Nietzsche, too, regarded nihilism as the "inner logic" of Western history, but he conceived this nihilism not as the self-concealment of being, but in- stead as the erosion of previously valid values. For Nietzsche, nihilism named not only something negative, i.e., the "no" to the old values, but also some- thing positive, i.e., the "yes" to new values to replace the dissolved ones. During the "in-between" condition which prevails after the collapse of the old values and the advent of the new ones, all sorts of half-baked doctrines (such as world happiness, socialism, Wagnerian music) come on the scene to stem the anarchic tide. [HW: 208/69] Nietzsche regarded such doctrines as instances of "incomplete" nihilism, the kind which still clung to the previous kind of suprasensory values. "Completed" or "positive" nihilism, by way of contrast, would posit values based on an entirely new principle, the principle of enhancing life in the form of the Will to Power.

According to Heidegger, the term "value" arose in connection with the transformation of humanity into the self-certain subject. The Greeks never spoke of "values." Values are, in effect, ways in which the human subject justifies and enhances its own Will to Power. A value is both a point of view and that on which the point of view is focused. Value is also the calculating involved in attempting to achieve a goal. All this, in Heidegger's view, was the culmination of Plato's idea of truth as correctness of vision, and the Roman imperial view of truth as the over-seeing which discloses things as penetrable by calculating intelligence and ripe for domination. With Nietzsche's vitalism, "value" became the posited point of view which both preserves and enhances the clever animal's Will to Power. The Will opens up a field of vision which discloses possibilities both for consolidating the power of Will and for expanding it out beyond itself. Power is power only insofar as it wills: more power. As a Will which always wills out beyond itself in an ever-expanding spiral, the Will to Power embodies Nietzsche's other crucial doctrine: the Eternal Return of the Same. [HW: 219–220/82] The eternally returning Will to Power is the Will to Will.

At work in the clever animal, this Will to Will drives the *subiectum* to conceive of truth as certainty. Certainty simply means that way of evaluating things which promotes security, power, and—still more power. The self-willing animal is Nietzsche's version of Descartes's self-certain, re-presenting ego. In justifying itself, in securing security for itself, in describing things as objects amenable to technological domination, in making itself the basis for all value and truth, the clever animal invents "Christian, humanistic, Enlightenment, bourgeois, and socialist morality." [HW: 228/92] Despite their differences, these different moralities agree that "justice" is simply the name for the enforcement of those values which promote the Will to Power of the value-positer. Nietzsche's conception of value "touches squarely the essence of the justice that at the beginning of the consummation of the modern age, amidst the struggle for mastery of the earth, is already historically true, and that therefore determines all human activity in this period. . . . " [Ibid.]

The embodiment of Nietzsche's this-worldly conception of justice is the Overman. The Overman is what humanity becomes at the culmination of productionist metaphysics: the half-steel, half-organic human machine which wills that the earth become a titanic foundry to promote ever greater instances of the Will to Will, the quest for power for its own sake. Reading Nietzsche's Overman through Jünger's writings, Heidegger claimed that today what is needed "is a humanity which is, from the ground up, adequate to the unique and basic essence of modern technology, i.e., lets itself be dominated wholly by the essence of technology. . . . Only the Overman is adequate to the unconditioned 'machine economy' and the other way around: the former needs the latter for arranging unconditioned domination over the earth." [GA, 48: 205]

For Heidegger, then, the Overman is demanded by the final stage of the history of metaphysics. If humanity did not allow itself to be stamped by modern technology, that history could not be brought to a completion—and a new beginning would not be possible. In light of Heidegger's claim that the "saving power" arises only within the "growing danger," we might say that the emergence of the Nietzschean-Jüngerian Overman is in and of itself a harbinger of the longed-for new world, despite the fact that the Overman is simultaneously the culmination of the nihilism which constitutes the "inner logic" of Western history. [GA, 48: 98]

In the Overman is incorporated the objectifying power of the human *subiectum*, the self-certain "ground" for all objects, i.e., for all things which can be re-presented, known, and controlled. The reflecting, representing ego bends everything back upon itself, opens up a sphere in which everything appears to us in terms of our projects which constitute that very sphere. Of this phenomenon, Nietzsche asked: "How were we able to drink up the sea? Who gave us the sponge to wipe away the entire horizon? What did we do when we unchained the earth from its sun?" [Cited in HW: 241/106] Modern man has wiped away the suprasensory world of what truly is; he has drunk up the sea of entities. Through man's insurrection, "all that is, is transformed into object. That which is, as the objective, is swallowed up into the immanence of subjectivity. The horizon no longer emits light of itself. It is nothing but the point of view posited in the value-positing of the Will to Power." [HW: 241/107]

In his lectures on Heraclitus, given during the same era as the Nietzsche lectures, Heidegger noted that for the Greeks, humans were embraced in a cosmos in which entities intuited humans; the entities shone forth on their own, gave themselves into shining appearance. For modern humanity, however, entities are objectified and represented by humans; the entities are allowed to appear only insofar as they conform to the expectations imposed upon them by the self-certain subject. Human reason portrays itself as the source of light in which entities can "be" as objects, but in fact "the primordial catching sight, through which the visible first comes into the open, and which itself remains the open, pre-readying of the capacity-for-appearing of the entity is not of human origin." [GA, 55: 351] For Heraclitus, "man in his essence belongs to being and is determined by being gathered upon this and from this receives the possibilities of himself." [GA, 55: 356] By way of contrast, every self-founding of man upon his capacities, a self-founding which has forgotten being, is *hubris*.

At first glance, Heidegger would seem to be contradicting himself by declaring that modern man is guilty of *hubris* and insurrection against being, on the one hand, while saying that his behavior is determined by the self-concealment of being, on the other. When we speak of an act of *hubris*, we often assign to it a voluntary quality, as if the hubristic person had some choice in the matter. For Heidegger, however, the arrogant acts of modern man were not really the result of free decision, but instead were provoked by the self-withdrawal of being. Nevertheless, while Western man was in some sense destined to behave in the way he did, he was also responsible at least for the stand he took in the face of that destiny. In more than one place, Heidegger insisted that the technological era was not imposed on us as an inexorable destiny. This era could undergo a "turning," depending in part on the stand we take within it—depending, in other words, on our capacity for "stepping back" from calculative thinking about entities into Heidegger's "meditative thinking" about being.

In the conclusion of this book, I offer some critical observations about Heidegger's controversial "history of being." Here, I would like to note that many scholars contest his interpretation of Nietzsche. They argue that Heidegger projected his own concepts onto Nietzsche's texts and forced them to speak an alien Heideggerean language.[12] Nietzsche's basic concepts, we are told, had nothing to do with Jünger's call for the technological domination of nature, but rather anticipated Heidegger's own notion of authenticity, understood as the courageous affirmation of mortality and finitude necessary for "letting entities be."[13] Heidegger's treatment of Nietzsche suggests that he owed more to his predecessor than he cared to admit.

Nevertheless, Heidegger insisted that he was not being arbitrary in reading Nietzsche in terms of Jünger's predictions about the future and in terms of Heidegger's own account of the history of being. Despite the fact that he had Zarathustra tell men to "remain faithful to the earth," Nietzsche also anticipated Jünger's credo that future humanity should carry out the technological mobilization of the earth. In *The Will to Power*, for example, we read:

> Once we possess that common economic management of the earth that will soon be inevitable, mankind will be able to find its best meaning as a machine in the service of this economy—as a tremendous clockwork, composed of ever smaller, ever more subtly "adapted" gears; as an ever-growing superfluity of all dominating and commanding elements; as a whole of tremendous force, whose individual factors represent *minimal forces, minimal values*.
>
> In opposition to this dwarfing and adaptation of man to a specialized utility, a reverse movement is needed—the production of a synthetic, summarizing, justifying man for whose existence this transformation of mankind into a machine is a precondition, as a base on which he can invent his *higher form of being*.[14]

Here, we find doctrines later to be appropriated by Jünger, including his elitist view that the industrial masses would serve as the basis for the arrival of the *Typus*, the new kind of man whose very existence would be so sublime and awesome that it would "justify" the technological world as an aesthetic phenomenon. [GA, 48: 185ff.] While conceding that such passages exist, critics would argue that they must be read in the larger context of Nietzsche's writings, a context allegedly ignored by Heidegger and Jünger. There is no doubt that Heidegger gave a violent reading to Nietzsche's writings. Heidegger claimed, however, that such a reading was justified by the fact that he alone discerned the "inner logic" of nihilism at work in Nietzsche's thought, the final stage in the nihilistic history of Western metaphysics. Spengler had provided the negative account of the "deep lawfulness" of Western nihilism, while Jünger conceived of the positive character of that nihilism, though without penetrating into the metaphysical essence of the Will to Power. [GA, 48: 28, 137] Heidegger made it his own task to disclose the metaphysical dimension of Nietzsche's thought and thereby to reveal the essence of the nihilism described by Spengler and Jünger.

Despite granting the importance of Jünger's contribution to a positive account and description of the technological age, however, Heidegger concluded that Jünger was not a real "thinker." Only a thinker could occupy the final spot in the historical sequence initiated by Greek metaphysicians. In Heidegger's view, Jünger's texts lacked the metaphysical power which resonated so clearly in Nietzsche's. What Jünger *described*, Nietzsche *thought*. And in thinking with and against Nietzsche, Heidegger believed he alone understood the metaphysical origins of—as well as the way beyond—the age of nihilism and technology.

As the metaphysician of the technological era, Nietzsche brought to a culminating point Aristotle's original metaphysical conception of man as a rational animal. Jünger described important aspects of this new kind of man, the *Typus*, but because Jünger was not privy to the requisite insight regarding being, its history, and its relation to humanity, he often spoke about the new man in terms of naturalistic and biological categories unacceptable to Heidegger, but all too close to the racist doctrines of various Nazi ideologues. Heidegger construed this new man as the courageous, noble, unflinching, and sacrificial Overman, destined itself to lead existing humanity through and beyond the nihilistic metaphysics of the technological era. National Socialist ideologues also used the language of Nietzsche and Jünger to describe the revolutionary new man. Heidegger maintained, however, that by using naturalistic-biologistic categories to describe the new man, those ideologues proved that the new Nazi man was "new" only in the sense of being the final form—that of the technological animal—assumed by humanity in the degenerate history of metaphysics.

The "new" man envisioned by the crude representatives of National Socialism was akin to Nietzsche's "last man," who seeks to gain happiness, security, and comfort by dominating the earth with industrial technology. From Heidegger's point of view, such a man was blind to being as such and thus incapable of receiving a new disclosure of being that would usher in a post-technological, post-metaphysical era. Heidegger believed that the "inner truth and greatness" of National Socialism lay precisely in its capacity for receiving such a disclosure and for bringing forth a new kind of humanity, one endowed with many of the qualities admired by Jünger, but without being burdened by the naturalistic self-understanding which was leading to the triumph of the producing-consuming animal. For Heidegger, then, the technological man of the future was Janus-faced: he could become the harbinger of a new era of Western history, or he could be the final representative of the West's endless decline into oblivion. In the following chapter, we focus on Heidegger's account of the latter type of technological man, the clever animal who seeks to control all things, but who ends up being enslaved by the drive to acquire more and more power for its own sake.

Chapter 12. Production Cycles of the "Laboring Animal": A Manifestation of the Will to Will

In the post-Darwinian era, technology has often been defined in terms of the adaptive behavior of the human organism. Without tools, we are told, early humans would scarcely have been able to survive in a world full of predators organically "equipped" with keen smell and hearing, powerful limbs, penetrating fangs, and tearing claws. The human animal's advantage over the other animals was its intelligence. The clever animal invented tools which were, in effect, extensions of its limbs and other organs. Over the millennia, especially after the introduction of agriculture made standing armies and big cities possible, many of the most important subsequent technical developments were spurred by military requirements. In the past few centuries, the quarrelsome but clever human animal has achieved technical feats that once would have been regarded as the stuff of dreams. World War I, however, made clear the double-edged character of modern technology: on the one hand, it made possible triumphs in medicine, communication, travel, scientific research, food production, manufacturing, and living conditions; on the other hand, it also made possible the implements of mass destruction used in modern warfare. About 1930, Freud remarked, "Men have gained control over the forces of nature to such an extent that with their help they would have no difficulty in exterminating one another to the last man. They know this, and hence comes a large part of their current unrest, their unhappiness and their mood of anxiety."[1]

A. CRITIQUE OF THE IDEA THAT HUMANS ARE RATIONAL ANIMALS

We have already noted how social Darwinism helped to promote among Europeans concerns about the "vitality" of the "white race." National Socialism in particular exploited racist doctrines. Heidegger, however, completely rejected the Darwinian, naturalistic view of humanity. He agreed that Western humanity had undergone a long decline, but it was *metaphysical*, not biological, in character. Nietzsche, the last metaphysician, conceived of man as the clever animal, thereby revealing the continuity between his own thought and that of Aristotle, who first defined man as the rational animal (*zoon logikon echon*). During the course of Western history, what the Greeks originally meant by *logos*—the power that gathers and distinguishes all things—had degenerated into "instrumental rationality." Whereas *logos* once meant the disclosive power that possessed humanity, in the technological era, *logos* came to be regarded as the instrument possessed by humans.

This modern view of rationality was closely related to the triumph of a paradoxical phenomenon, which I call "naturalistic humanism." The paradox of such humanism lies in the fact that it combines the view that humans are *like* other animal species struggling for survival, with the view that humans are *unlike* other animals in being endowed with the rationality which gives them the "right" to define, evaluate, and use things in any way they choose. Naturalistic humanism, in Heidegger's view, represented the self-elevation of the human animal, in the form of the self-certain subject, to the level of the deceased Christian God. Conceiving of itself as the foundation for all meaning, purpose, and truth, and driven by the animalistic craving for power and control, the humanistic animal set out to dominate the entire planet. Heidegger's famous critique of humanism was directed at this naturalistic humanism. In effect, he called for a "higher humanism," one which would free humanity from the animal compulsions present in the incessant cycle of production and consumption, hot wars and cold wars. Presumably, such a higher humanism would also make possible a different relationship between humanity and other entities, a relationship in which things appeared as other than standing-reserve.

Hence, one of Heidegger's major tasks was to develop an alternative conception of humanity, one which would avoid the naturalistic traps which ensnared many philosophical anthropologists. Time and again he proclaimed that *the* question for Europeans was: Who is man? The wrong answer was: An intelligent animal. He did concede that there seems to be a relationship between people and animals. For example, in his "Letter on Humanism" (1946), we read: "on the one hand [animals] are in a certain way the most closely related to us, and on the other they are at the same time separated from our ek-sistent essence by an abyss." [WGM: 157/206] This "abyss" is opened up by the fact that humans can understand the being of entities, but animals cannot. Heidegger explored the difference between humans and animals in great detail in his 1929–30 lecture course *Basic Concepts of Metaphysics*. There, he defined animals as *zoon*, which means not merely "animal" or "life" but "that which rests within itself and which spontaneously brings itself to presence." While capable of making themselves present, however, animals cannot apprehend their presencing or that of other entities. Because humans can understand such presencing, Heidegger described humans as "world-building." Animals, which are not "open" to things in the way humans are, are "world-poor." Inanimate things, by way of contrast, are entirely "world-less." As world-poor, animals are in some sense open to things, but they are "benumbed" and captivated by the self-enclosing structure of their behavior which is strictly limited by their environment. Although animals behave according to various temporal structures, and even possess a measure of "selfhood," they can never confront entities as "objects" over against them. Deprived of true language, they cannot understand that and what things *are*. For this reason Heidegger claimed that the human hand is capable of "work," i.e., the activity of disclosing things, in a way that the animal claw or paw is not.

Heidegger cautioned, however, that his description of animals as world-poor did not mean that he regarded them as "lower" kinds of entities. Animals are whole and complete just as they are:

> We are indeed accustomed to speaking of higher and lower animals, and yet it is a basic error to suppose that amoeba and infusoria are less complete animals than elephants and apes. Every animal and every kind of animal is just as complete as every other. With all that has been said, it becomes clear that the talk of world-poverty and world-building is from the start not to be taken in the sense of an evaluated rank-ordering. [GA, 29/30: 287]

Heidegger resisted the customary rank-ordering of animals because he rejected the modern scientific view that humans are the "highest animal." Humans are not animals at all, according to Heidegger. Even eminent thinkers such as Max Scheler, however, had fallen victim to this error by arguing that human life emerged from animal life, and that the latter had evolved from inorganic reality. [GA, 29/30: 283][2] Scheler conceived of humanity as the microcosm which contains all essential features of the macrocosm. Interpreting Scheler rather ungenerously, however, Heidegger maintained that such a concept was too close to Darwin's idea that humans are descended from apes.

Years later, in his "Letter on Humanism," Heidegger reiterated that humans are *not* merely more complex or higher forms of animal life:

> The human body is something essentially other than an animal organism. . . . Just as little as the essence of man consists in being an animal organism can this insufficient definition of man's essence be overcome or offset by outfitting man with an immortal soul, the power of reason, or the character of a person [cf. Scheler]. In each instance essence is passed over, and passed over on the basis of the same metaphysical projection. [WGM: 155–156/204–205]

The metaphysical "projection" governing the metaphysical conception of humanity is that of the rational animal: "Metaphysics thinks of man on the basis of *animalitas* and does not think in the direction of his *humanitas*." [WGM: 155/204] Heidegger argued that naturalism flawed the writings of Nietzsche and Rilke, who suggested that humanity could become more authentic by regressing to the supposedly "free" and "spontaneous" level of the animal. Heidegger regarded psychoanalysis, with its emphasis on "unconscious drives," as yet another example of naturalistic reductionism. [GA, 54: 182ff.; 226–235] Heidegger bemoaned the fact that in the modern world everything non-rational (including being) was pushed either into the swamp of the merely irrational, or into the unconscious or instinctual realm. [GA, 53: 115ff.] His resistance to naturalistic interpretations of human existence was shared by many of his contemporaries. The Christian doctrine that only man is created in God's image, and is thus radically other than animals, continued to influence students of humanity, including atheists such as Marx. Although a great admirer of Darwin, Marx insisted that more than "natural" evolution was involved in the emergence of the human species. Endowed with the capacity for consciousness, humanity began to influence its own evolution within the domain of "history" in a way radically different from what takes place in the merely animal realm: "Men can be distinguished from animals by consciousness, by religion or anything else you like. They themselves begin to distinguish themselves from animals as soon as they begin to *produce* their means of subsistence, a step which is conditioned by their physical organization."³

In *Capital* Marx also claimed that even the worst architect is better than the bee, no matter how complex a honeycomb the bee might build, for "the architect builds the cell in his mind before he constructs it in wax."⁴ Engels once remarked that man can never regress to the level of the beast: "The lowest savages, even those in whom regression to a more animal-like condition with a simultaneous physical degeneration can be assumed, are nevertheless far superior to these transitional beings [apes]."⁵

Despite the protestations of Marx, Engels, and Heidegger, contemporary ethologists emphasize the similarities and continuities—social, emotional, even linguistic—between human behavior and that of other primates.⁶ From the ethological viewpoint, Heidegger went astray in lumping *all* animals into the same "world-poor" category. Surely a chimp is more open to things, more capable of distinguishing between objects and itself, than is a caterpillar. While attempts to teach non-human primates to become linguistically competent have produced controversial results, many people regard human language as a complex natural phenomenon, which is a development of capacities shared by animals, including dolphins and whales.

For Heidegger, however, all such explanations of human existence fail to understand themselves as *interpretations*, i.e., as language-based attempts to describe things as they appear within the world opened up by language. He insisted that language cannot be reduced to a physical event, but instead constitutes the ontological disclosedness in terms of which "events" can first appear. Language is not one thing among others but an ontological domain of unconcealment in which things come to "be."

Just as it is wrong to interpret human existence and language naturalistically, it is wrong to interpret animals mechanistically. For example, Heidegger contended, it is a categorical mistake to conceive of animal organs as "tools." Tools have the character of "readiness" (*Fertigkeit*); they are separable from the user, and can be used by anyone at any time. Organs, such as the eye, however, are not separable from the organism; organs are not external "parts" with which the organism is somehow "equipped." Instead, organs are brought forth by the "capacity" (*Fähigkeit*) of the organism itself. We should not say that the eye has the capacity for seeing, but instead that the organism has a capacity for seeing which is realized by the organ called the eye. [GA, 29–30: 323–324] Consider the amoeba; it has no fixed organs. Any part of the amoeba can become the mouth, the gut, and so on. Hence, *"the capcity for eating, for digesting is earlier than the actual organs."* [GA, 29–30: 327] The organism's capacity for growing its own organs must not be confused with the "in order to" or instrumentality that defines tools. *"The capacity takes the organ into its service. . . . [T]he character of the in-order-to, which we observe in* every tool, work-tool, and in every machine, is fundamentally different in the organ and in the tool. The eye is not in service for seeing, as is the pen-holder for writing, but instead the organ stands in the service of the capacity forming it." [GA, 29–30: 330]

While both the organ and the tool are useful, the organ is intrinsically related to the very *being* of the organism that grows it in a way that the tool is not related to the human who uses or makes it. A hammer does not "develop" toward its "end" as does the eye of a growing animal guided by its own *telos*. A tool lacks intrinsic *telos* and is defined instead by the humans who make and use it. Moreover, unlike an organ, a tool has no temporal dimension; it does not "develop." Hence, it is temporal only insofar as it is a human product. The tool and the organ are superficially similar in that neither can operate on its own. A pen, for example, is *ready* for writing but is not itself *capable* of writing. Similarly, the eye cannot see on its own, but only by virtue of the seeing capacity of the organism. Here, however, the similarity ends. Unlike organs, tools are *external* to the one who uses them.

To this claim, the following objection might be made: A blind person using a cane reports experiencing the cane as an organ of touch; the cane user "feels" at the tip of the cane. There are other tools that people seem to experience as extensions of their sensory-organic existence.[7] Heidegger insisted that when we are fully involved in work with tools, the tools themselves disappear; somehow, they become incorporated into our work intentions and goals. Nevertheless, we do not *grow* tools in the way that we grow a hand or heart. Whereas a sociobiologist might explain human tool use as an instance of evolutionary adaptation, Heidegger would argue that tool-making and tool-using are possible only within the linguistic-historical world which cannot be explained in terms of "natural" evolution.

Because only humans exist within a world, only humans have hands, properly speaking.[8] The human hand, then, cannot be conceived as an animal organ. Human hands are essential ingredients in humanity's activity (*Handlung*) of opening up a world and disclosing entities within it. Mediating between the entity and the human individual, the hand is ontologically disclosive.

> Through the hand happens prayer and murder, greetings and thanks, the oath and the wink, but also the "work" of the hand, the "handiwork" and the implement [*Gerät*]. . . . No animal has a hand, and never does there arise a hand from a paw or a claw or a talon. Even the despairing hand is never and in the last a "talon," with which the man "grasps." *Only from and with the word has the hand arisen.* Man does not "have" hands, but instead the hand has in it the essence of man [*hat das Wesen des Menschen inne*], because the word as the essential realm of the hand is the essential basis of man. [GA, 54: 118–119; my emphasis]

Heidegger went so far as to define *thinking* as handiwork: "All the work of the hand is rooted in thinking. Therefore, *thinking itself is man's simplest, and for that reason hardest, hand-work*, if it would be accomplished at its proper time." [WHD: 51/16–17; my emphasis] In 1930, he used agrarian metaphors to describe the hand-work of the thinker. Plato and Aristotle, who stood in the brightness of originary presencing, brought in the first "harvest," by interpreting presencing as *eidos* and *energeia*. Hence, they initiated the history of productionist metaphysics. At the end of the history of metaphysics, we have been reduced to threshing empty straw. Hence, "we must once again venture out and bring in the harvest. . . . We must learn tilling and plowing once again, that is our fate, to learn to dig up the ground, so that the blackness and darkness of the ground comes to the sun. . . . " [GA, 31: 47–48]

Heidegger's agrarian metaphors aimed to show humanity's relationship to the "earth," but also to underscore the difference between humanity and nature. Plants and animals are driven by instincts, but only humans can differentiate themselves from things in a way which makes agriculture and animal husbandry possible. While usually insisting that the "drives" of animal life differ from the "comportment" of human existence, he did concede that an instinctual survival drive may be present in humanity. He insisted, however, that if human existence is defined *solely* in terms of such a drive, then humanity forsakes its primary obligation and possibility—to preserve entities and to guard the self-disclosure of being—for utilitarian considerations. By letting himself be guided by such animal impulses, Heidegger argued, modern man becomes Nietzsche's loathsome "last man."

Behind the wishes and drives of these "last men," according to Heidegger, is a profound unrest for which every "enough" becomes a "not-enough." The unrest involved in the constantly mounting and expanding "interests" and "uses" is not an artificially induced greed (*Raffgier*), but instead a kind of eagerness (*Gier*) resulting from the pressing-forth and drives of the merely living. [GA, 51: 4] The limitless Will to Will of modern technology is directly related to the reduction of humanity to the status of an animal with infinite craving.

> Living things essentially remain compelled and forced into the proper drive. Certainly, "the living," which we know as plants and animals, appear to find and contain in this drive their fixed form, over against which man can elevate the living and its drive purposely to the leading measure and can make "progress" into a "principle." If we notice only that which we use, then we are yoked into the compulsion of the unrest of mere life. This living awakens the appearance of moving and self-movement and, thus, of freedom. [GA, 51: 4–5]

While within the sphere of drives and wants, needs and interests, there may be some room to calculate and plan, this is not genuine freedom. Instead, in such circumstances people are "unbound in the sphere of compulsion"; there is "slavery under the domination of the constantly 'useful.' " [Ibid.] Blindly confident, technological humanity believed that "progress" was the road forward to a better future. In the middle of World War II, however, Heidegger asked: "Forward? Where to, please? To the smashed cities on the Rhine and the Ruhr?" [GA, 55: 78] Compelled by the drive for life, "the laboring animal is left to the giddy whirl of its products so that it may tear itself to pieces and annihilate itself in empty nothingness." [VA I: 65/87]

B. HANNAH ARENDT'S CONCEPTION OF THE "LABORING ANIMAL"

The writings of Hannah Arendt, Heidegger's student, help to explain Heidegger's aversion to the modern definition of humans as clever, laboring animals. In *The Human Condition* she argued that there are three basic kinds of human activity: *labor*, which produces goods that are immediately consumed to sustain life; *work*, which produces enduring structures in which human behavior can take place; and *action*, which opens up and preserves the historico-political world through language.[9] According to Arendt, action was regarded by the Greeks as the highest human possibility, while mere labor was scorned as something fit only for slaves. This negative evaluation of labor stemmed from the Greek view that any trait that people shared with other living things was subhuman. Since labor was necessary to provide the human body with the goods needed for life, and since animals also had to forage in order to feed themselves, labor itself was somehow subhuman and suitable only for slaves. Over the centuries, however, especially in modern times as the economic conception of human life came into prominence, labor increasingly gained respect, while attention turned away from civic "action." Hence, while the Greeks regarded participation in political affairs to be one of man's highest possibilities, modern *homo economicus* largely eschewed the duties of "public life" in order to pursue his "private interests" in the marketplace. Indeed, liberal societies claim that the state has no business promoting any particular "good" for people, including the "good" of political involvement; rather, each person must be allowed to discover and to pursue his or her own good.

As Arendt noted, both Adam Smith and Karl Marx defined humanity in terms of its capacity for productive labor. However, "it is indeed the mark of all laboring that it leaves nothing behind, that the result of its effort is almost as quickly consumed as the effort is spent. And yet this effort, despite its futility, is born of a great urgency and motivated by a more powerful drive than anything else, because life itself depended on it."[10] Arendt argued that the process of secularization in Europe, and the corresponding loss of belief in an afterlife, led people to become preoccupied with earthly life, particularly "appetites and desires, the senseless urges of his body which [they] mistook for passion. . . . "[11] While the self-interested ego at work in *homo economicus* may have retained a slim measure of the *vita activa*, socialism promised a world in which only one interest remained: that of re-producing the human species. Hence,

> What was left was a "natural force," the force of the life process itself, to which all men and all human activities were equally submitted ("the thought process itself is a natural process" [Marx]) and whose only aim, if it had an aim at all, was survival of the animal species man. None of the higher capacities of man was any longer necessary to connect individual life with the life of the species; individual

life became part of the life process, and to labor, to assure the continuity of one's own life and the life of his family, was all that was needed.[12]

In 1941, Heidegger said that humanity had long been classified as a "living being," one that "discovers and builds and can use the machine, a living being which can *reckon* with things, a living being which *in general* can place *everything* into his reckoning and calculating. . . . " [GA, 51: 90] Seventeen years later, Arendt observed that Western people, by defining themselves not in terms of world-engendering linguistic actions but instead in terms of natural "processes," ended up in a technical-mechanical world of labor that was far from the *vita activa* crucial to the Greek *polis*. The Greeks were struck by the overpowering surge of being and were compelled to bring it to a stand through language. In bringing being to a stand, they opened up a world in which entities could manifest themselves in their diverse ways. Western humanity has lost all contact with the original power and instead directs all its energies toward controlling the entities originally disclosed by the Greeks. The ever-expanding cycle of production and consumption, which discloses everything only as raw material for fueling the technological system, resembles all too closely the life of the animal, benumbed by its environment, living within a self-circumscribing circle of instinctual behavior, incapable of encountering entities as such. The cycle of technological humanity, however, is not completely circumscribed, but instead is (at least in principle) capable of infinite expansion: the technological world can suck all things into its labor processes.

In a manner reminiscent of Jünger, Arendt remarked that, seen from the viewpoint of an extraterrestrial intelligence, "modern motorization would appear like a process of biological mutation in which human bodies gradually begin to be covered by shells of steel."[13] Dominated by the never-ending organic cycle of production and consumption, humanity was declining to the condition of Nietzsche's "herd," which was concerned only with a full belly and a warm house. The consumerist society brought forth by the laboring animal "may end in the deadliest, most sterile passivity history has ever known."[14] Similarly, Heidegger believed that both human obliviousness to being and "the destruction of the earth can easily go hand in hand with a guaranteed supreme living standard for man, and just as easily with the organized establishment of a uniform state of happiness for all men." [WHD: 11/30]

In the postwar era, much has been heard about the economic "promise" of the atomic age. What does it mean, asked Heidegger, that we lived in the "atomic age"? "For the first time in his history, man explains an epoch of his historical *Dasein* from the pressure and the preparation of an energy of nature." [SG: 199] Just think: "The *Dasein* of man imprinted by atomic energy!" [Ibid.] Even being thus "imprinted" by a force of nature, however, humanity can never fall to the level of a mere animal. No matter how naturalistic and materialistic the atomic age might be, "its materialism is nothing at all material. It is instead a *Gestalt* of *Geist*." [SG: 199–200] Having been stamped by the *Gestalt* of the atomic age, humanity has been compelled to unleash the monstrous energies of the atom, and then compelled to develop ever greater methods for its control: "For the self-establishing of life, however, it must constantly secure itself anew." [SG: 202]

C. THE CYBERNETIC CHARACTER OF MODERN TECHNOLOGY: WILL WILLS ITSELF

Proponents of technological development proclaim that it ends human slavery to natural forces and makes possible a new era of freedom. Technological devices, from this point of view, are instruments for human ends. While conceding that it may still make *some* sense to speak of technical devices as human "means," Heidegger argued that for the most part "man is the slave of the power which governs all technological producing." [ZFB: 14] Industrial society is sometimes spoken of as the new "subject" which dominates the natural "object," but in fact "industrial society is neither subject nor object. Rather, as opposed to the semblance of its independence which alone is authoritative and placed upon itself, it is placed into subjection [*Botmässigkeit*] by the same power of the positing which challenges-forth that has also transformed the former objectivity of objects to mere disposability [*bestellbarkeit*] of standing-reserve." [ZFB: 12]

The "power" spoken of here refers to the being of entities in the technological age, i.e., to the "presencing" which compels humanity to organize everything in an endless quest for power for its own sake. That the technological system is not under human control, Heidegger argued, can be discerned in self-referential, cybernetic systems. The cybernetic character of modern technology distinguishes it from the Machine Age. The great iron works and mills of the Industrial Revolution were still owned and controlled by self-interested human subjects striving for power. In the twentieth century, however, the technological disclosure of entities mobilizes everything—including people—into the project of increasing the power of the technological system itself, all under the guise of "improving" the human estate. In his *Spiegel* interview in 1966, when asked what has taken the place of philosophy, Heidegger replied: "Cybernetics." [Sp: 212/279] A bit earlier, he was asked why he thought modern technology should be "overcome," when in fact everything was functioning: power plants are being built, production is at a peak, and people in the industrialized world have a high standard of living. "What is missing here?" Heidegger replied: "Everything is functioning. This is exactly what is so uncanny, that everything is functioning and that the functioning drives us more and more to even further functioning, and that technology tears men loose from the earth and uproots them." [Sp: 206/277] Elsewhere, he argued that even philosophy had become nothing but a servant for total mobilization. The cybernetic-computer revolution was a sign that primordial *logos* had been reduced to logistics:

> In the West, thought about thinking has flourished as "logic." Logic has gathered special knowledge concerning a special kind of thinking. This knowledge concerning logic has been made scientifically fruitful only quite recently, in a special

science that calls itself "logistics." It is the most specialized of all specialized sciences. In many places, above all in the Anglo-Saxon countries, logistics is today considered the only possible *Gestalt* of strict philosophy, because its results and procedures *yield an assured profit for the construction of the technological universe*. In America and elsewhere, logistics as the only proper philosophy of the future is thus beginning today to seize power over the spirit. [WHD: 10/21; my emphasis]

In a little-known essay on art, given during a trip to Greece in 1967, Heidegger expanded upon the cybernetic character of modern technology. He cited Nietzsche's remark from *The Will to Power*: "The victory of *science* is not what distinguishes our 19th century, but instead the victory of scientific *method* over science." [D: 140] "Method" here means the way in which the sphere of the objects to be investigated gets demarcated in advance. Scientific method is a projection of the world which determines that the real is what can be calculated in mathematical terms. Hence, "method is the victorious challenging of the world upon a thoroughgoing availability for man." [D: 141] This victory of method achieves its uttermost possibility in cybernetics. The word "cybernetics" is taken from a Greek word meaning "steersman." According to the cybernetic world projection, the basic trait of all calculable world processes is steering. Information provides the mediation necessary for one process to be steered through another. To the extent that the steered process provides information which affects the steering process, steering has the character of informational feedback. Cybernetic processes, then, have a circular dimension, an ordering circuit:

Upon [this ordering circuit] rests the possibility of self-ordering, the automation of a system of movement. In the cybernetically represented world, the difference between automatic machines and living things disappears. It becomes neutralized by the distinctionless process of information. The cybernetic world project, "the victory of method over science," makes possible a completely homogeneous and in this sense universal calculability, i.e., controllability of the lifeless and the living world. In this uniformity of the cybernetic world, even man gets trained [*eingewiesen*]. [D: 142]

Indeed, humanity gets taken into the cybernetic project in a particularly powerful way. Having become the subject for whom the entire world is its object, technological humanity becomes an element in the gigantic feedback circuit in which information about the object alters humanity. For as science discovers more about the genetic structure of the human organism, this information can alter the way in which humanity treats its own organic body. In this way, humanity becomes its own object: biochemical researchers define human life in terms of the genetic structure of the germ cell. Learning the "alphabet" of the genetic code may eventually enable scientists and engineers to produce and breed humans. Reflecting on these trends in 1966, Heidegger suggested that technological man has been inspired by Nietzsche's remark that "man is the as yet uncompleted animal." Guided by the technological principle of total self-control, an American genetic researcher concluded, "Man will be the only animal which is able to guide its own evolution." [D: 143] While humanity cannot yet manufacture itself in factories, it is moving in that direction. Certainly, Heidegger claimed, "futurology" represents the impulse to planning and control which foresees a totally managed world in the not-too-distant future.[15] Seen in such terms, humanity has been reduced to the status of a merely social being (*gesellschaftliche Wesen*) in industrial society.

The self-organizing, self-perpetuating character of modern technology was profoundly disturbing for Heidegger. Today, he claimed, technology commands ever greater production and control—for their own sake. No human "purpose," such as a higher living standard, motivates the technological system; rather, it drives itself onward for ever greater power. In analyzing Nietzsche's doctrine of the Will to Power, Heidegger argued that the Will does not will something it does not have, but instead *the Will wills itself.*

> All is the same thing as to will to become *stronger*, to will to grow . . . [. . . .] "Stronger" here means "more power," and that means: only power. For the essence of power lies in being master over the level of power attained at any time. Power is power only when and only so long as it remains power-enhancement and commands for itself "more power." . . .
> . . . Hence, also, Will and power are, in the Will to Power, not merely linked together; but rather the Will, as the Will to Will, is itself the Will to Power in the sense of the empowering of power. [HW: 216–217/77–78]

In the technological era, the Will to Will mobilizes humanity to transform the planet into a titanic factory for no other end than to magnify power. The usual "means-end" interpretation of instruments, then, makes no sense when applied to the technological era: humanity has become the "means" for an end that lies beyond human control or ken. The technological system may be likened to an organic process moving in a circle, but the difference is that there are self-limiting factors in organic cycles that are not discernible in the infinitely expanding spiral of the technological system. Once again calling on Nietzsche, Heidegger asked: "What is the essence of the modern power-machine other than *one* expression [*Ausformung*] of the eternal return of the same? But the essence of that machine is not anything machine-like or even mechanical." [VA I: 118/431]

One may understandably be skeptical of the claim that the repetitive cycles of the internal combustion engine are stamped by the metaphysical structure of the eternal return of the same. Nevertheless, Heidegger was by no means alone in advancing the claim that the system of industrial technology had in effect become self-organizing and autonomous, such that it could no longer be conceived as a human "means." Such a claim was a commonplace in early twentieth-century Germany. In 1939, Friedrich Georg Jünger, brother of Ernst, described how technological processes, in describing everything in causal-mechanical and economic terms, lays waste to the natural and social order.[16] Jacques Ellul has argued that "technique has become autonomous; it has fashioned an omnivorous world which obeys its own laws and which has renounced all tradition."[17]

More recently, Langdon Winner also maintains that technical innovations are no longer neutral means but instead help to shape the ends of society. Because technology radically alters the arena for human activity, *techne* has become *politeia*. This alteration of the socio-political domain usually occurs without those transformations ever being being subjected to public debate. According to Winner, there are at least five ways in which technology alters the socio-political terrain: (1) technological advances in transportation and communication facilitate and encourage centralized control in virtually all institutions; (2) the "economies of scale" associated with new techniques call for gigantic social structures which were unheard of in previous centuries; (3) the rational arrangement of technological systems produces its own forms of hierarchical authority independent of the political realm; (4) centralized, hierarchical technological systems tend to crowd out other types of human activity (e.g., agribusiness displaces family farming); (5) large socio-technological organizations exercise undue influence on the very political structures designed to "control" those organizations.[18]

Time and again, Heidegger insisted that "technology is in its essence something which man cannot master by himself." [Sp: 206/276] Indeed, instead of mastering technology, technological humanity has developed remarkable disciplines designed to master humanity itself, so that individuals will be amenable to the demands of technological society.[19] Once again, this theme has been voiced by a number of other prominent twentieth-century authors. C. S. Lewis, for example, in "The Abolition of Man," remarks that in the quest to gain total control over nature, humanity forgets what it means to be genuinely human and ends up becoming the object of technological domination:

> The price of conquest is to treat a thing as mere nature. . . . But as soon as we take the final step of reducing our own species to the level of mere nature, the whole process is stultified, for this time the being who stood to gain and the being who has been sacrificed are one and the same. . . . But once our souls—that is, our selves—have been given up, the power thus conferred will not belong to us. We shall, in fact, be the slaves and puppets of that to which we have given our souls.[20]

D. MICHEL FOUCAULT AND THE QUEST FOR BIO-POWER

Another proponent of the idea that people themselves have become the objects of technical control is Michel Foucault, who acknowledged that Heidegger was the most important influence on his thinking. Foucault worked out in detail many of the various "disciplinary" practices associated with the rise of the technological era. Dreyfus and Rabinow have suggested that while Heidegger focused on the transformation of the object in the technological era, Foucault focused on the transformation of the "subject."[21] While this may be an overstatement, Dreyfus and Rabinow are certainly correct in arguing that Foucault's examination of the disciplines imposed in such modern institutions as factories, asylums, hospitals, barracks, as well as in the domain of sexuality, goes far beyond anything attempted by Heidegger. For Foucault, modern institutions subject people to constant surveillance and disciplines designed to achieve "normalcy," i.e., designed to turn human beings into the "bio-power" suitable for the needs of the totalizing aims of the technological system. In the technological era, an important change in the character of power takes place:

> In traditional forms of power, like that of the sovereign, power itself is made visible, brought out into the open, put constantly on display. The multitudes are kept in the shadows, appearing only at the edges of power's brilliant glow. Disciplinary power reverses these relations. Now it is power itself which seeks invisibility and the objects of power—those on whom it operates—are made the most visible. It is this fact of surveillance, constant visibility, which is the key to disciplinary technology.[22]

For Heidegger, the technological understanding of being attempts to make entities wholly present as standing-reserve; for Foucault, the technological system attempts to make human beings wholly present as bio-power. The "human sciences," including sociology and psychology, are enlisted in the effort to disclose everything possible about the "normal" man and woman, so that they can be properly adjusted to the imperatives of the technological system, which itself lies hidden in the apparently beneficial institutions carrying out the disciplinary training needed to keep the system going. Foucault said that the power relations at work in the institutions of modernity are "intentional and non-subjective."[23] But how can we speak of intentionality without a subject? Somehow the disciplinary practices themselves embody the power that was the object of Foucault's investigations. "There is a push towards a strategic objective, but no one is pushing. The objective emerged historically, taking particular forms and encountering specific obstacles, conditions and resistances. Will and calculation were involved. The overall effect, however, escaped the actors' intentions, as well as those of anybody else."[24]

Heidegger claimed that the technological disclosure of entities as raw material compelled humanity to erect a world consistent with that disclosure: the world of total mobilization. Analogously, Foucault explained that the technological era occurred because the human subject revealed itself as capable of being made completely present for study (surveillance) and control (disciplinary practices). The power at work in this new disclosure of humanity conceals itself; made manifest by the hidden workings of power are the disciplines and institutions whose aim is to normalize human life. Foucault went much further than did Heidegger in describing how the body is subjected to disciplinary controls. Nevertheless, as we saw in chapter eleven, Heidegger sketched out how various disciplinary practices in Roman, medieval, and modern times helped to foster the being of the technological subject. It was this power-oriented, self-actualizing, anthropocentric subject which Heidegger believed was the image of humanity under "humanism." His famous critique of humanism was not a call for barbarism, then, but instead an attempt to point out that humanism was internally related to the ills of industrial technology and modernity.

During the twentieth century, many students of human behavior have discussed the relation between personal and social pathology, on the one hand, and the practices, discourses, and institutions of industrial modernity, on the other. Recently, for example, David Michael Levin has argued that three different and widespread pathologies—depression, schizophrenia, and narcissistic character disorder—are in some sense metaphors of the nihilism of the post-modern situation. According to Heidegger, as we have seen, this nihilism is describable in terms of the grandiose but ultimately meaningless Will to Will. Levin concludes that depression, schizophrenia, and narcissism are "historically conditioned pathologies" of this peculiar mode of the Will to Power:

> These pathologies develop out of particular dispositions that are characteristic of the individual will in modern times: pride, delusions of omnipotence, apathy, despair, dependencies, self-alienation, low self-esteem. And they are produced collectively in particular formations of the will to power: social and political institutions, dominant systems of discourse, symbolic rituals. Psychopathology is as much a question of political economy and conceptual paradigm hegemony as it is a question for biology, medicine, and psychiatry.[25]

Heidegger's own interest in pathological dimensions of the modern technological subject was revealed in the lectures and seminars he gave for Medard Boss's psychiatry students and faculty in Zollikon, Switzerland, from 1959 to 1969. Early Heidegger argued that the dualistic experience of the modern subject was a central feature of inauthenticity and suffering. By identifying itself with an empirical entity, the ego, *Dasein* conceals its own no-thingness, i.e., its mortal openness for the self-manifesting of entities *including* the ego. Having misidentified itself with a particular thing, the ego, *Dasein* sets out to defend itself against all other things. The insecurity of the ego's position in a threatening world leads it to become paranoid regarding others. Such collective insecurity is behind the never-ending arms race betwen the two present-day superpowers.[26] In his later writings, Heidegger concluded that this egocentrism was not a function of individual weakness, but instead was a feature of humanity in the age of the Will to Will. Like Freud, Heidegger concluded that pathology can be a cultural, not merely a personal, phenomenon.

In his meditations on the nature of art as authentic disclosiveness, Heidegger was seeking to move beyond the anthropocentrism or cultural narcissism of "secular humanism" toward what he regarded as a "higher humanism," one that would free humanity from its bondage to incessant industrial production for its own sake and would free it for its authentic capacity for disclosing things appropriately. Before considering such authentic producing, however, let us ponder for a few moments Heidegger's account of how the industrial mode of producing at the end of the history of productionist metaphysics transforms the everyday world in a way that makes any authentic encounter with things impossible.

Chapter 13. How Modern Technology Transforms the
                 Everyday World—and Points to a New One

Two decades ago, Alvin Toffler coined the term "future shock" to describe the fact that technological innovations are occurring so rapidly that they are transforming the socio-political realm in a dizzying, even dangerous way.[1] Toffler asked: How can existing institutions, social practices, values, and expectations for the future keep pace with the accelerating flood of new devices and systems? Heidegger, too, experienced a kind of "future shock" in the face of Germany's industrial transformation. Reflecting on this transformation, he concluded that "improved" technical devices did not merely serve the same function, only more efficiently, than did the outmoded devices. Rather, he argued, new devices alter the nature of the function or practice in question. In this chapter, we first consider how novel technical devices transform everyday practices. Then we turn to a larger but related issue: how the global system of modern technology, which produces those new devices, has altered both earth and world. Heidegger's name for the technological disclosure of things was *Gestell*. Finally, we examine Heidegger's reflections on whether and how humanity might be "released" from the "stamp" of *Gestell*.

A. HOW NEW TECHNICAL DEVICES ALTER EVERYDAY LIFE

Consider a device such as the typewriter. Surely, we may think, this device is simply a more efficient means for writing. According to Heidegger, however, efficiency was the wrong measure to use in evaluating writing. For him, writing was essentially *handwriting*. It was his experience that his own thinking occurred through his hand while writing. Typewriting undermined both thought and language, he argued, because in typewriting the word no longer "comes and goes through the writing and authentically acting hand, but rather through its mechanical pressure [on the typewriter keys]. The typewriter tears the writing from the essential realm of the hand and, that is, of the word." [GA, 54: 123] Typewriters have their place, for example, in reproducing and preserving things already written by hand. But there is no place for typing in personal correspondence! When typewriters first appeared, Heidegger remarked, people were affronted to receive letters written on them: "Today, [however,] a handwritten letter interrupts quick reading and is thus outmoded and unwanted." [Ibid.] Typescript conceals the personal character of a letter's author, thereby contributing to the homogenization of modern humanity.

Heidegger's objection to the practice of typing personal letters was shared by many of his contemporaries. And like many conservatives, he maintained that it was no accident that the printing press was invented at the dawn of the modern age. Letters, once drawn by hand, were reduced to characters that could be set and pressed by machine. The successor of the printing press, the typewriter (*Schreibsmaschine*) reveals the intrusion of the machine into the domain of language. Although the typewriter is a symptom of the technological age, Heidegger conceded that it is "still not a machine in the strict sense of modern technology, but is an in-between thing, between a tool and a machine, a mechanism. Its production, however, is conditioned by machine technology." [GA, 54: 126]

Today, the typewriter is giving way to the word processor. Apart from the keyboard needed for giving instructions to the computer, there are few similarities between the typewriter and the word processor—aside from the fact that both serve the same "function." Recently, people have argued that the change involved in moving from typing to word processing may be as great as the change involved in moving from handwriting to typewriting. Even the term "word processing" suggests a massaging, shaping, and managing of words as if they were plastic raw material. The advantages of word processing are undeniable; it is far more efficient than handwriting and typewriting. Moreover, because the words one produces on the monitor are not yet permanent, but instead are merely patterns of electrons, some users report that they feel less anxiety about writing on the word processor, because they are less committed to the words which they "process."[2] Nevertheless, it remains to be seen what long-term impact the advent of word processing will have on language and writing.

Albert Borgmann has elaborated on Heidegger's view that changes in devices alter important life practices.[3] The phonograph, for example, is a remarkable invention which makes great (and mediocre) music available to millions who would not otherwise hear it. On the other hand, the phonograph has also made it unnecessary for people to make their own music, which often pales in comparison with what is available on records. The phonograph still requires some mechanical manipulation on the part of its user. One must, for example, dust the record, clean the needle, and so on. But technical progress has led to a new device, the compact disc player, which is so easy to use and whose discs are so indestructible that it has become almost transparent as a device.[4]

Reactionary Europeans of Heidegger's generation feared such new devices, for they fostered the rise of mass culture. Mass culture, so they believed, threatened the viability of "high culture," supported by those elites sufficiently educated and suitably reared to be able to appreciate the nobler forms of art. Even Herbert Marcuse, who was no reactionary, argued that mass culture threatens to engulf previously independent spheres of life, including the world of art. Today, "the music of the soul is also the music of salesmanship. Exchange value, not truth value counts."[5] Heidegger argued that, in the technological epoch, the poet stands in the same relationship to the publishing business as does the stockboy who stacks copies of the poet's verses in a warehouse. [VA I: 87–106] While critical of Marxism in many ways, Heidegger agreed with his claim that the fetishism of commodities transformed the world by conceiving of everything in terms of commodities, i.e., in terms of monetary values. As Heidegger maintained,

> Self-willing man everywhere reckons with things and men as with objects. What is so reckoned becomes merchandise. Everything is constantly changed about into new orders. . . . [Commercial] man moves within the medium of business dealings and "exchange." Self-assertive man lives by staking his will. He lives essentially by risking his nature [*Wesen*] in the vibration of money and in the currency of values. As this constant trader and middleman, man is the "merchant." He weighs and measures constantly, but does not know the real weight of things. Nor does he ever know what in himself is truly weighty and preponderant. [HW: 289/135]

For modern commercial-technological humanity, nothing is "sacred." Everything has its price; everything can be calculated and evaluated according to the economic interests of someone or other. In the pre-technological era, when humanity still felt itself to be a part of the world instead of its master, people had to adapt themselves to the natural order as best they could. Even medieval humanity, to be sure, projected a certain order onto the world, but at least that "order" was believed to have been created and sustained by God— not by humans. The old-fashioned view that people must adapt themselves to the pre-existing order of things may be discerned in the objection which many people made in regard to the first airplanes: "If God had intended us to fly, He would have given us wings!" In the technological age, however, instead of conforming to the natural order, people force nature to conform to their needs and expectations. Whenever nature proves unsatisfactory for human purposes, people reframe it as they see fit. For Heidegger, such technological "reframing" compels entities to be revealed in inappropriate ways. The "factory farm," for example, treats corn and cattle as if they were merely complex machines, not living things. Such reframing, however, is a necessary consequence of the economic imperatives of the food industry.

An analogous kind of reframing occurs in social practices when new devices are introduced to replace ones that are inefficient in one way or another. Albert Borgmann offers the example of what happened when the gas-fired furnace replaced the wood-burning stove.[6] In an earlier age, if the home was heated by hearth or by Franklin stove, some family members would be responsible for finding and chopping wood. This practice required familiarity with the countryside and an appreciation of the value and relative scarcity of good wood-burning timber. Someone else would have the responsibility of lighting and tending the fire. And the whole family would gather around the fire for social interaction without the benefit of radio, television, or phonograph. The same heating "function," as Albert Borgmann has explained, is accomplished in a way that is cleaner, safer, and more efficient with the forced-air furnace.

But by replacing hearths and wood-burning stoves, forced-air furnaces

have also forever altered the everyday practices of an earlier age. No one
gathers around the gas or electric furnace in the basement of the house. Moreover, the electricity required for one such furnace may be generated by a coal-
burning power plant, the coal for which is extracted from monstrous strip
mines and the smoke from which promotes both the greenhouse effect and
acid rain. If the electricity is generated by a nuclear power plant, there is no
carbon-based pollution, but the plant's radioactive waste is incredibly poison-
ous and will remain so for thousands of years. Because the coal-fired power
plant, the strip mines from which the coal is extracted, and the forests and
lakes dying from acid rain are many miles from the comfortably heated subur-
ban home, and because the radioactive threat to life seems to be far in the
future, the home's occupants may be unaware of the devastation wreaked for
the sake of "clean" and "efficient" heat. The disappearance of the family
hearth, then, not only undermined important social practices but also contributed to industrial practices which are destroying the natural world.

     There is no denying, of course, that heating one's home by firewood has
become impossible for most people in the industrialized world. Moreover, in
those areas of the United States where wood-burning stoves have made a
comeback (such as in New England), significant air pollution has often resulted. Heidegger's point, however, was that technological innovations are
portrayed as "solving" problems, but in fact frequently give rise to new
problems that may be greater and more intractable than the problems they
originally "solved." Once again, nuclear power is a good example of how a
"solution" turned out to be an enormous problem requiring a "solution" that
has yet to be discovered: safe disposal of vast amounts of hazardous radioactive waste.

     Technological innovations have often been depicted as "labor-saving,"
but in Heidegger's view the more time "saved" by a new device, the more
time "available" for production. Does the computer really save time and labor, or does it generate the possibility (and thus the necessity) of far more
paperwork than ever before? How about the photocopying machine? When
limited to typewriter and carbon paper, people thought twice before sending
memos to five hundred people.

     Moreover, as we noted earlier, new devices cannot always be said to
accomplish the same thing as the devices they replace. Heidegger argued that

> it is not at all the case that the same processes that were formerly begun and
> accomplished with the help of land letter-carriers and mail-coach are fulfilled by the
> use of other means. Rather, airplanes and radios from out of themselves, that means,
> from out of their machine-essence and from the wide extension of their essence,
> determine the new arena [*Spielraum*] of possibilities, which become plannable and
> capable of being carried out by human willing and for its acting. [GA, 53: 53]

     In 1943, Heidegger addressed the fact that the invention of new technol-
ogies often leads to paradoxical situations in everyday practices. Such new
technologies are said to promote a better life by making possible more intense

"lived experience" (*Erlebnis*). But consider, Heidegger argued, the case of a well-known researcher who must fly from Berlin to Oslo to give a public address:

> One finds the "experience" [*Erlebnis*] splendid. But one does not give the least thought to the fact that this experience is the purest affirmation of the Will to Power, the essence of which bears the possibility of an airplane and a ride in it. From out of one's Christian experiences, one finds the Nietzschean doctrine of the Will to Power detestable and [nevertheless] cheerfully flies over the Norwegian fjord. Perhaps one gives an address, laden with the history of spirit, against "nihilism" and flies around in the airplane, uses the auto and razor-blade—and finds the Will to Power detestable. Why is this grandiose mendacity possible? Because one, neither with his Christian standpoint nor in his ride in the airplane, still does not think of being [*Sein*] even once and gets driven about by the oblivion of being [*Seinsvergessenheit*] in the purest oblivion. [GA, 55: 106]

Passages such as these reveal in a striking way Heidegger's disdain for all aspects of the technological era. After the war, however, even he learned to appreciate flying. In a letter to Medard Boss in 1963, for example, he remarked, "My wife and I, after the unusually beautiful flight to Zurich [from Sicily], arrived punctually at 8 P.M. in Freiburg." [ZSB: 329][7] Right after speaking of his pleasant flight (made possible, of course, by the world-tourism industry as well as by the factories of world capitalism), Heidegger went on to say that in Freiburg "the apple and pear trees stand all about in full bloom. The native land [*heimatliche Land*] appears new in contrast with the gazed-upon sea and island and its inhabitants." In a world in which space and time were being compressed by electronic media, jet airplanes, and multinational corporations, Heidegger could still experience something of a "home" in Freiburg. Clearly, however, he believed that his homeland was threatened by the advances of modern technology, which cannot make important ontological distinctions regarding time, place, things, and people.

B. THE TECHNOLOGICAL TRANSFORMATION OF SPACE AND TIME

In 1950, Heidegger spoke out prophetically against television, which combined the capacities of film and radio in order to make everything immediately present for inspection:

> All distances in time and space are shrinking. Man now reaches overnight, by plane, places which formerly took weeks and months of travel. He now receives instant information, by radio, of events which he formerly learned about only years later, if at all. The germination and growth of plants, which remained hidden throughout the seasons, is now exhibited publicly in a minute, on film. Distant sites of the most ancient cultures are shown on film as if they stood this very moment amidst today's street traffic. Moreover, the film attests to what it shows by presenting also the camera and its operators at work. The peak of this abolition of every possibility of remoteness is reached by television, which will soon pervade and dominate the whole machinery of communication. [VA II: 37/165]

In the technological era, space and time are no longer understood in terms of what Heidegger defined as an authentic homeland, a place in which the destiny of a people can work itself out within a familiar natural context. His personal attachment to his own Swabian soil was widely known, as was his antipathy toward big cities. His close friend Heinrich Petzet noted that Heidegger would become almost physically ill when approaching a big city, so upset was he by the combined effects of urban pollution and social disloca-tion.[8] Profoundly affected by his own sense of place, Heidegger complained bitterly about the universalizing impulse of modernity to destroy everything unique and irreplaceable. In 1966, he expressed reservations about the root-lessness of modern people in a way which made him seem hopelessly out of touch with the nature of the modern world: "I do not know whether you were frightened, but I at any rate was frightened when I saw pictures coming from the moon to the earth. We don't need any atom bomb. The uprooting of man has already taken place. The only thing left is purely technological relation-ships. This is no longer the earth on which man lives." [Sp: 206/277]

Numerous people, many of whom cannot be labeled "reactionaries," have expressed similar sentiments about the rootlessness of the technological world. Simone Weil, for example, in *The Need for Roots*, and Wendell Berry, in *The Unsettling of America: Culture and Agriculture*, have argued that an authen-tic human life requires a sense of stability and identity—a sense of rooted-ness.[9] People raised in cities may be puzzled by the extraordinary attachment to "place" exhibited by traditionally agricultural or peasant people, or even by people from small towns. From the viewpoint of "progressive" ideologies, including Marxism and liberalism, such attachments are in fact impediments to improving the human estate. Marx, for example, spoke more than once about the "idiocy" of rural life. And he insisted that only a romantic fool would spend time mourning the eradication of traditional places and ways of life by the capitalist juggernaut that (despite its own intentions) was preparing the way for a universal workers' paradise. In the Grundrisse, Marx stated that in the industrial age

> for the first time, nature becomes purely an object for humankind, purely a matter of utility; ceases to be recognized as a power for itself; and the theoretical discov-ery of its autonomous laws appears merely as a ruse so as to subjugate it under human needs, whether as an object of consumption or as a means of production. In accordance with this tendency, capital drives beyond national barriers and prejudices as much as beyond nature worship, as well as all traditional, confined, complacent, encrusted satisfactions of present needs, and reproductions of old ways of life. It is destructive towards all of this, and constantly revolutionizes it, tearing down all the barriers which hem in the development of the forces of production, the expansion of needs, the all-sided development of production, and the exploitation and exchange of natural and mental forces.[10]

Even the leftist Herbert Marcuse, however, mourned the passing of the backward, pre-technological world, in which labor was still a "fated misfor-tune," but also a world in which people and nature were not yet reduced to instruments. With its manners, style, and literature, "this past culture expressed the rhythm and content of a universe in which valleys and forests, villages and inns, nobles and villains, salons and courts were a part of the experienced reality. In the verse and prose of this pre-technological culture is the rhythm of those who wander or ride in carriages, who have the time and pleasure to think, contemplate, feel, and narrate."[11]

For technological humanity, space is no longer the beloved ancient landscape which provides the familiar context for everyday life; and time is no longer the historical realm in which human deeds and events occur in a way that is linked to past and future. Heidegger used the term "dwelling" (*wohnen*) to describe this lost way of life, rooted in a particular earth and world. Such dwelling requires a different sense of space and time from the abstract one imposed by the technological epoch. In Hölderlin's hymns about the rivers Ister and Rhine, Heidegger observed, one reads of "space" in terms of local place (*Ortschaft*) and of "time" as the historical wandering (*Wanderschaft*) of a particular people in that local place. Heidegger hoped that Hölderlin's poetry could found a historical world in which the earth could once again become native soil, the context in which place and history, space and time, could be unified in a way that bestowed meaning, purpose, and continuity.

Hölderlin's unity of earth and world, place and history, space and time, however, must not be confused with the technological concept of the "space-time-world." As opposed to a mode of dwelling which occurs in the native world and on its earth, a dwelling which discloses things appropriately, living in the technological space-time-world involves a "grasping and snatching" kind of disclosure which seeks to dominate not only entities but space and time as well. [GA, 53: 47] Einstein's unification of space and time does not "unify" those realms in the way that the poet's language does. Heidegger remarked that "from out of the mathematical-technological projection of lifeless nature, the character of reality becomes 'ordering.' " [Ibid.] Such "ordering" was initiated by Descartes, who conceived of space algebraically, as a system of linear coordinates. Space was thereby reduced to number. Later, when space was conceived in its relation to time and to moving things, a fourth coordinate was needed. This coordinate was, in Einstein's view, time itself. The mathematical conception of space-time gave rise, in Einstein's thought, to the conception of the "world line" in which the empirical experience of the "passage" of time, and hence the entire domain of "history" and "change," are revealed as illusory.

The conceptions of space and time developed by early modern physics also made it possible to conceive of "work" in a new way: as the product of force and direction divided by time. Hence, "reality is nothing but the quantum of working." [GA, 33: 48] Working, in turn, became conceived in the technological era in terms of ordering for its own sake, work for its own sake. Human working and the forces of nature became projected onto each other:

> The name for human doing and striving, "work," was carried over to the performance of mechanical force. Work is equated with mechanical energy. Conversely,

the predominance of the physicalistic concept of work in its essentially technological meaning took effect upon the determination of human work as "performance" [*Leistung*]. The performance principle is an essential principle of human acting and behaving. [Ibid.]

The ancient Greeks conceived of human performing not as a way of "producing" something in the sense of manufacturing it, but rather as a way of disclosing something—letting it be. The Greek concept of performing was transformed by its translation into the Latin term *fungere*, whence we get "function," in the sense of a performing which involves a process which produces a result. The reality of the real becomes, thereby, reduced to its "working," its "performing," its "functioning." Terms that once depicted human disclosive activity became transformed into descriptions of the "activity" of natural processes. In turn, human activity was conceived as a special case of these processes.

Moreover, the ancient definition of reality as "substance," that independent mode of presencing which defines and "produces" a living thing, was transformed by modern science into the "function" of a thing. An animal is not a self-presencing entity, then, but rather an organic "function" comprehensible within the matrix of space-time coordinates. The modern conception of work as function lets entities be conceived as abstract quanta of energy. Such a conception ignores the outward appearance of things which gives them substantial "form." In the technological era, then, space and time have become a system of mathematical coordinates, the framing structure (*Rahmenbau*), which organizes in the most efficient way possible the industrial functions needed to work up standing-reserve into finished commodities. [GA, 53: 46ff.] Individual entities thereby disappear into quanta of energy which can be stored and switched about at will: "Grasping and seizing are the names for a unified will to domination of space and time." [GA, 53: 47] The technological disclosure of things, then, reduces them to functional means for the purpose of ever greater productivity.

C. THE FUNCTIONALIZATION OF WORK, WORKER, AND EQUIPMENT

The functionalization of time and work becomes apparent in the modern industrial plant. First of all, industrial workers are bound by the time clock, which has no relationship to the normal rhythms of life or the seasons. Several years ago, a senior manager at a major plastics plant near Baton Rouge told me that out of every hundred workers who were hired, only about thirty were able to adapt successfully to the temporal regime of industrial work. Many of the "unsuccessful" workers were hunters, trappers, and small farmers who lived in a world where punctuality and rigid work habits were simply unknown. Some workers would quit after receiving their first paycheck and would return a few weeks later when their money had run out—not understanding the continuity of work required in the industrial plant.

Work in such plants is organized according to functional processes, whether workers are on an assembly line or organized into teams of "production units." The function or role required by the assembly line or production unit is, in principle, "fillable" by any worker. Often, those functions have been analyzed and rationalized by industrial social engineers who, in the tradition of the founder of scientific management, Frederick Taylor, compute the time needed to perform a particular action. Human work, in other words, becomes conceived in terms of abstract mechanical forces and activities. Industrial production processes call for repetitive work actions which *intentionally* undermine the all-around skill of the kind known to a master mechanic or tradesman. As one critic has noted, "The physical processes of production are now carried out more or less blindly, not only by the workers who perform them, but often by lower ranks of supervisory employees as well. The production units operate like a hand, watched, corrected, and controlled by a distant brain."[12] By reducing human work to the status of a function within a production unit, which itself is but a functional unit within the factory, which in turn is a functional unit of a multinational corporation, work is systematically degraded. Heidegger argued that such degradation was not the result of a particular economic system, such as capitalism, but rather was systemic to production in the technological era.

Heidegger believed that the impulse to functionalism is also reflected in the history of equipment: from tools to implements to machines to machine-making machines to cybernetic technological systems. Ancient Greek craftsmen had an authentic relationship both to their tools and to the materials with which they worked. Even during the first years of the Industrial Revolution, at least some workers retained a measure of understanding and respect for their equipment and materials. With the advent of large-scale industrial plants, assembly lines, and offices staffed with thousands of white-collar workers, both the understanding of the nature of one's work and respect for one's equipment deteriorated. Increasingly, everything has become regarded as plastic material which can be shaped, used, and discarded at will.[13]

In a number of places, Heidegger reflected upon the differences among tools, machines, and modern technological systems. A tool is useful for something, and so is a machine. But a tool does not have to be a machine. "Indeed, on the contrary, a tool for travel *can* be a machine, such as a car or an airplane, but it does not have to be. Indeed, a writing tool can be a machine (a typewriter), but it does not have to be." [GA, 29/30: 314] Like tools, machines are built according to plans. The plan for machine-building is guided by the serviceability of the machine. And this serviceability is governed by the purpose it is to serve. In the design for the machine, there is a preparatory ordering of the fittings (*Gefügen*) in which the individual parts of the machine move "here and there" against each other. [GA, 29/30: 315] Decisive for the machine-character of the machine, then, is "the independent course of the fittings regulated for determinate movements." [GA, 29/30: 51] At a certain stage in the history of productionist metaphysics, the machine could still be regarded as a means for human ends, but in the technological era this has changed:

The outstanding feature of modern technology lies in the fact that it is not at all any longer merely "means" and no longer merely stands in "service" for others, but instead [in the fact] that it itself unfolds a specific character of domination. Technology itself demands [fordert] from out of itself and for itself and develops in itself a specific kind of discipline and a unique kind of consciousness of conquest. Hence, for example, the fabrication of factories for fabricating fabricated articles, namely machines, which [in turn] fabricate other machines. Hence, the founding of a factory of tool machines is a unique triumph that is in itself [characterized] by graduated stages. [GA, 53: 53–54]

Power machinery, especially in the form of an assembly line, may well seem autonomous by comparison with hand tools. And the phenomenon of machines producing other machines gives the impression of an "autonomous technology." Workers on assembly lines report feeling subordinated to the rhythms and patterns of the highly complex machines: workers seem to be the means to the ends of the machinery. Marx spoke of the machine as an "animated monster" which "dominates and pumps dry living labor power." The modern machine mutilates the worker, turning him into a "fragment of a man," and destroys "every remnant of charm in his work."[14] Marx warned, however, that machines must not be "fetishized," i.e., depicted as independent agencies. Rather, they must be understood as what they really are: the complex means of production ordered up by the capitalist economic system. In Marx's view, the technological means of production are not in and of themselves alienating, for they are ultimately nothing but means for human ends. By changing the ownership of the means of production, workers would supposedly develop authentic social relations of production, relations that would permit the use of machinery without the alienation experienced under capitalism.

Some critics, including Heidegger, challenge this point of view by saying that industrial production methods are *intrinsically* alienating and exploitative, no matter who "owns" the means of production. Are workers in a Russian steel mill any less alienated than their counterparts in a liberal democracy? Does a woman working on an assembly line in Hong Kong experience herself as any less a "means" to an external "end" than a man working at a similar job in North Korea? Heidegger agreed that machines must not be viewed as autonomous agents, but neither must they be understood merely as "means" belonging either to the bourgeoisie or to the working class. Machines are, Heidegger argued, part of a global system of production that is no longer "owned" by anyone or any class; rather, humanity itself is in the service of a self-governing technological system. Seen from this point of view, capitalism and communism are merely two sides of the same system. Marxist critics reply that Heidegger is, in effect, a "technological determinist," someone who believes that technological systems have evolved with a momentum of their own which transcends human control. Such a determinist fails to see that human agency gave rise to the technological system and can thus also transform that system. As we shall see in our conclusion, however, Marx himself seems to have been a historical determinist in important ways.[15]

No matter how complex and "autonomous" machinery may appear, Heidegger argued, it is in every case an instance of standing-reserve within the global technological system. An airplane ready for takeoff is in fact capable of flying off on its own, but the fact that it can become detached from the ground says nothing about its essential imbeddedness within the technological system. This imbeddedness is also not influenced by the fact that the pilot may experience the aircraft as a "means" to the end of performing his or her job. Correctly revealed, the aircraft

> stands on the taxi strip only as standing-reserve, inasmuch as it is ordered to ensure the possibility of transportation. For this it must be in its whole structure and in every one of its constituent parts, on call for duty, i.e., ready for takeoff. (Here it would be appropriate to discuss Hegel's definition of the machine as an autonomous tool. When applied to the tools of the craftsman, his characterization is correct. Characterized in this way, however, the machine is not thought at all from out of the essence of technology within which it belongs. Seen in terms of the standing-reserve, the machine is completely unautonomous, for it has its standing only from the ordering of the orderable.) [VA I: 16–17/17]

In the technological world, even language becomes an instrument serving the production process. Heidegger argued not only that German dialects are being pushed aside by standardized German (promoted by ratio and television, as well as by schools), but that the German language itself is being replaced by Anglo-American—the universal language of modern technology. [GA, 53: 79–80] Instead of being the servants of the language at work through them, i.e., instead of letting entities show themselves appropriately through language, people in the technological era are compelled to use language to disclose things one-dimensionally. Many critics other than Heidegger have argued that in a "totally administered society" (Marcuse), language becomes distorted in such a way that it actually conceals more than it reveals. Such distorted language usually promotes an ideological goal.

For example, it is sometimes claimed that a new technological breakthrough is important because it will mean a step forward in human progress. Indeed, a major U.S. corporation once used the slogan "Progress is our most important product." What is revealed here is an image of the human community working together on projects that will benefit the whole species. What is concealed becomes clear only when we ask such questions as: Progress for whom? At whose expense? Moreover, when we speak of "human progress," Heidegger maintained, we talk as if humans were actually in charge of things—but, he believed, they are not. The technological way of disclosing things has reached such a point that humanity itself is no longer the "subject" spoken of by Enlightenment thinkers, nor are there any longer "objects" in the way things were experienced a century ago. Instead, "whatever stands by in the sense of standing-reserve no longer stands over against us as object." [VA I: 16/17] While humanity itself can never be transformed completely into standing-reserve [VA I: 18/18], technological humanity has become in effect the most important raw material in a process which no longer makes basic ontological distinctions among different kinds of entities.

D. *GESTELL*: THE "ESSENCE" OF MODERN TECHNOLOGY

Heidegger used the term *Gestell* to describe the historical "stamping" which compels humanity to disclose everything one-dimensionally, as standing-reserve. Heidegger's choice of this term was motivated, in part, by its proximity to what Jünger called the *Gestalt* of the worker which now "stamps" humanity. The prefix *ge-* in German is used to indicate the linking together of elements referred to by a particular noun. For example, *Gebirg* means a mountain chain. Similarly, *Gestell* means the linking together of what is "posited" (*stellt*) by the technological subject: in other words, *Gestell* means positing that all things are basically the same—raw material. When Heidegger called *Gestell* the "essence" (*Wesen*) of modern technology, he played on the term *Wesen*. Ordinarily, *Wesen* means "essence." Hence, the *Wesen* of modern technology would mean what is essential or basic to it. But Heidegger interpreted *Wesen* to mean "presencing" or "being," the manifesting of a thing. Hence, the *Wesen* of modern technology—*Gestell*—means the way in which things are disclosed or manifested in the technological era. In a striking passage, Heidegger described how *Gestell* posits or sets (*stellt*) everything into presence in the same way:

> The revealing that rules throughout modern technology has the character of a setting-upon [*Stellens*], in the sense of a challenging forth [*Herausforderung*]. That challenging happens in that the energy concealed in nature is unlocked, what is unlocked is transformed, what is transformed is stored up, what is stored up is, in turn, distributed, and what is distributed is switched about ever anew. Unlocking, transforming, storing, distributing, and switching about are ways of revealing. But the revealing never simply comes to an end. [VA I: 16/16]

The revealing never ceases because modern humanity "puts to nature the unreasonable demand" that it supply energy which can be not only extracted but stored as well. [VA I: 14/14] Modern technology extracts energy from water, coal, or uranium, stores that energy for future use, and distributes it at will. In reply to the question whether the old windmill did much the same thing, Heidegger said: "No. Its sails do indeed turn in the wind; they are left entirely to the wind's blowing. But the windmill does not unlock energy from the air currents in order to store it." [Ibid.] Presumably, Heidegger's point here is that windmills somehow "let the wind be wind," since wind cannot be forced to blow except where it does blow. While he may have been correct in some ways about this matter, he was clearly naive in his use of the windmill as an instance of pre-technological equipment. Forty miles east of San Francisco in Livermore, California, thousands of turbo-driven windmills cover the hills for as far as the eye can see. The wind's energy is, in fact, being harnessed and stored to produce electricity.

Heidegger's view that the technological age compulsively discloses things as raw material to be transformed into products by the most efficient technical means was similar in many respects to the views of members of the Frankfurt Critical School, including Max Horkheimer, Theodor W. Adorno, and Herbert Marcuse. This school was originally founded in the late 1920s by disaffected leftists who wanted to explain why the German communist revolution failed in 1920–21, especially in light of the fact that Germany was an industrialized country supposedly ripe for revolution. One of the major conclusions of the Frankfurt School was that the very system of industrial production that was supposed to liberate humanity not only had altered the institutions of society, but also had changed the character structure of the members of the working class. Industrial modes of production tend to make people more authoritarian, more willing to accept a passive position in a hierarchical society. Instead of being an emancipatory factor in human history, then, industrialism threatens to become a more complex form of enslavement, one legitimated by the "dictates" of efficiency, rationality, and technical expertise.

In addition to "stamping" the workforce with the imprint of the authoritarian personality, industrial technology also wraps itself in the garb of "reason" in order to be able to dismiss all opposition to it as "irrational." What the Frankfurt School called the "logic of domination" refers in part to industrial technology's capacity for using the "rationality" of science and technology as means for containing the liberating social forces which technology makes possible. In the 1960s, Jürgen Habermas summarized this curious "logic": "What is singular about the 'rationality' of science and technology is that it characterized the growing potential of self-surpassing productive forces, which continually threaten the institutional framework *and at the same time*, set the standard of legitimation for the production relations that restrict this potential."[16]

Similarly, Herbert Marcuse argued that in advanced industrial society, science and technology become a kind of ideology. Such a society seems to be run by economic and social technocrats who use the latest scientific findings both to maintain economic growth and to "prove" that there is no acceptable alternative to the existing form of society. The growing importance of science and ideology, according to Habermas, leads to

> a perspective in which the development of the social system *seems* to be determined by the logic of scientific-technical progress. The immanent law of this progress seems to produce objective exigencies, which must be obeyed by *any* politics oriented toward functional needs. But when this semblance has taken root effectively, then propaganda can refer to the role of technology and science in order to explain and legitimate why in modern societies the process of democratic decision-making about practical problems loses its function and "must" be replaced by plebiscitary decisions about alternative sets of leaders of administrative personnel.[17]

Despite agreeing with the first-generation members of the Frankfurt School that science and technology had undercut the possibility of democratic discourse about political options, however, Habermas increasingly distanced himself from his predecessors' gloomy view that rationality itself was intrinsically domineering. Like Heidegger, both Horkheimer and Adorno believed that the Enlightenment marked not a high point but a low point in Western history. Enlightenment rationality was a far cry from the ideal of "objective reason" or cosmic *logos* which guided Greek thinking. Through the course of history, according to Horkheimer and Adorno, such reason had become reduced to "subjective reason" or "instrumental rationality." The Enlightenment's "rationality," then, was not disinterested and impartial, but in fact was motivated by hidden power interests which became clear only during the nineteenth and twentieth centuries. Habermas, however, has resisted the claim—shared by Horkheimer, Adorno, and Heidegger—that the rational ideal of the Enlightenment was nothing but an ideological screen for the "logic of domination."[18] We ought to give up rationalist pretensions to absolute knowledge about nature or society, Habermas counsels, but we ought not to forget the genuinely emancipatory intention of the Enlightenment, nor ought we to condemn reason as being intrinsically power-motivated. Rather, we ought to commit ourselves to establishing societies where people can engage in the kinds of rational, non-coerced, and undistorted communication required if we are to arrive at a consensus about how to live in the technological era.

Of course, from Heidegger's point of view, no such human discourse could have any real effect on the outcome of the technological era. Modern technology was not the consequence of human action, and thus could not be essentially changed by such action. Especially during the late 1930s and up until about 1945, Heidegger often spoke about modern technology—its practices and its consequences—in the darkest possible terms. After World War II, however, when he realized that the German "revolution" had completely failed, he had to come to terms with the fact that the technological epoch was not nearly at its end, but instead was only just beginning. Given that this epoch might last five hundred years, Heidegger's later remarks about modern technology became more temperate and reserved. For example, in a television interview given in 1969, he maintained, "I am not *against* technology. I have never spoken *against* technology, also not against the so-called demonic [nature] of technology. But instead I attempt: to understand the essence [*Wesen*] of technology."[19]

Despite Heidegger's protestations to the contrary, he did indeed speak out against modern technology. Yet, he also said that we must learn to move beyond our usual evaluation of technology in terms of optimism vs. pessimism. Insofar as modern technology is the final phase in the history of being that shapes the West, we must learn how to cope with that destiny. Resisting it is foolish, as Heidegger observed in 1955:

> No one can see the radical changes to come. But technological advance will move faster and faster and can never be stopped. In all areas of his existence [*Dasein*], man will be encircled ever more tightly by the forces of technical apparatuses and automatic devices. These forces, which everywhere and every minute claim, en-

chain, drag along, press and impose upon man under the *Gestalt* of technological installations and arrangements—these forces, since man has not made them, have moved long since beyond his will and have outgrown his capacity for decision. [G: 19/51]

After stating that humanity will become ever more shackled to the technological system and its devices, however, Heidegger conceded that that system has become indispensable to us. Hence, we must not blindly attack technology or consider it the work of the devil, since the technological disclosure of entities drives us onward toward ever greater improvements. [G: 24/53] And, even if we have become enslaved to technology, we do not have to remain enslaved. But precisely how are we to be freed from the stamp of *Gestell*?

E. RELEASEMENT FROM MODERN TECHNOLOGY

Heidegger formulated what is probably his most famous answer to this crucial question in 1955:

> We can indeed use technological objects, and yet at the same time with all correct use keep ourselves free from them, so that we can let go of them at any time. We can thus take technological objects into use, as they must be taken. But we can at the same time let these objects remain with themselves as something which does not concern us in the innermost and authentic [ways]. We can say "yes" to the unavoidable use of technological objects, and we can at the same time say "no," insofar as we do not permit them to claim us exclusively and thus to warp, confuse, and finally lay waste to our essence [*Wesen*]. [G: 22–23/54]

It is difficult to see how Heidegger hoped to reconcile what he predicted as humanity's total enslavement to technology with his claim that humanity can be freed from such enslavement simply by learning to use technical objects appropriately, i.e., in a way that lets people avoid being turned into automatons. Present-day humanity is certainly not characterized by the "releasement" (*Gelassenheit*) necessary to "let technological devices be." Perhaps Heidegger was holding open a vision of what humanity might become, centuries from now, if being revealed itself once again in such a way that a new relationship to things became possible for Western humanity.

In important respects, Heidegger's account of the person "released" from the "claim" of *Gestell* is reminiscent of what Eastern thinkers have described as an "enlightened" person, someone no longer driven by the compulsion to control and to master everything.[20] Such a person no longer conceives of himself or herself as a self-contained ego that is constantly threatened by and thus compelled to dominate surrounding objects, but instead as a receptive openness for all things. How to arrive at this "released" condition, however? In effect, by taking our present-day situation as a *koan* to be studied. In the Rinzai sect of Zen Buddhism, a *koan* is an apparently contradictory saying or expression which a Zen student is required to "solve." The *koan* cannot be solved by merely rational means, however, but requires an existential break-down of the rational ego's rational way of framing things—and thus a break-through to a less constricted, more expansive way of being in the world. This expansive, open way of being is sometimes described as a "still mind" in the Zen tradition. During World War II, Heidegger argued that "We all . . . must really arrive at the point that understanding stands still for us. For only when this understanding—which is always busy and crowded with the expressions 'logical' and 'illogical,' but which is at the same time 'normal'—stands still can the other essential sort of thinking move into play." [GA, 55: 6]

Such "essential thinking" is neither rational nor irrational, but a medita-tive openness for the being of entities. Such openness precedes the distinction between the rational and irrational. But does the technological era not result from the complete self-concealment of being? How then can we become open to it, especially if its self-concealment has stamped us with the one-dimen-sional rationality of *Gestell*? And if being has in fact completely concealed itself from us, how is it that we nevertheless encounter entities *as* entities, even if merely as "standing-reserve"? These questions compose the *koan* for the West. Heidegger formulated the *koan* in the following terms:

> *Being has already cast itself upon us and has cast itself away from us.* . . . This ap-pears to be a "contradiction." Only we do not want to snatch what is disclosed there [and put it into] a formal scheme of formal thinking. In this way, everything becomes merely weakened in its essence and becomes essence-less under the appearance of a "paradoxical" formulation. As opposed to this, we must attempt to experience [the fact] that we—placed between the two limits—are transferred into a unique abode from which there is no exit. Yet since we find ourselves transferred into this situation of no exit, we will notice that perhaps even this uttermost situation without exit might arise from being itself. . . . [GA, 51: 80–81]

Instead of trying to "solve" the problem of modern technology by furious actions and schemes produced by the rational ego, then, Heidegger counseled that people learn that there is no exit from that "problem." We are cast into the technological world. Insight into the fact that there is no exit from it may, in and of itself, help to free us from the compulsion which characterizes all attempts to become "masters" of technology—for technology cannot be mas-tered. Instead, it is the destiny of the West. We can be "released" from its grip only to the extent that we recognize that we are in its grip: this is the paradox.

Wolfgang Schirmacher has offered a related but somewhat different inter-pretation of what it means to "let technical devices be." Such letting-be means to integrate machines into our existence as mortals who must make things not only in order to survive, but also in order to respond to the ontolog-ical claim made upon us. In some sense, the sometimes perverted purposes to which we put technical devices are a sign of our failure to understand what it means to be human. All attempts to use technology to "dominate" nature arise from the illusion that we are separate from the natural order of things. Instead of dreaming either of technological domination or of a return to a pre-technological era, we should "learn to live with the realization that *artificiality is the nature of man*, and that technology, in all its uncommonly diverse forms, is the realized, cosmic mode of being peculiar to our nature and which must be further perfected. Technology is the human way of corresponding to the universe. . . . "[21]

While Heidegger would have been suspicious of Schirmacher's attempt to interpret human existence in "cosmic" terms, since such talk comes too close to the "naturalism" he abhorred, nevertheless, Heidegger would agree that making and producing ("artificiality") are universal conditions of human existence, and that such "technology" is the basic human way of responding to the claim made upon us. We are called on to "let things be," to bear witness to them. Making and doing—these are how humans can "let things be." The way beyond the technological era, then, would involve restoring the authentic meaning of technology: *techne*, the kind of "making" which discloses things appropriately. Such an authentic mode of *techne*, however, is nothing that humanity could "will," but instead could arise only in connection with a radical "turning" within the history of being. Despite the catastrophic consequences of his first attempt to prepare the way for this "turning," the postwar Heidegger continued to attend to the "traces" and "winks" left behind by the vanished gods and made himself ready for the advent of new gods, whose arrival would mean that being had manifested itself once again. Heidegger believed that his own meditations on being, his own attempt to recall what had been forgotten, were of world-historical significance for the following reason: Being has concealed itself from technological humanity. Thus deprived of a relationship with the originary ontological source, humanity can experience only entities, and then only in a constricted way: as standing-reserve. If and when a new encounter with being were to occur, if being showed itself once again, this event *in and of itself* would indicate that the reign of the technological era was coming to a close.

Modern technology was for Heidegger the greatest danger for humanity, not because industrial technology threatens to destroy the biosphere, but rather because technology reduces humanity to the status of a clever animal with no insight into its own authentic possibility and obligation: to disclose things and to shelter their being. Yet in this very danger, so he believed, there is harbored the possibility of a turning in which the self-concealment of being would be overcome. Heidegger prepared for this turning in his meditations on the nature of art, *techne*, as authentic producing, authentic letting-be. In our final chapter, we shall examine Heidegger's return to the ancient Greek conception of *techne* as a possible guide for initiating a new, non-domineering mode of making and producing.

Chapter 14. Authentic Production: Techne as the Art of Ontological Disclosure

Heidegger believed that his task was not only to deconstruct the history of productionist metaphysics, but also to prepare the way for an alternative to that history. That is, he was not satisfied merely to reveal how the productionist understanding of being had stamped contemporary humanity with *Gestell*, his version of Jünger's *Gestalt* of the worker. He also hoped that in meditating on the nature of *Gestell*, in taking a step back from its impulse toward total mobilization, he would be granted insight into the new beginning needed to save humanity and the earth from destruction. In this chapter, we examine Heidegger's claim that what the Greeks meant by *techne* offers a clue for redefining the nature of working and producing. Authentic *techne* is a way of disclosing that unites art and production. In the 1930s, Heidegger had hoped that National Socialism would institute a post-industrial "technology" modeled on his vision of the unity of art and production. Postwar Heidegger continued to meditate on the possibility that art, particularly poetry, could help to initiate the "turning" necessary for moving beyond the one-dimensional technological understanding of things.

As we have already seen, early Heidegger concerned himself with the nature of handicraft production. Later Heidegger modified his earlier idea that the production of works is an *a priori* feature of human existence. It is true that all people produce works, he concluded, but productive activity which takes place within the sway of productionist metaphysics is different from productive activity that occurs outside of it. Early Heidegger's example of the workshop as an element with the referential totality is ambiguous. On the one hand, it suggests that handiwork is an *a priori* feature of authentic production; on the other hand, it also suggests the extent to which handiwork may be drawn into and shaped by the technological imperative. Despite the difference between Greek craft technology and industrial technology, even the Greeks, at least during and after the time of Plato, were beginning to be influenced by productionist metaphysics. While Plato may have drawn the model for his productionist metaphysics from his experience with artisans of various kinds, he in turn helped to initiate the metaphysics which culminated in the triumph of industrial modes of production. Heidegger concluded that because the grip of industrial technology is so strong, the mere fact that a contemporary person is engaged in handiwork does not in and of itself protect him or her from producing in accordance with the demands of productionist metaphysics.

Hence, it would appear, handiwork is a necessary but not sufficient condition for authentic producing.

While Heidegger concluded that Greek existence, too, was shaped by a productionistic attitude toward things, he also believed that he had found resources in the Greek tradition for what he regarded as an ontologically appropriate alternative to that metaphysics and to its concept of production. Productionist metaphysics conceived of making in terms of "actualizing" or "effecting" a thing, in the sense of "causing" it to be present. The pre-metaphysical conception of production, which Heidegger "discovered" by a generative reading of ancient Greek thinkers such as Heraclitus, spoke of producing as a "letting-be" or a "freeing" which enabled an entity to come into presence, to show itself, to emerge, analogously to how a living thing comes forth into presence. Heidegger's constructive "retrieval" of this alternative conception of producing was central to his political effort to save the West.

In this chapter, we shall first examine a crucial ambiguity in Heidegger's notion of "producing." We may pose this ambiguity, which we discussed in a preliminary way in chapter eight, in the form of a question: What is the relation between producing defined as "letting something be disclosed," on the one hand, and producing defined as the emergence of a thing, as in the blooming of a flower, on the other? Next, keeping this ambiguity in mind, we examine Heidegger's concept of *techne* as authentic disclosing or producing. The highest form of *techne* is the work of art, which—as we have seen—founds a world wherein things may present themselves in a particular way. In the postwar years, Heidegger suggested that the work of art could refine our ontological understanding in such a way that we could learn to "free" things from their captivity in the matrix of instrumental dealings associated with the industrialism spawned by productionist metaphysics. In what may be regarded as an attempt to move beyond the anthropocentrism associated with his claim about the saving power of art, Heidegger went so far as to say that virtually *any* thing could play the world-gathering role formerly assigned primarily to the work of art. We conclude with the question of whether the work of art, or any thing at all, has the power necessary to transform the technological world.

A. A PUZZLING DUALITY: BEING AS SELF-EMERGENCE AND BEING AS APPEARING

The earliest Greek thinkers—the so-called pre-Socratics—achieved so much, Heidegger believed, because they dwelt so close to the ontological difference between being and entities. Although they were unable to experience this difference as such, they experienced the being of entities as presencing, as rising into appearance, as unconcealment. Parmenides, "the true founder of ancient ontology," came close to Heidegger's basic insight when he said that "being and thinking are the same." [GA, 24: 154/110] Heidegger interpreted this phrase to mean that there is an internal relationship between the temporal transcendence of human existence, on the one hand, and the presencing of entities, on the other. With the advent of Plato's productionist metaphysics, this relationship became obscured, and being itself was reduced to the status of a superior kind of entity. In the nineteenth century, in a late stage of the history of productionist metaphysics, people said that for something "to be" meant for it to be composed of matter-energy, exist in space and time, be publicly observable, be amenable to quantifiable measurement, and be in principle predictable. Since there is no "is-meter," no way to take physical measurements of the being of an entity, positivists concluded that apart from such physical attributes the term "being" was meaningless. [GA, 51: 26] "To be" means to be an object in "nature," the all-encompassing system of matter-energy which causes everything to be the thing that it actually is. In the technological era, of course, "nature" reveals itself less as an object for scientific investigation and more as a commodity for consumption.

Heidegger argued that the productionist view of nature, while correct in certain ways, was a constricted interpretation of the early Greek experience of nature as *physis*. The word "nature" derives from the Latin *natura*, itself a translation of the Greek word *physis*. *Natura* is related to the Latin *nascor*, "to be born." *Nascor* contains a faint echo of the Greek word *physis*, which came closest to naming what Heidegger called "being." Meditation on the hidden implications of *physis*, he believed, would help in developing a new understanding of the "nature" spoken of in Hölderlin's poetry. This new understanding would presumably make possible a new, non-domineering relationship with things. Heidegger's controversial "etymology" of the term *physis* led him to give it two meanings. First, by linking *physis* to *phyein*, he defined *physis* as the event of self-emergence, as when a bud bursts forth into a flower. Second, by linking *physis* to *phainesthai*, he defined *physis* as appearing, shining, showing forth. For Heidegger, then, *physis* meant both self-emergence and appearing.

The first aspect of *physis*, an entity's self-emerging, would seem to be in some measure independent of the second aspect of *physis*, the appearing or presencing of an entity within a historical world. Heidegger seemed to give priority to this second aspect, however. Hence, he rejected the Roman notion that *physis* named a totality of entities that are present-at-hand as a cosmos governed by natural law, to which humans are also subject. Instead, he argued that *physis* meant the disclosive event which first makes possible the appearing of entities, and hence every possible human encounter with them. The faint echo of "presencing" found in the Latin *natura* was grounded in its stem *nascor*, "to be born." This dimension of presencing, however, concerned the self-emergency of things, not their "being" in the sense of their appearing within a world. Moreover, the Romans conceived of the self-emergency of entities in terms of the metaphysical conception that "to be" means to be produced, caused, or created.

According to Heidegger, however, for an entity "to be" primarily means for it to manifest itself *as an entity*. Hence, unlike for the Romans, "for the Greeks standing-in-itself was nothing other than standing-there, standing-in-the-light. Being means appearing. Appearing is not something subsequent that sometimes happens to being. Being presences [*west*] *as* appearing." [GA, 40: 108/86] The fact that for the Greeks "to be" meant to appear or to shine forth "punctures the empty construction of Greek philosophy as a 'realistic' philosophy which, unlike modern subjectivism, was a doctrine of objective being." [Ibid.] Nevertheless, almost overwhelmed by the presencing of entities, the Greeks were forced to bring them to a stand by establishing limits for them. Gradually, they transformed the presencing of things into an underlying and constantly present metaphysical "ground." This, as we have seen, was the start of productionist metaphysics.

In defining *physis* (being) primarily as presencing-appearing, however, and in failing to emphasize adequately the second aspect of *physis*, the entity's self-emerging, Heidegger ran the risk of defining the "being" of entities as a phenomenal event wholly dependent on the fact that humans happen to exist. As we saw in chapter eight, these two dimensions of *physis* may not be compatible.[1] On the one hand, the being of an entity means its "presencing" in the sense of its "appearing" or its "self-manifesting." This conception of being was Heidegger's most important contribution to the history of ontology. The event of such presencing requires a clearing or an absencing. This clearing is constituted by human existence. Without human existence, an entity cannot "be" in the sense that it cannot show itself or present itself. We may call this the *aletheia*-logical, or truth-like, conception of being. Heidegger translated the Greek term *aletheia*, usually interpreted as "truth," as "unconcealment."

On the other hand, Heidegger defined being as the ontological power somehow at work in the self-emerging or self-producing of living things, such as plants and animals. Presumably, the ontological organizing principle of an elk is in no way dependent on the fact that the elk "appears" within the clearing of human existence. In lectures from 1929 to 1930, Heidegger analyzed the being of animals, by which he seems to have meant the ontological principle which explained their morphology and behavior. [GA, 29/30] Moreover, in *Being and Time* he described the ontological structure of the human entity, *Dasein*. This structure, care, is itself articulated in terms of three-dimensional, ecstatic temporality. In speaking of "the being of *Dasein*," Heidegger often seems to have meant not only the "appearing" of the site needed for things to show themselves, but also the metaphysical structure of a particular kind of entity, a structure that is analogous to the metaphysical structure that makes an animal the living entity that it is. We may call this the *ousia*-logical, or substance-like, dimension of being. The *ousia*-logical dimension of an entity, for example, a fish, is not self-evidently related to the *aletheia*-logical dimension of the fish. The fish may manifest itself only within a historical world, but the fish spontaneously grows quite independently of such an event.

According to William F. Vallicella, in the early 1930s Heidegger "solved" the conflict between these two conceptions of being—one dependent on human existence, the other not—by eliminating the *ousia*-logical conception of being.[2] Henceforth, the "being" of an entity meant the event of its appearing.

In effect, Vallicella argues, Heidegger concluded that the quest to discover the *ousia*-logical dimension of being was tied up with the history of productionist metaphysics. Was not the search for the "being" of an entity in the sense of its ground or principle of self-producing and self-organizing a typical example of productionist metaphysics?

Despite such reservations, in my view, Heidegger never completely abandoned the *ousia*-logical conception of being, although he increasingly turned his attention away from it toward being understood as the changing ways in which things manifest themselves or appear historically. The turn away from the *ousia*-logical conception of being paralleled Heidegger's turn away from transcendental philosophy and *Dasein* analysis. He no longer conceived of presencing and unconcealment primarily in terms of the transcendental being-structure of human *Dasein*, but instead defined the openness of human *Dasein* as being "appropriated" (*ereignet*) as the site through which presencing occurs.

Had Heidegger concluded that the "being" of an entity is really nothing more than its appearing, he would have been open to the charge that he was a subjective idealist. He always insisted, however, that entities "give themselves" to us; they are *discovered*, not "projected" or "invented" by us. The question facing Heidegger, then, was this: how to describe the relative spontaneity and autonomy of entities without resorting to metaphysical categories? In the mid-1930s, he attempted to answer this question by developing the distinction between "earth" and "world." As we saw earlier, he argued that entities cannot be reduced to the event of their appearing within a historical "world," for entities belong to the "earth" which can never be made fully present in any world. Two basic features of earthly things are that they bring themselves forth (as when a plant blooms) while at the same time they conceal crucial aspects of themselves (as when the later stages of the plant's growth are hidden from view). The growth of a plant or animal can be *disclosed* only within a historical world; nevertheless, there is an intrinsic "innerness" (*Innigkeit*) to living things, and in fact to every entity, which resists disclosure. Conceptually, this idea of "innerness," drawn from Hölderlin, seems close to the *ousia*-logical dimension of being discussed above, although without the notion of "ground" or "foundation" often associated with *ousia*. If we try to penetrate into the innerness of an entity, Heidegger claimed, we destroy the entity in question. The technological era, driven to make everything wholly present and to eliminate all secrets, ends by pushing entities beyond their own appropriate limits.

In an important statement about the intrinsic limits which govern earthly things, Heidegger said:

> The unnoticeable law of the earth preserves the earth in the sufficiency of the emerging and perishing of all things in the allotted sphere of the possible which everything follows, and yet nothing knows. The birch tree never oversteps its possibility. The colony of bees dwells in its possibility. It is first the will which arranges itself everywhere in technology that devours the earth in the exhaustion

and consumption and change of what is artificial. Technology drives the earth beyond the developed sphere of its possibility into such things which are no longer a possibility and are thus the impossible. The fact that technological plans and measures succeed a great deal in inventions and novelties, piling upon each other, by no means yields the proof that the conquest of technology even makes the impossible possible. [VA I: 90/108]

Note that Heidegger concluded that the intrinsic "possibilities" of living things are *discovered, not created*, in the historical world. If living things did not somehow contain their own intrinsic limit and measure, there would be no basis for objecting to the technological disclosure of things. If for something "to be" simply meant for it to present itself in accordance with the categories imposed by a historical epoch, then there would be nothing problematic, for example, about animals presenting themselves merely as objects in the laboratory or as commodities on the factory farm. Nor would there be any reason even to pause in the face of the possibility of storing aborted human fetuses in cold-storage warehouses as a spare-parts "resource" for infants with genetic problems. Certainly, animals and fetuses *can* present themselves as objects and commodities, but Heidegger would have said that such a disclosure is inappropriate, indeed "impossible." It is impossible to treat an animal like a machine, because under such an arrangement the animal would have been destroyed as an animal and transformed into something monstrous, something which oversteps the *limits* proper to animals. The technological disclosure of things was so horrifying in Heidegger's eyes because it lacked insight into the intrinsic limits of things; without such insight, there cannot be a genuine historical world. Hence, he regarded the technological world as an un-world.

Heidegger's difficulty, however, as we have noted, was how to speak of the intrinsic measure and limit of living things ("the hidden law of the earth") without resorting to one of the foundationalist doctrines of productionist metaphysics: that entities are "grounded" upon their "essence." Equally unpalatable to Heidegger were scientific explanations for the enduring structure of living things, e.g., the role played by DNA in determining cell reproduction in plants and animals. Unwilling to appeal either to metaphysics or to science, Heidegger concluded that the "hidden law of the earth" is an impenetrable mystery.

Such a conclusion may seem obscurantist, but there is a constructive element to it. Heidegger criticized both metaphysics and the science which arose from it because they presupposed both that there is an enduring, objective "structure" to things, and that reason is capable of discovering this structure. Today, this view of "reality" and "reason" has been called into question, partly because of Heidegger's own "deconstruction" of metaphysical foundationalism, partly because of developments within science and other disciplines. Many people no longer accept two views that were popular in the first part of the twentieth century: the claims that absolute truth is attainable, and that science is the only valid methodology for attaining that truth. Hesitation to accept these absolutizing claims is not unrelated to the modern political consequences of totalitarian regimes which claim to be founded on the truths of science. Moreover, the unexpected and sometimes horrifying consequences of the technical application of scientific discoveries have made many people suspicious (rightly or wrongly) about the Enlightenment's confidence that science would make possible utopian social conditions.

The target of Heidegger's sometimes violent criticism of science was primarily its claim to being the only valid way of understanding things. He counseled that a wise humanity would discern the *hubris* involved in saying that things are "nothing but" matter in motion, or "nothing but" commodities. Freed from one-dimensional understanding, presumably a post-technological humanity would make more variegated and possibly more appropriate ontological distinctions about entities. Thus, Heidegger's idea of the self-sufficient "emerging and perishing of all things," the jutting forth of the earth which resists all attempts at total disclosure and total control, was a way of highlighting the limitations of rationality, especially the kind of rationality which imposes its own measure on things instead of being open to the measure internal to things themselves.[3] Science, in his view, was genuine only if it succeeded in taking the measure from things, instead of imposing measure upon them.[4] Moreover, because he insisted that human existence is radically finite, he believed that it was impossible in principle for humanity ever to gain an "absolute" or "total" understanding of things.

The impossibility of certainty regarding our understanding of things does not justify despair, but rather a healthy skepticism regarding all-embracing or totalizing attitudes and the projects springing from them. If we are to treat things appropriately, then, we must at least become attuned and receptive to those things as much as we can within the limits of the conceptual grid which helps to determine how things can show themselves. For Heidegger, such a receptive approach to things was central to all authentic producing, understood as the disclosive act of letting something be.

Letting something be, taking its measure—these activities can occur only within a world. As we saw earlier, "world" names the historical clearing opened up by a work of art, of statesmanship, or of thought. The world enables entities to show themselves in various ways, to be distinguished from other entities. But there is much that cannot appear within a finite world. Moreover, the very event of presencing which makes it possible for things to manifest themselves conceals itself. Hence, in a given world, people do not notice presencing, but instead the things which are present. At first, Heidegger spoke as if the dimension of concealment involved in a historical world were *identical* with the self-concealment or self-withdrawing dimension of the earth. Later, however, he emphasized that the self-concealing dimension of the earth is *different* from the self-concealment involved in the disclosive event called world. The self-concealment pertaining to the earth lies forever beyond the play of unconcealment-concealment that is constitutive of worldhood.

For the most part, Heidegger focused not on the self-concealment of the earth, but on the self-concealment of the presencing involved in worldhood.

He frequently cited Heraclitus's saying *physis kryptesthai philei*, "*physis* loves to conceal itself." Here, *physis* means the worldly appearing of entities, not their non-historical, spontaneous rising-into-presence. What conceals itself above all, then, is the being of entities, the event of their unconcealment and appearance. Heidegger also cited Parmenides' saying *gar auto noein estin to kai einai*: "there is a reciprocal relation between apprehending and being." Ordinarily, *einai* is translated as "being," but Heidegger interpreted it as meaning much the same as *physis*. Further, he interpreted *noein* not as "thinking" but instead as "apprehending" (*Vernehmen*). To apprehend means to receive, to let come to oneself that which presents itself and appears. *Noein* apprehends *physis*: there is a reciprocal relationship or bond between presencing and human existence. The former claims the latter as the site for its own occurrence: "Apprehension is not a mode of comportment which man has as an attribute, but rather the other way around: apprehension is the happening that has man." [GA, 40: 150/119] Human apprehension, *noein*, is inextricably related to *logos*. Usually translated by a variety of terms, including "language," "logic," "reason," and "rationality," *logos* is derived from the term *legein*, meaning to gather or to assemble. As we saw in chapter seven, the *logos* at work in human existence enables humanity to bring the presencing of entities to a stand, and thereby to open up a historical world in which entities can appear in a differentiated way. In 1935, Heidegger stated: "Standing and active in the *logos*, which is ingathering, man is the gatherer. He undertakes to administer and succeeds in administering the governing power of the over-powering." [GA, 40: 181/144] Elsewhere, he made clear that "man" does not undertake this task by and for himself, but instead is "appropriated" to become the site through which entities can disclose themselves.

B. AUTHENTIC PRODUCING AS *TECHNE* AND *POIESIS*

One of the most important ways in which such self-manifesting of entities can occur is by virtue of a work of art, the highest form of *techne*. *Techne* is often translated as both "art" (in the sense of "fine art") and "handicraft," but Heidegger insisted that Greek *techne* had "absolutely nothing to do with a 'primitive' understanding of the world in the horizon of skilled handiwork, in distinction to our supposedly higher mathematical-physical [horizon]." [GA, 33: 131] For Heidegger, moreover, the art work cannot be understood as arising from handicraft; instead, handicraft and equipment can attain their ontologically disclosive power only *within* the world opened up by the work of art. The primary meaning of *techne*, then, is "art," defined as the capacity for disclosing something, for bringing it forth, for letting it be seen. Hence, even modern *techne* or "technology" is a mode of artistic disclosure, though a highly constricted one.

To emphasize that genuine *techne* involves a disclosing that preserves and guards things, instead of exploiting and dominating them, Heidegger claimed to discern a link between *techne* and the word *techo*, that which involves procreating and engendering, as in "raising" children, "bringing them into the world." [GA, 55: 201] For the Greeks, *techne* meant both the event of bringing something into the open, and the know-how required for accomplishing that disclosure. Authentic producing, then, understood in terms of the Greek insight, involves disclosing something appropriately, letting it come forth into its own, bringing it into the arena of accessibility, letting it lie forth as something established stably for itself. To "pro-duce" something means to lead it forth (*pro-ducere*), to release it so that it can manifest itself and linger in presence in its own way.

Authentic producing, *techne*, then, is not a matter of an "agent" using "force" to push material together into a specific form. Rather, it is a disclosure of entities for their own sakes. This conception of *techne* is consistent with Heidegger's contention that the very being of human *Dasein* is "care." To exist authentically means to care for oneself, for others, and for things in an appropriate way. The destructive aspect of modern technology, then, is directly related to the constriction of human *Dasein*'s capacity for genuine caring. Alienation from this capacity was, for Heidegger, the worst kind of "dehumanization" and led to the most terrible crimes against humans and non-humans alike.

As we saw in chapter ten, the early Greek capacity for caring was already stamped by an instrumentalist attitude, so that for them to "free" something meant primarily to let it become "involved" as an instrument for human ends. [GA, 24: 162–163/115] Looking beyond such instrumentalism, however, Heidegger discerned what he regarded as a deeper meaning of producing hidden in the Greek term *techne*. The great work of art, especially poetry, is the *techne* which enables people to be at home with things, to understand in advance what things are, so that within this articulated and intelligible matrix of entities people can pro-duce things, bring them forth, let them be. The primordial disclosure of the world-founding work of art makes possible the productive disclosure of things *within* the world. Obviously, though so far as I know Heidegger himself never addressed this issue, people were producing things *before* the Greeks were granted their primal encounter with being. Such pre-Greek producing, however, was allegedly not comparable with the producing which occurred in the Greek world. For Greek craftsmen existed in the mysterious glow of the *presencing* or *being* of entities. Greek producing, then, even though quickly affected by the advent of productionist metaphysics, shared in an ontological disclosiveness that was unavailable to earlier people. In making something, in other words, the Greek artisan knew that he was *letting it be*. The decline of the West occurred as human *Dasein* lost touch with the awesomeness of the gift and responsibility of this ontologically disclosive capacity.

One might raise the following objection to Heidegger's notion that *techne* really means art as ontological disclosure. If in fact *techne* means "art," why is it that our own term "technology" has at best only an indirect connection with art? We speak of the "techniques" of an artist, for example, but we do not regard technology as an art, except perhaps in a derivative sense. For example, the word "artisan" contains the word "art." Understood in one way, an artisan is a skilled craftsman, sometimes a "technician," someone whose

abilities have reached such a point that his or her work is regarded as requiring talent analogous to that of a fine artist. The artisan, however, is not an
artist, because the artisan produces things for use, while the artist does not. In
Heidegger's view, of course, the most important figure in society is not the
artisan, no matter how many "useful" things he or she produces, but instead
the artist, for the latter founds the world in which the activity or pro-ducing,
letting things be, can take place.

Despite this distinction between art and artisan, however, Heidegger was
concerned to demonstrate the inner relationship between poetry or art, on the
one hand, and producing—especially handicraft—on the other. Indeed, his
engagement with National Socialism was an attempt to realize in the political
domain his own conviction about the need to reunite the artistic and productive realms. He hoped that producing and working would no longer be alienated from their ontological roots in the disclosive activity of poetizing, so that
"technology" would no longer be separated from "art." Of course, in many
respects Heidegger's vision of the unity of technology and art was a far cry
from what Lacoue-Labarthe has described as the "national aestheticism" of
the Nazis. Nevertheless, Heidegger—in a way analogous to, but incomparably
more sophisticated than, that of Hitler—called for a workers' state in which
social fragmentation, nihilism, and the evils of industrialism would all be overcome in the transformation of producing into a disclosive activity both akin to
and grounded in Hölderlin's poetry.

To stress the profound relation between poetizing and producing, between "art" and "technology," Heidegger said that *techne* is related both to
poiesis and to *episteme*. The term *poiesis* is ambiguous. It means both "poetry"
and "producing." The ambiguity of this term is addressed in Plato's *Symposium*, where we read: "so that the fabrications [*ergasiai*] by all the arts [*technai*]
are productions [*poieseis*] and all the fabricators [*demiourgoi*] of these are artisans [*poeietai*]."[5] The one who engages in world-founding *poiesis* is an artist;
the one who engages in producing things within the world is an artisan.
Notice, however, that both poetry and producing share this basic trait: both
are *modes of disclosing*. Poetry discloses the gods needed to order and found
the world; genuine producing discloses things respectfully, in accordance with
the vision of the poet.

Techne is also related to *episteme*. Both terms mean knowing something in
the sense of being completely at home with it, understanding it, being an
"expert" with it. Heidegger emphasized that *episteme* must not be translated
as "theoretical" knowing which is somehow opposed to the allegedly "practical" knowing of *techne*. We must not become stuck in the old distinction
between theory and practice. Heidegger once remarked, "What the Greek
thinker already knew, Goethe grasped in the statement: 'The highest would
be: to conceive that everything factical is already theory.' " [ZSB: 328]
Whereas early Heidegger emphasized the priority of practice over theory, here
we are told that the practical is already the theoretical, i.e., that the practical
presupposes the disclosure of what things are.

Heidegger, following Aristotle, did indicate that while both *techne* and *episteme* are modes of knowing, *techne* involves a particular kind of knowing. *Techne* is a dimension of truth (*aletheia*): unconcealment, unhiddenness, disclosedness. *Techne* reveals in advance what does not yet show itself and thus does not yet stand forth:

> Whoever builds a house or a ship or forges a sacrificial chalice reveals what is to be brought forth. . . . This revealing gathers together in advance the aspect and the matter of ship or house, with a view to the finished thing envisioned as completed, and from this gathering determines the manner of its construction. Thus what is decisive about *techne does not lie at all in making and manipulating or in the using of means, but rather in the aforementioned revealing* [*Entbergen*]. It is as revealing, and not as manufacturing [*Verfertigen*], that *techne* is a bringing-forth [*Her-vor-bringen*]. [VA I: 13/13]

Techne as ontological revealing makes possible "production" as we ordinarily understand it. Only because artisans are capable of understanding and disclosing in advance what the envisioned product *is*, can they do the things needed to gather the thing together, draw it forth, let it be. Heidegger clarified the disclosive character of *techne* with the example of the process involved in making a chalice. To understand the point of that example, we must first consider how we ordinarily conceive of "making." Usually, we think of "making" as a type of causal activity, for example, "causing" matter to assume a new shape or to stand in a new relation to other matter. The term "cause" is a Latin translation of the Greek term *aition*. Aristotle provided the most famous philosophical analysis of "cause" in his study of the "four causes": formal, efficient, material, and final. Interpreting these four causes in the usual way, we arrive at the following account of the making of a chalice. The silversmith makes a chalice (final cause) by working up (efficient cause) silver (material cause) according to a pattern or model (formal cause). Notice that the efficient cause or "agency" of the silversmith is the crucial ingredient in this account. Efficient cause, especially since the rise of modern science, has displaced the other three kinds of cause. "Final cause," for example, the purpose of a thing, was rejected by scientists as an unnecessary and misleading factor in explaining the workings of the universe. Efficient causation became the dominant way of explaining how things come "to be."

Heidegger rendered Aristotle's term *aition* not as "cause" but instead as "that to which something is indebted." In addition, instead of speaking about "causing" something to happen, Heidegger used the term "occasioning." He then showed how the four causes, now understood as the "four occasionings of indebtedness," may be used to describe the production of a silver chalice. The chalice is "indebted" for its coming-into-being to each of the four "occasionings." [VA I: 8–9/7] First of all, the chalice is indebted to the silver for the matter (*hyle*) which constitutes it, and to the aspect (*eidos*) or form which gives it shape. Above all, however, it is indebted to the *final* cause, the *telos* of the chalice as a sacrificial vessel. *Telos* means the defining boundaries and limits within which a thing can gather itself together and come to a stand. The silversmith is also responsible for the chalice, but not as an "efficient cause," which corresponds (in Heidegger's view) to no Greek idea at all. Rather, the silversmith

> considers carefully and gathers together the three aforementioned ways of being responsible and indebted. To consider carefully [*überlegen*] is in Greek *legein, logos*. *Legen* is rooted in *apophainesthai*, to bring forward into appearance. The silver-smith is co-responsible as that whence the sacrificial vessel's bringing forth and resting-in-itself take and retain their first departure. The three previously mentioned ways of being responsible owe thanks to the pondering of the silversmith for the "that" and the "how" of their coming into appearance and into play for the production of the sacrificial vessel. [VA I: 9/8]

These four ways of being responsible bring the thing into appearance, they let it come forth into presencing (*An-wesen*). "They set it free to that place and so set it on its way, namely, into its complete arrival. . . . It is in the sense of such a starting something on its way [*anlassen*] into arrival that being re-sponsible is an occasioning to go forward [*Ver-an-lassen*]." [VA I: 10/9] The mystery is how these four "occasionings" are gathered together in such a way that the artifact is "freed" to become itself. Heidegger suggested that *logos* played this gathering role, but *logos* must be understood not as an overarching causal agent but instead as somehow analogous to the *Tao*, that "pathless way" which makes it possible for entities both to gather themselves into a stable presence (as in the case of living things) and to be gathered into presence through an artisan attuned to *logos*. [US: 198/92]

We may summarize Heidegger's concept of *techne*, or authentic produc-ing, as "the disclosive occasioning that makes presencing and bringing-forth possible." These two aspects of *techne*, presencing and bringing-forth, corre-spond to the dual nature of *poiesis* as art and producing. *Poiesis* belongs not only to the work of art and to producing, however, but also to *physis*:

> Not only handicraft manufacture, not only artistic and poetical bringing into ap-pearance and concrete imagery, is a bringing-forth, *poiesis*. *Physis*, also, the arising of something from out of itself, is a bringing-forth, *poiesis*. *Physis* is indeed *poeisis* in the highest sense. For what presences by means of *physis* has the bursting open belonging to bringing-forth, i.e., the bursting of a blossom into bloom, in itself (*en heuatoi*). In contrast, what is brought forth by the artisan or the artist, e.g., the silver chalice, has the bursting open belonging to bringing-forth not in itself, but in another (*en alloi*), in the craftsman or artist. [VA I: 11/10–11]

In this discussion of *physis* as the highest instance of *poiesis*, we should keep in mind the conflict discussed in the first part of this chapter, namely, the conflict which obtains between *physis* understood as the event of dis-closedness and *physis* understood as the event of self-emergence. These two dimensions of *physis* seem to correspond to the two aspects of *poiesis*, art (disclosedness) and production (bringing-forth). In the quoted passage, *physis* seems to mean self-emergent coming-forth, as when a plant "produces" itself by coming into bloom. Because the plant produces itself, it is a higher mode of *poiesis* than an artifact, which is produced through another. But what relation is there between the self-producing dimension of the plant (*poiesis* in one sense) and the presencing or disclosedness of the plant (*poiesis* in the other sense)? In his various accounts of *physis*, Heidegger often spoke of "presencing" and "bringing-forth" in the same breath, thereby tending to blur the distinction we mentioned earlier between self-generating or self-producing, on the one hand, and presencing or appearing, on the other. Moreover, he frequently emphasized that the latter was more primordial than the former. By doing so, however, he seems to neglect the earthly dimension of living things, i.e., the fact that their "self-producing bringing-forth" cannot be reduced to the "presencing" that occurs within any historical world. It is true that a living thing which has brought itself forth (one sense of *poiesis*) could never *appear* or *be* outside the world opened up by the work of art (the other sense of *poiesis*). World-founding poetry (*techne*, *poiesis*) does not create but only discloses living things.

To these objections, Heidegger might reply in the following way. It is wrongheaded to separate earth and world, as if one could speak about earthly things (such as animals or plants) wholly apart from any consideration about their relation to a historical world. Earth and world are different and irreducible, but they are internally related, always contending with one another. Rightly understood, *physis* names both the earthly and the worldly dimension of things. *Physis* names the self-generating bringing-forth of living things, but also names the presencing by virtue of which such things come into appearance within a world. *Physis*, then, is not only generative but also disclosive. *Physis* brings forth the humans necessary to disclose what *physis* brings forth. The self-producing dimension of *physis* at work in living things resists the intrusions of the disclosive dimension of *physis* at work through human existence. The producing done by humans is radically dependent on both aspects of *physis*: the self-producing aspect involved in the emergence of things of the earth, on the one hand, and the disclosive aspect involved in letting those things appear with a world. The name for *physis* in human existence is *poiesis*: the disclosiveness (art in its broadest sense) which makes bringing-forth (producing of all kinds) possible.

Techne as emergent artwork and *physis* as self-emergent entity are similar in that both works of art and things are ultimately purposeless, despite the fact that paintings of the great masters are now worth millions of dollars, and despite the fact that trees and animals are also highly valuable in the marketplace. Their monetary value and utility, however, are incidental to their being. Neither a poem nor a rose has any "reason" for being. Commercial rationality wants to discover the utility of all things, while scientific rationality wants to adduce a reason for them. Recalling the mystical poet Angelius Silesius, however, Heidegger remarked that the rose is "without why." It blooms because it blooms.[6] It has no "purpose" apart from one arbitrarily assigned to it. By way of contrast, Heidegger claimed that the very being of equipment is constituted by its usefulness. In this respect, equipment is radically different from works of art and "natural" things.[7] Treating works of art and living things as commodities betrays a basic misunderstanding of what they are. If the work becomes a commodity on the art market, for example, it cannot "be" what it is; it cannot do its "work." The great work of art makes no practical contribution to a given world, but instead opens the way to a new one. In this respect, all great art is truly revolutionary.

Neither works of art nor natural things require any metaphysical "ground" on which to stand. The art work is not "based" on something external to it, such as a Platonic form; rather, emerging from the abyss (*Abgrund*) of "holy wildness," it provides the grounds and limits for things *within* the world it founds. Even in opening up a new world, the work of art does not serve a "purpose"—as if inaugurating worlds were the "ultimate purpose" of "reality." Moreover, living things are not "founded" either on the will of the Creator or on the principle of sufficient reason; they are because they are. According to Heidegger, there is no "reason" why there is something rather than nothing. Worlds world; entities appear. Ultimately, that is all he could say about the *fact* that things are.

Technological humanity, however, conceives of everything in terms of purposive or instrumental rationality. Talk of producing as a "freeing" or as a "letting-be" makes no sense. Instead, to make something means to *cause* it to happen, to *will* that it come forth, in order to serve some purpose within the technological system. The all-embracing purposiveness governing the technological system, however, conceals a deeper purposelessness in at least two respects. First, at least in Heidegger's view, the technological system has left behind the purposiveness characteristic of merely *human* projects. As the manifestation of the Will to Will, the technological system strives for ever greater production for its own sake. Ultimately, then, modern technology is purposeless, which is why critics such as Marcuse have accused it of being "irrational." Second, Heidegger argued that the purposelessness of modern technology is not "irrational" but instead a-rational, i.e., the technological world is but another historical world, none of which have any ultimate "purpose," none of which point beyond themselves to some ultimate "ground" or "foundation." The distinction between rational and irrational can arise only by virtue of the ontological disclosure of entities within a historical world. Worlds are not themselves grounded on anything. There is no "reason" why things "are" at all. To overcome our terror and awe in the face of the sheer presencing of things, we make up stories, myths, fables, and scientific theories to "explain" what and why and how things are. Productionist metaphysics constitutes the "story" told by Western man to explain—and thereby to control—the cosmos.

Heidegger maintained that even though modern technology was so very dangerous, it was also—paradoxically—the way beyond that danger. If we could take the "step back" from the constant purposiveness demanded by modern technology, we could suddenly encounter the purposelessness of it, i.e., we could discover that the technological world is analogous to a work of art in the sense of being a disclosure of entities simply for the sake of that disclosure, not for any "higher" or "ulterior" purpose. Heidegger owed this insight, at least in part, to Jünger, as well as to Nietzsche and Hölderlin. Discovering the purposeless and "artistic" character of modern technology is

tantamount to revealing that it has no rational basis, no foundation, no ultimate "justification," such as the "iron laws of social evolution." Heidegger
maintained that, in fact, the quest for the metaphysical foundation necessary
to justify taking total control over things has led not to control but instead to
the threat of complete destruction. The point of Heidegger's meditation on the
nature of modern technology is to reveal that it has no justification. It is only
one more historical world that is "without why," like the rose and like the
work of art. In other words, because there is no rational basis for the technological way of life, things *could be otherwise*. Discovering the groundlessness of
the technological era makes possible the openness—and the anxiety—necessary for the arrival of a new, post-modern era.

C. THE ART WORK, THE THING, AND AUTHENTIC PRODUCING

Heidegger made it very clear that there was nothing people could "do" to
inaugurate a new age. In fact, more "doing" would only exacerbate the grip of
the technological system, since "doing" for the sake of controlling is central
to productionist metaphysics. Are there not, however, non-controlling ways of
behaving which might help to elicit the change in attitude necessary for the
advent of the new gods necessary for a post-technological world? In chapter
eight, we saw that Heidegger urged people to allow Hölderlin's poetry to elicit
the mood necessary for historical transformation. Postwar Heidegger continued to believe that meditation on the nature of the work of art was an appropriate way of "stepping back" from the technological impulse and thereby
becoming prepared for the dawn of something new.

He warned, however, that works of art can scarcely "work" in the technological era. In a late essay, "Art and Space" (1969), he described how the
technological alteration of space demanded had changed the character of
works of art. Ordinarily, thinking in terms of the scientific conception of
homogeneous, isotropic space, we say that a work of art "occupies" space in a
way analogous to how a power plant "occupies" space. Heidegger insisted,
however, that in addition to neutral scientific space which can be seized for
some purpose, there are at least two other kinds of space: first, the space of
everyday activity; second, the space of the work of art. Indeed, he contended
that neutral, profane space is itself derivative from the "place" (*Ort*) opened
up by the work of art. [AED: 208] Instead of "occupying" a pre-given space,
then, the work of art "embodies" (*verkörpert*) a place and opens up the arena
in which entities can encounter each other. Does the place embodied by the
work come before the clearing away (*Einraumen*) constituted by the work?
The question admits of no easy answer. We can say, however, that a work of
art embodies a place within a pre-existing historical world, while simultaneously clearing a new place for itself out of which a new world can unfold.

Kathleen Wright has elaborated upon the theme that in the technological
age the work of art embodies a neutral "space," hardly a "place" at all.[8]
Nevertheless, she argues, even in such a placeless space the work of art can
reveal something crucial about the being of entities. Consider once again van

Gogh's portrait of the peasant's shoes. The painted shoes stand alone in an empty space, surrounded by a picture frame, hung on a museum wall. According to Heidegger, however, if we allow ourselves to let the painting do its "work," we suddenly discover that from out of the painting there pours the "place" of the peasant woman: her fields and hills, her cares and joys, her toil and rest. Such a revelation discloses that the undifferentiated space in which the museum stands as a segment of the culture industry is not primal or "real" space in which the peasant woman also stands, but instead the undifferentiated technological space is a denuded, impoverished version of the lived place disclosed by works of art.

In a dialogue composed in 1944–45, "Toward the Situating [*Erörterung*] of *Gelassenheit*," Heidegger indicated that we must no longer conceive of space in terms of Kant's transcendental philosophy, namely, as the spatial horizon—projected by the subject—in which objects can appear. Instead, we must learn to see that human understanding is itself "appropriated" (*ereignet*) by a "region" (*Gegend*) that transcends the merely human. [G: 29–71/58–90] Human openness arises within and from a greater, non-human "opening" or "region." This "region" makes possible a non-anthropocentric clearing in which entities may gather each other into a mutual play that constitutes the "worldhood" of the world.

Heidegger regarded most twentieth-century art as cut off from this transcendent region. Hence, such art is trapped within a post-romantic, expressivistic, and hence anthropocentric and subjectivistic conception of art. Abstract expressionism, he noted, is consistent with the character of the technological age: "That in such an age art becomes object-less testifies to its historical legitimacy, and this above all when the object-less art conceives of itself such that its productions can no longer be works, but instead something for which the suitable word is still lacking." [SG: 66]

Far from being a "work" which helps to found a new world, modern art is not in any sense a "work" at all. While in the 1930s Heidegger had hoped that the work of art would show the way beyond the technological age, in 1966 he stated, "I do not see how modern art shows the way, especially since we are left in the dark as to how modern art perceives or tries to perceive what is most proper to art." [Sp: 219/283] For one thing, great art had been undermined by subjectivism, commercialism, and other aspects of the technological epoch. In a way, however, the condition of the work of art in the technological age still reveals something about that age: namely, that there are no things or even objects left. The modern work of art reveals that everything has been reduced to undifferentiable standing-reserve. What the Nazis had described as "degenerate art"(*entartete Kunst*), then, could nevertheless play a significant role in disclosing the nihilism of the present age.

D. RILKE, CÉZANNE, AND THE DISCLOSURE OF THE "THING"

Later Heidegger continued to look for those works of art which could help disclose the being of entities, i.e., which could reveal things *as things*, in their individuality, their depth, their being—other than as commodities. Through his reading of Rilke's letters, he became fascinated with Cézanne's still lifes and his late landscape paintings.[9] In the autumn of 1907, during his sojourn in France, Rilke spent many hours at a retrospective exhibition of the paintings of Cézanne, who had died the previous year. Gazing upon Cézanne's extraordinary paintings of ordinary things, Rilke came to understand the vocation of an artist: to let things show themselves as what they are, apart from our instrumental projections. Cézanne himself described his work as an attempt to bring about the "*réalisation*" of the things he painted. Rilke translated this term as *Dingwerdung*, "thing-becoming," the process by which the thing was allowed to become the thing it was.[10] Cézanne was not a representational painter, in Heidegger's view. Had he been so, perhaps he would have become an abstract expressionist or a cubist, in order to represent the being of the thing in the industrial age. By transforming apples on the table into "art apples," Cézanne was attempting both to free them from their status as commodities and to free the viewer from the instrumental way of perceiving things.[11] Considering Cézanne's paintings of pieces of fruit, Rilke observed: "In Cézanne, they cease to be edible altogether, that's how thinglike and real [*dinghaft wirklich*] they become, how simply unconsumable in their obstinate presence."[12] Like Cézanne, Rilke believed that the task of the artist was *to let the thing be*:

> First, artistic perception [has] to overcome itself to the point of realizing that even something horrible, something that seems no more than disgusting *is*, and shares the truth of its being with everything else that exists. Just as the creative artist is not allowed to choose, neither is he permitted to turn his back on anything: a single refusal, and he is cast out of the state of grace and becomes sinful all the way through.[13]

For Rilke, Cézanne (and van Gogh) had moved far beyond the conception of art as the depiction of the "beautiful." Such great painters sought to disclose the sheer *isness* of things, in such a way that everyday life itself might be transformed by a new mode of ontological perception. Unlike aesthetes, such as Jünger, who were attracted to the sensational and mind-altering effect of the horrifying almost as an end in itself, Rilke believed that the artist was obligated to disclose the horrifying and the ugly simply because they, too, are things to which one ought to bear witness. In lectures from 1927, Heidegger cited a long description of the wall of a demolished house from Rilke's *The Notebooks of Malte Laurids Brigge*. Rilke's words, Heidegger said at the time, make possible the "elementary emergence" of human existence and its world. [GA, 24: 244–247/171–173] Rilke's depiction is possible only as an articulation of what is already "actual" (*wirklich*) in the wall, "which leaps forth from it in our natural comportmental relationship to it." [Ibid.] To some extent, later Heidegger found this account of the nature of art—as disclosing things in terms of their relation to human existence—too anthropocentric. Hence, his later interest in Cézanne's paintings, which sought to reveal things apart from everyday concerns which were increasingly instrumental and commercial.

Despite his belief that Rilke's poetry was caught within the metaphysical tradition, Heidegger was influenced by Rilke's meditations on the nature of the "thing." For example, Heidegger's essays "The Thing" (1951) and "Building Dwelling Thinking " (1952) explore the possibility for a non-subjectivistic, non-anthropocentric encounter with things. He argued that the abstract "space" of the subject must give way in favor of the concrete "place" of things." By "things," however, he meant not only the pre-eminent thing, the work of art, but also everyday things, such as bench, bridge, plow, pond, tree, hill, heron, deer, bull, mirror, book, crown, and cross. [VA II: 55/182] Virtually anything, he maintained, can play the role of opening up the place in which things can encounter each other. A footbridge, for example, is not set into a pre-existing space, but rather helps to gather into a shared place the constituents of the fourfold: earth and sky, gods and mortals. [VA II: 27–29/ 152–154]

For another example of an everyday thing which gathers together the elements of a world, Heidegger turned to the jug. The jug "works" by holding open a void in which wine may be held. For Heidegger, of course, this "void" is not an empty nothingness but instead the self-concealing source in terms of which all things appear. Meditating on the curious character of the jug discloses that the jug somehow draws upon the same "nothingness" which constitutes the temporal-linguistic transcendence of human existence. The jug can play the ontologically disclosive role once assigned almost exclusively to the work of art. As the bearer of wine, the jug gathers the elements of the world— earth and sky, gods and mortals. The jug reveals that the wine comes from grapes nourished by the warmth of the sun, and that the wine enlivens the mortals who imbibe it and provides a libation for the gods. The jug allows earth and sky, gods and mortals to come near to each other, to manifest themselves to each other in a mirroring play which lets each come into its own. [VA II: 51–52/179] Rational cognition, Heidegger maintained, is simply incapable of understanding the appropriating event (*Ereignis*) which makes possible the "worlding" of the world constituted by the interplay of the fourfold. [Ibid.] The mutual appropriation of earth and sky, gods and mortals both takes place within and gives rise to the "region" which constitutes a particular world. Heidegger used the term "dwelling" (*wohnen*) to describe the mode of existing involved in the mutual play in which everything is allowed to show itself appropriately.

Heidegger's account of the world as fourfold was borrowed from Hölderlin, but also may have been indebted to Rilke's reflections on Cézanne's portrait of Madame Cézanne. Describing Cézanne's astonishing sense of color, Rilke remarked:

> In the brightness of the face, the proximity of all these colors has been exploited
> for a simple modeling of form and features: even the brown of the hair roundly
> pinned up above the temples and the smooth brown in the eyes has to express

itself against its surroundings. *It's as if every place were aware of all the other places*—it participates that much; that much adjustment and rejection is happening; that's how each daub plays its part in maintaining equilibrium and in producing: *just as the whole picture finally keeps reality in equilibrium.*[14]

The self-organizing, internally gathering dance of colors which Rilke perceived in Cézanne's paintings would appear to be a basic analogue to the mirror play of the elements of Heidegger's fourfold. One critic has remarked that Rilke conceived of the work of art itself as a symbolic *thing* "which has as its aim the presentation of a work of art, autonomous, consistent within itself and no longer subject to the laws of reality."[15] For Heidegger, the art-thing did not symbolize anything, but instead disclosed the elements in a world which could not be comprehended in terms of the categories applying to technological "reality." The work of art, according to Heidegger and Rilke, transforms our ontological perceptiveness: it enables us to enter into a new relationship with things. The work of art reveals to us what things *already are*, but which we have been too blind to see.

Heidegger's account of the world as fourfold differs in many ways from his early account of world as referential totality of human *Dasein*. In world as fourfold, ordinary things (both natural and artifactual) provide the focal point around which the world "worlds." *Dasein*, instead of being like the subject around which everything revolves, becomes instead a vital element within a larger context that in some sense gives rise to its own turning. As the entity endowed with language, *Dasein*'s role is to be the medium through which there may occur the "naming" needed for worldhood. The poet's saying "entrusts world to the things and simultaneously keeps the things in the splendor of the world. It grants [*gönnt*] to things their presence. Things bear world. World grants things." [US: 24/201–202] While still insisting that only linguistic "mortals" attain to the world *as world*, Heidegger nevertheless conceded that the world is sustained only in the context of a dance with non-human partners. Only in surrendering to or complying with the spontaneous movements of those partners, it would appear, is anything like an authentic world possible. Hence, later Heidegger stressed the importance of paying attention to "little things" in our everyday practices in order to counter the homogenizing, technological glance which compels things to conform to our expectations.

One reason he kept returning to consider the nature of handiwork was that it involved micro-practices that were not totally obliterated by the technological understanding of being. The master craftworker, in contrast to the factory worker, at least has the opportunity to respect the material with which he or she deals. If there are craftworkers who have not been "stamped" by modern technology, they provide the limits which enable both the artifact and its material to manifest themselves, to come forth appropriately. As we saw in our consideration of the chalice, we must not conceive of an artifact in terms suitable only for productionist metaphysics, i.e., as something composed of malleable "stuff" which has been "shaped" by the act of an agent. For Heidegger, authentic producing requires behavior consistent with letting a thing present *itself*. The material from which a thing is made—rock, wood, leather—can be allowed to manifest itself in a way appropriate to that material, or it can be dominated, treated indiscriminately by the industrial "worker." A great woodworker spends a long time with wood; he or she takes grain, texture, color, density into account so that they may provide outstanding "occasions" for the chair which is in the process of "arriving." The assembly-line worker, by contrast, who exists in a world where efficiency is the governing determinate for producing, must ignore the qualities of material and must treat everything as interchangeable parts.

We might ask how, in a technological world, great artisans can continue to exist. Certainly, their numbers have dwindled. The degradation of work in the twentieth century has reduced the time and skill needed for authentic craftwork, except for those who "drop out" of the social mainstream in order to pursue what they consider to be authentic producing. Nevertheless, great craftworkers remain. Perhaps the attraction such craftspeople have for us today lies in our awareness that they are attuned to things in a way in which most of us are not. Consider, however, the admiration many people display for the intricate circuitry of a computer or the engine of a Mercedes-Benz. We often express amazement at the precision and beauty of such products. For the most part, they have been produced in gigantic factories in which much of the work is done by robots attended by human personnel.[16] Yet are not robots and computers themselves human inventions? Even graphic artists who use computer imagery remind us that the artist is still responsible for the images executed by the computer program. Technological products, too, are in some sense human crafts—but not handicrafts. Because Heidegger believed that the human hand was essentially linked to our awareness of being, he maintained that authentic "producing" had to involve work with the hands. Skilled hands "know" the materials with which they work. But those hands have always used tools. At what point does a tool escape the play of the hand and become master of it? Is it then that the potential for truly human acting and producing vanishes? If so, is that why Heidegger regarded the technological world as an "un-world," a placeless place? Heidegger believed that technological humanity no longer "dwells" upon the earth, but instead regards it as Le Corbusier suggested that we regard a house: as a machine to live in.

E. HEIDEGGER AND "DEEP ECOLOGY"

Heidegger's criticism of the tendency of modern technology to treat the earth as a machine or as raw material for exploitation has led some people to interpret his thinking as being consistent with contemporary environmentalism. In this section, I want to examine briefly the extent to which Heidegger's thought may be "applied" to the environmental movement called "deep ecology."

The consequences of treating the earth as a machine, or as an infinitely exploitable commodity, have become increasingly clear in the form of acid rain, polluted rivers and lakes, the "greenhouse effect," deforestation, desertification, soil erosion, thinning of the ozone layer, mass extinction of species, and a myriad of other environmental problems. Environmental reformers want to curb the destructive practices of industrialism in order to prevent the collapse of the biosphere and, thereby, to save humanity from a form of suicide. Those who call themselves "deep ecologists," however, maintain that reformism will succeed only in postponing the inevitable catastrophe.[17] Deep ecology is "deep" in that it asks deeper questions about the origins of our ecological crisis. Deep ecologists agree that the real explanation for the environmental crisis is the radically anthropocentric character of Western culture. Such anthropocentrism stems from at least two sources, including Christianity and Greek philosophy.

In a famous essay, Lynn White, Jr. called Christianity "the most anthropocentric religion the world has known" because it overemphasizes the importance of humans in creation, while downplaying the glory of the rest of the creation.[18] Important strands of Greek philosophy, especially its Stoic offshoots, also promoted the idea that humans are at the top of the "great chain of being" and that therefore all other things are for man's use.[19] The integration of Christian theology and Greek metaphysics helped to make possible the scientific revolution. Centuries later, Francis Bacon championed the view that scientific knowledge made it possible for man to gain power over nature.[20] Descartes's division of reality into thinking minds and extended matter promoted the dualism which stripped the natural world of beauty and intrinsic value. Later, the Enlightenment proclaimed that rationality would make possible a utopian life for humanity, by making possible not only the conquest of nature but also the reorganization of human society. The application of scientific findings to industrial processes enabled Enlightenment ideologies—including capitalism and later socialism—to initiate the extraordinary global industrialization process that has wreaked such havoc on the natural environment, as well as on many aspects of human society.

Hence, for deep ecologists the real roots of our ecological crisis lie in the anthropocentric humanism which portrays all non-human entities as raw materials that are useful in enhancing human projects. Deep ecologists argue that only if humanity overcomes its anthropocentric bias and moves toward a position of biocentric egalitarianism will it be possible to avoid environmental catastrophe. Only by learning to love and to respect *all* things, not only people, will humanity develop the alternative technologies and the simpler ways of life necessary to preserve the planet from the dire consequences of unchecked population growth and rampant industrialism, not to mention the nuclear holocaust that may result when the two superpowers go to war over whose version of "humanism"—Marxism or capitalism—will prevail.

In light of our examination of Heidegger's critique of the anthropocentric humanism so central to modern technology, the reader will readily notice the significant similarities between the deep ecological and the Heideggerean accounts of Western history. Inspired by such similarities, a French environmental theorist has asked: "In the final accounting, will Heidegger have been the first theoretician in the ecological struggle?"[21] As I have argued in detail elsewhere, Heidegger's critique of anthropocentric humanism, his call for humanity to learn to "let things be," his notion that humanity is involved in a "play" or "dance" with earth, sky, and gods, his meditation on the possibility of an authentic mode of "dwelling" on the earth, his complaint that industrial technology is laying waste to the earth, his emphasis on the importance of local place and "homeland," his claim that humanity should guard and preserve things, instead of dominating them—all these aspects of Heidegger's thought help to support the claim that he is a major deep ecological theorist.[22]

While it is tempting to "apply" Heidegger's thought in this way, there are several problems which should give deep ecologists pause before they adopt Heidegger as one of their own. These problems include (1) residual anthropocentrism, (2) the reactionary dimension to his critique of industrialism and modernity, and (3) his antipathy toward science. Earlier in this essay, we noted that despite his critique of anthropocentric humanism, Heidegger remained part of the Western tradition insofar as he regarded humanity as radically other than all other things. Early Heidegger in particular emphasized that human *Dasein* is essentially different from all other things. Not only different but also more important: other things can manifest themselves and thus "be" only insofar as human *Dasein* constitutes the clearing in which that manifesting can occur. Later Heidegger tried to temper this anthropocentrism, but not by integrating humanity into what deep ecologists regard as the "seamless web" of life. Recall that Heidegger was a severe critic of all "naturalism," of which he might have regarded deep ecology as a spiritually elevated but nevertheless misguided example. Instead of overcoming anthropocentrism by integrating humanity into the natural nexus, Heidegger overcame it by saying that humanity is not autonomous and central, but instead exists in the service of what transcends all entities: being as such.

Second, deep ecologists must examine seriously the implications of Heidegger's involvement with National Socialism.[23] His willingness to support an authoritarian regime to "solve" the problems posed by modernity and industrialism, the ease with which he abandoned the principles of respect for the rights of others, his talk about a mystical "union" between *Volk* and earth, and his hierarchical views about those "gifted" with insight about the meaning of history—all this must give pause to those deep ecologists, most of whom recognize that authoritarianism, hierarchism, and communitarianism without respect for individual freedom are by no means "solutions" to the environmental crisis, but instead are important causes for it. Heidegger may be defended from the charge of racism, for he rejected Nazi prattle about "blood ties" with the land. But this defense is not of much help for those concerned to use Heidegger in support of deep ecology. Deep ecologists want to be able to speak about the organic relatedness of all life on earth without being accused of reverting to fascist mythologizing.[24] Heidegger "solved" this problem by redefining "rootedness in the earth" in a way that virtually eliminated the organic dimension of this notion. But deep ecologists rightly hesitate to adopt this solution, since they have a more positive, less suspicious view of the organic domain than did Heidegger, who was such an enemy of all "naturalism."

Heidegger's suspicion of naturalism leads us to the third problem regarding the Heidegger–deep ecology relation. Deep ecologists portray humanity as a highly intelligent animal which has arisen by virtue of billions of years of terrestrial evolution. According to deep ecologists, only by understanding the extent to which human life is inextricably involved with other forms of life, and indeed with virtually all the processes—organic and inorganic—taking place on earth, can humanity understand that by destroying the planet, humanity is destroying itself. The science of ecology, despite being to some extent implicated in the scientific view of all things as mere "objects," is regarded by deep ecologists as a basic source of information concerning how to structure an appropriate mode of human "dwelling" on earth. Attempts by humans to deny their animal origins and their dependence on the rest of life have led, in the view of deep ecologists, precisely to that arrogant anthropocentric humanism which treats all non-human life merely as a commodity for human ends. Insofar as Heidegger refused to take seriously the organic dimension of human existence, he may well be accused of having remained in a curious way tied to the human-centered, dualistic metaphysical tradition of which he was so critical.

Keeping in mind these significant caveats, I believe that Heidegger's writings offer much food for thought regarding the environmental crisis. Whether we finally accept or reject his analyses of Western history and his alternative to it, the process of arriving at such an evaluation will have forced us to examine critically the undeniably anthropocentric attitude of much of Western history and culture. Perhaps elements of Heidegger's notion of a "higher humanism" can contribute to the quest by deep ecologists to define humanity anew. The key here may be Heidegger's conception of human existence as "care." Arne Naess, one of the leading figures in deep ecology, seems to echo this conception when he says of humanity: "Up to this point, we know of no other life form in the universe whose nature is such that, under favorable circumstances, it more or less inevitably would develop a broad and deep concern for life conditions in general."[25] For both Naess and Heidegger, humanity becomes what it truly is only when it learns to let other entities "be" what they are. In such caring, humanity becomes what it most authentically can be.

F. ON LETTING THE HUMAN BODY "BE"

Heidegger equated "letting things be" with the act of "producing" things. To produce something in the right way means to disclose it in a way that shows concern for the thing's own character and possibility. Heidegger emphasized the role of the human hand in such authentic "letting-be." Although he insisted that the human hand was radically other than the ape's hand, because humans are open to the being of entities while apes are not, Heidegger's focus on the disclosive and caring dimension of the human hand indicates that he understood the importance of "embodiment" for human existence. Unfortunately, his discussions of embodiment are limited, perhaps because of his uncertainty about how to define the body without lapsing into naturalistic categories. A more extended analysis of such embodiment would certainly help to make clearer the relevance of Heidegger's thought for offering an alternative to the technological way of understanding things, an alternative that may be consistent with important principles of deep ecology.

Fortunately, David Michael Levin has taken important steps toward such an analysis. He argues persuasively that only by exploring how modern technology has transformed and perverted bodily experience can we move in the direction of the "new beginning" for which Heidegger longed. In two remarkable books, *The Body's Recollection of Being* and *The Opening of Vision*, Levin draws upon thinkers such as Maurice Merleau-Ponty, C. G. Jung, Medard Boss, and Eugene Gendlin to show how in the technological age we have become cut off from the wisdom of the body.[26] When Heidegger spoke of the "rootlessness" of modernity, he should have emphasized that it is from our own corporeality that we have become uprooted and alienated. Developing Heidegger's hints, especially in light of Merleau-Ponty's work, Levin shows that bodily motility, gestures, and feeling are crucial factors in opening up the "clearing" of human existence. Humanity may contribute to the "gathering" of entities in the fourfold primarily through the language of bodily movement of all sorts, including dance, the basis for all religious ritual and mythology. Emphasizing the deep relationship between human embodiment and experience of the divine, Levin sides with Nietzsche against those patriarchal religious traditions which castigate the body. Only the ontological wisdom of the lived body may bring us out of the abstract, dualistic egocentrism that produces so much suffering and destruction.

Attempts to conceive of Heidegger as the ecological thinker of our age must recall the fact that our primary encounter with "nature"—with spontaneous arising into presence—takes place through our own bodily awareness. Paul Shepard, Norman O. Brown, Herbert Marcuse, Michel Foucault, Max Horkheimer, T. W. Adorno, Rosemary Radford Ruether, and a host of other twentieth-century writers have argued that the so-called domination of external nature is an outgrowth of the domination, discipline, and revilement of our own bodily existence. Concern about the "exploitation" of nature through industrial technology may be regarded as displaced concern about the exploitation and repression of our own feelings, desires, wishes, moods, and fantasies. Such repression manifests itself in many different ways. To take only one example: *anorexia nervosa*. Surely there can be no better symbol of the nature of the modern age than young women who starve themselves in order to assert absolute control over their own bodies.

One may argue that such bizarre attitudes on the part of young women toward their bodies are the consequence of a process that began long before the Greeks: the process of the denial of the body, which Plato spoke of as a "prison" for the soul and which became a symbol for human finitude and dependence. Western man came to identify the body with the female and the natural, i.e., with what is "lower" than the soul. In denying his finitude, patriarchal man attempted to make himself into a god: pure, immortal intellect. In the technological age, man seeks to gain immortality by conquering the body, the female, the whole of "inferior" nature. Hence, there arises the illusion of a purely ideational existence in which the body becomes something to be pushed around like a box of corn flakes. In rejecting and starving her own body, the young woman has internalized patriarchal contempt for the female and for the body. She will kill what she finds so loathsome, in order thereby to attain the control promised at the level of pure "spirit."

While Heidegger may have devoted insufficient attention to the phenomenon of the body, not to mention patriarchalism, he certainly emphasized the importance of recognizing human finitude. Like Nietzsche, he believed that the technological destruction of the earth was an expression of the Will to Power's hatred of everything that reminded it of its finitude, limitation, and time. Taking revenge on the earth meant subjugating and dominating it, instead of preserving and caring for it. The highest revenge, in Nietzsche's view, however, was positing that the merely finite, temporal world was "non-being" when compared with the eternal, unchanging realm. [VA I: 109/423] To become the Overman, to relinquish the spirit of revenge, Nietzsche maintained, man as he is now must affirm the most horrifying thought: the eternal recurrence of the merely transient and temporal.

For Heidegger, however, the transformation of the temporal into the eternal, even if for the sake of affirming the temporal, was nevertheless a metaphysical deed. To move beyond the metaphysical history of the West would require that there arise a non-metaphysical, non-vengeful relationship with finitude and temporality. Only by acknowledging its mortality could humanity be freed from the delusions of grandeur which foster the titanic projects of the technological age: "The self-assertion of technological objectification is the constant negation of death." [HW: 279/125] Only by experiencing the unsettling truth about our mortality and by surrendering to the healing power of pain will we awaken from our technological compulsions. Death anxiety and pain reveal to us that we are the ontological difference which both divides the world and things, and gathers them together. [US: 26–27/204] When we harden ourselves to pain and death, we separate ourselves from things and from other people so that they appear as threatening objects to be dominated or as pleasing objects to be exploited. Mortals cannot truly be mortals until they surrender to and affirm their mortality; only then can the "dance" of the fourfold and true "dwelling" on the earth begin.

Some people have argued that a central feature of human evil is the attempt to become immortal, perfect, wholly independent, all-powerful, and completely defended. Evil erupts when appropriate limits and measures are lacking. The disappearance of such limits is what Heidegger had in mind by the death of God. Where, in a post-metaphysical age, are we to find appropriate measures for our behavior? At least in part, as Werner Marx has observed, through what is revealed to us by the love, compassion, and respect (for other people and for the earth) which are the grace-granting companions of anxiety and pain.[27] Human "salvation," then, can come only when "there is a turn with mortals in their natures." [HW: 273/118]

For a long time, Heidegger hoped that Hölderlin's poetry would help to prepare humanity for that turn. That hope was—and remains—a vain one. Indeed, in light of the dynamics at work in the technological world, Heidegger's hope for the saving power of art seems almost absurd. Perhaps Hegel was right after all in saying that in the modern age the spiritual power of art has disappeared. But is there an alternative way to understand the disclosive power of *techne*? What if the *techne* needed by technological humanity were the art of disclosing, encountering, and affirming our own mortality, corporeality, and suffering? Through such *techne* perhaps we could learn to "let ourselves be." Certainly this is the insight recommended by many of the world's great spiritual traditions. In contemplating the notion of *techne* as the disclosure of ourselves as mortals, we move away from the idea of the work of art as a separate entity opening up new possibilities. Heidegger himself turned in this direction when he concluded that virtually any thing can play the role of revealing the self-gathering worldhood of the world. Perhaps what is needed is a new *techne* of the body, a new way of being embodied, a redemption of the body. But in an age so pervaded by a schizophrenic attitude toward the body, toward feelings, toward woman, toward nature, and so bereft of any sense of the immanent presence of the divine, we may rightly wonder whence such redemption might come.

Conclusion: Critical Reflections on Heidegger's Concept of Modern Technology

Heidegger's most important contribution to the philosophy of technology was his claim that modern technology is the final stage in the history of the self-concealment of being. Other authors to whom Heidegger was indebted, such as Spengler and Jünger, addressed the all-embracing character of modern technology, its uncheckable tendency to transform both social institutions and the natural environment. What Heidegger added was the claim that the "essence" of modern technology is *Gestell*, a one-dimensional way of disclosing entities as raw material, a disclosure which provokes humanity to behave in accordance with the technological imperative of infinitely expanding production for its own sake. As we have seen, even for the ontological dimension of Heidegger's account of modern technology, he was indebted in many ways to Jünger's idea of the *Gestalt* of the worker-soldier which "stamps" humanity in the technological era.

Heidegger's thought is controversial not only because of its links with reactionary writers and Nazi politics, but also because it explains modern technology in seemingly idiosyncratic terms: as the outcome of the "play" of being, rather than as the outcome of the judgments, decisions, and actions of determinate human agents. Nevertheless, his analysis has proved appealing to many people precisely because they share his intuitions: that in the technological age no one seems to be in control of the "system," that enhancing the power of the "system" has become more important than contributing to the well-being of the people who work in it, that technical innovation and industrial development have a "logic" and "purpose" that transcends the intentions of individuals, that the spread of the industrial worldview (whether socialist or capitalist) replaces traditional values with largely economic or utilitarian considerations which are blind to personal, social, religious, and environmental concerns, and that the likely consequence for the global technological impulse is dystopian, not utopian. This skeptical, negative interpretation of modern technology stands in stark contrast to the meliorist vision that was popular not so long ago.

To be sure, since the beginning of the Industrial Revolution, there have always been strident—if usually marginalized—critics of the Enlightenment confidence in the power of reason, industry, science, and technology to improve the human estate by making "man" the master of his fate. Today, critics of modern technology are no longer marginal; indeed, they are major participants in a critical debate about the presuppositions of modernity and industrialization. The prominence of contemporary critiques of technology is not surprising at the end of a century which has suffered through two world wars and countless smaller ones, totalitarianism of the left and right, genocide, a nuclear arms race, the ongoing destruction of the biosphere, and other social disorders of unparalleled magnitude. If this bloody history and an arguably suicidal future are the results of modern technology, so some critics maintain, then it is obviously time to subject to critical scrutiny the presuppositions of the rationalistic and modernistic attitudes so essential to it.

In this work, I have attempted to explain how Heidegger's conception of modern technology resulted from his own critique of modernity, rationalism, and industrialism. Division One hypothesized that his critique was profoundly shaped by his reactionary political views. While the political analysis in Division One was necessary for understanding Heidegger's conception of modern technology, equally necessary was Division Two's examination of his deconstruction of metaphysical foundationalism and his related analysis of the origin and character of modern technology. Together, Divisions One and Two have constituted a hermeneutic circle in which our understanding of Heidegger's political views was deepened by our appreciation of his critique of metaphysics, and in which our insight into his interpretation of the history of productionist metaphysics helped us to see why Heidegger was attracted to the National Socialist promise of a new world of working and producing.

In dividing this book into these two divisions, I stepped outside of the one-dimensional hermeneutic circle that is typical of the way in which most of Heidegger's commentators have explained his concept of modern technology. Virtually ignoring the political implications of that concept, such commentators have maintained that a continual reading and rereading of early and later Heidegger's own critical reflections about foundationalist metaphysics and his poetic meditations on the new beginning for the West constitutes the *only* hermeneutic circle needed for understanding his concept of modern technology. By adding a division largely devoted to examining the political context, origins, and implications of Heidegger's concept of modern technology, and by making this the first division in this book, I intended to de-center Heidegger's thought, to displace is from what some have regarded as its suprahistorical position, and to reveal that it was not so unique as it might appear, but was, rather, one important voice in a cultural conversation into which Heidegger himself had been "thrown."

Here, I wish to further this decentering process by amplifying some of the critical themes which I introduced in Divisions One and Two. Had I considered the work complete when I had indicated the ways in which Heidegger's reactionary politics both influenced and were influenced by his critique of the metaphysical decline of the West, the hermeneutic circle constituted by the two divisions would have been self-reinforcing and closed. Resisting such closure from the start, however, I have suggested throughout this book some of the limitations of Heidegger's political and philosophical views. In what follows, I expand these earlier reservations. My aim is to reveal possible alternative ways of interpreting Western history, modern technology, human agency, and politics, alternatives which Heidegger systematically concealed, marginalized, or distorted in his surprisingly one-dimensional reading of these phenomena.

When Heidegger's text is de-centered, a whole new series of questions emerges: Is human autonomy largely illusory, and does human freedom simply amount to responding appropriately to the historical claim made by being upon humanity? Was Heidegger correct in arguing that modern technology is the inevitable outcome of certain developments within the "history of being"? On what basis did he maintain that his own controversial reading of Western history was privileged and prophetic? In the face of the increasingly serious threats to the biosphere posed by planetary technology, is there no alternative to Heidegger's recommendation that we cultivate the meditative openness necessary for a new revelation of being? Are there not plausible but different accounts of the origins of modern technology and of how to transform or limit its destructive features? This conclusion can only outline certain approaches which might be taken in exploring such questions. More important than attaining some ultimate "closure" is the need to keep them open as questions.

To approach these questions, I first examine the extent to which Heidegger's deterministic conception of history is indebted to Hegel's thought and thus has much in common with Marxism. Such an examination is important in order to show that Heidegger's account of the determinism at work in the "history of being" is not so idiosyncratic as it might first appear, but rather constitutes one voice in a long-standing "conversation" in German philosophy, a conversation to which Scheler, Spengler, and Jünger were all indebted. These thinkers explained the historical developments of the West in terms of forces or movements at work "behind the scenes" of the drama being played out by human actors. More primordial than the intentions, self-understanding, and agency of humans, in other words, were the striving for freedom by *Geist*, the development and overthrow of capitalism, the emergence of the *Gestalt* of the worker, and the history of being. Heidegger differed from Hegel and Marx in important respects, however. He maintained that Western history had no "purpose," "foundation," or "goal," and that whatever "direction" could be discerned in that history was a movement of decline away from the primal origins. Like Nietzsche and Jünger, Heidegger came to view history as an artistic—purposeless—phenomenon. Hence, he insisted that humans could do nothing to alter the technological destiny into which they had been "thrown." Only a new appropriation of humanity by being itself could free humanity from the grip of nihilistic modernity.

Second, I examine what some regard as the political danger involved in Heidegger's determinism and in his deconstruction of the foundations of Western metaphysics, including the foundations of the humanistic, emancipatory project of the Enlightenment. A number of contemporary pragmatists argue that while Heidegger and his deconstructionist followers are correct in criticizing the foundationalist pretensions at work in much of Western metaphysics and in pointing out the dark side of the Enlightenment's push for technological power over nature, those deconstructionist critics are ahistorical and politically naive in their failure to see that the Enlightenment promoted genuine political, economic, and civil liberties which may be ignored and condemned only at great risk. Heidegger took such a risk by aligning his own critique of Enlightenment metaphysics with the Nazi attack on the "alien" (French, British, American, and German!) Enlightenment commitment to individual liberty, toleration, rationality, and universal human solidarity. Pragmatists argue that the critique of foundationalism does not go hand in hand with a reactionary determinism which effaces human freedom. From this perspective, modern technology is not a destiny imposed upon humanity, but rather a manifestation of the effort by humanity to gain a measure of control over the forces of nature. Even if humanity is in important respects capable of self-determination, the question remains open whether humanity can direct the developments of modern technology in a way that avoids the nightmarish alternatives of nuclear war, environmental catastrophe, or new forms of totalitarianism.

Heidegger's account of the development and character of modern technology has been widely influential, but is only one of many such accounts. Marxists, Hegelians, liberals, theologians, anthropologists, and feminists (among others) offer competing accounts. Heideggereans have accused such alternative accounts of being trapped within metaphysical discourse, but such an argument presupposes the validity of Heidegger's own critique of the history of productionist-foundationalist metaphysics. Many of these alternative accounts question the validity of Heidegger's own "meta-narrative" of Western history. Here, I shall address only one of the above-mentioned alternatives, namely, the feminist one. Many feminists agree with Heidegger regarding the domineering character of modern technology, but they argue that this character stems from blindness not to the ontological difference but rather to the sexual difference. The contemporary drive for total control, then, may be regarded as a late stage in the patriarchal quest to exclude, repress, and deny all difference, all otherness which threatens the security of the masculinist ego. By de-centering Heidegger's self-enclosed narrative, the feminist narrative reveals that there may be a totalizing and thus a metaphysical impulse at work in the thinker most famous for his critique of metaphysics.

A. IN THE SHADOW OF HEGEL: HEIDEGGER'S HISTORICAL DETERMINISM

Marxist critics have argued that Heidegger is guilty of mystification not only for saying that the exploitative character of modern technology is shaped by the "self-concealment of being," but also for claiming that there is no alternative to this alleged "destiny." For Marxists, technology is not in and of itself alienating or destructive, but becomes so under the aegis of capitalism. As opposed to Heidegger's despair about the possibility of change being initiated by human action, Marxists continue to assert the melioristic Enlightenment view that rationality is a liberating force which not only improves humanity's material circumstances but also provides the critique needed to transform the socio-economic arrangements responsible for alienation and deprivation. Rejecting Heidegger's determinism, Marxists proclaim that humanity can liberate itself from the fetters forged by the capitalist misuse of technology.

From the Marxist perspective, then, Heidegger may have been brilliant, but he was nevertheless a reactionary thinker who believed that world history was shaped by mysterious forces inaccessible to rational analysis. Heidegger's reactionary attitude was shaped in part by his conservative religious orientation, which helped to foster his distrust of the self-assertiveness implicit in rationalism, his historical determinism, his condemnation of the human self-worship involved in socialism and liberalism, and his belief that technological nihilism is the inevitable outcome of the interplay between the self-concealment of being and the *hubris* of modern humanity. Marxist critics are correct that there are important analogies between Heidegger's view of history and the view offered by some theologians. Consider, for example, the following passage from Reinhold Niebuhr:

> Moral or historical evil is the consequence of man's abortive effort to overcome his insecurity by his own power, to hide the finiteness of his intelligence by professions of omniscience and to seek for emancipation from his ambiguous position by his own resources. Sin is, in short, the consequence of man's inclination to usurp the prerogative of God, to think more highly of himself than he ought to think, thus making destructive use of his freedom by not observing the limits to which a creaturely freedom is bound.[1]

Similarly, Heidegger maintained that the destructive consequences of modernity and industrial technology are the result of mortal humanity's hubristic attempt to make itself the God-like ground for all things, to compel all things to bend to human demands, and to act as if there were no transcendent dimension that is more originary than human reason. Marxists maintain that Heidegger's concept that technological nihilism results from the self-concealment of being is a secularized version of the claim that the world's suffering results from the self-concealment of God from a sinful humanity. Heidegger's appeal to "being," then, is allegedly simply a variation of the mystifying appeal of Jews and Christians to "God" as a device to explain the origins and to justify the preservation of exploitative social conditions. Yet Marxism itself has often been accused of being a Judeo-Christian heresy, proclaiming that the "New Jerusalem" will be achieved not by divine intervention but by the efforts of humanity. For, while Marx and his followers debunked the Biblical vision of history and the history of being, they still spoke in quasi-religious and mythical terms about the history of economic formations which have led to capitalism and will eventually lead to a socialist and even possibly to a utopian future.

Marx transformed the conflict between good and evil depicted in the "history of salvation"(*Heilsgeschichte*) into his own secular mythology involving the cosmic struggle between the greedy, blood-sucking werewolf, "Lord Capital," and the creative but suffering proletariat. In speaking of money as the "divine power" that overturns all things, and in describing greed as the "utterly alien power" or "inhuman force" holding sway over human existence, Marx revealed his belief that the behavior of capitalists was a function not of their individual "decisions," but rather of a power that transcends human control. As Robert Tucker has explained in reference to Marx's writings,

> Man as such has vanished, and in his place two personalities, the worker and the capitalist, stand in hostile confrontation. Neither is fully human. The worker is a non-man, a dehumanization, a proletarian devoid of every human attribute save the essential one—creativity. On the other side, the insatiable, acquisitive urge, the alien power within man that transforms his productive activity into alienated labour, has been detached and housed in "another man outside the worker." It has assumed flesh and blood, semi-human form, in the capitalist.[2]

Tucker's comment suggests how Marx, not unlike Heidegger, tended to conceive of humans—capitalists and proletarians alike—as functions of the hidden movement of capital, a movement which Marx believed he was the first to discern. The movement of capital stamped humanity in Marx's industrial era in much the same way that Jünger's *Gestalt* and Heidegger's *Gestell* stamped humanity in the technological era. Consider the following passage from *Capital*:

> As the conscious bearer [*Träger*] of this movement [of capital], the possessor of money becomes a capitalist. His person, or rather his pocket, is the point from which the money starts, and to which it returns. The objective content of the circulation we have been discussing—the valorization of value—is his subjective purpose, and it is only in so far as the appropriation of ever more wealth in the abstract is the sole driving force behind his operations that he functions as a capitalist, i.e., *as capital personified and endowed with consciousness and will*.[3]

While Marxists have accused Heidegger of being an idealist who explained history not in terms of practical human action but in terms of the processes of mystical ontological movements, Marx himself explained history in terms of the development of economic formations, including the movement of capital, which also undermined the importance of human agency. Despite the conflict between Marxists and Heideggereans, then, they share a hidden harmony because both Marx and Heidegger stand in the shadow of Hegel.[4] Like Hegel, they believed that while human actors did the work of producing the machinery and institutions of the industrial and technological eras, those same actors were not fully aware that they were playing roles in a drama which they themselves did not compose. Hegel maintained that the "cunning of Reason" harnessed human desire for Reason's own purposes.[5] Marx and Heidegger added their own unique twists to Hegel's belief that history possesses a determinate direction, a goal, a *telos* which shapes human affairs. Moreover, Marx and Heidegger, again like Hegel, each believed that he was the first to discover the secret laws that determine the course of history.

Whereas Marx, for example, assured the proletariat that its active participation was necessary to defeat capitalism, he nevertheless believed that social
evolution—including the rise of capitalism—was determined by the "iron
necessity" of the *telos* driving humanity to actualize its divine capacity for
creating and producing. Like Hegel, Marx shared the Biblical view that history
has a reason, purpose, basis, meaning, foundation, or goal. Neither thinker
provided many details about what would happen *after* the attainment of the
telos of history, but both agreed there would be no "new beginning" in the
sense of the initiation of a new historical sequence by being. Whatever was to
happen *after* the attainment of the "Absolute" (Hegel) or of "authentic communism" (Marx) would be events consistent with a completed and relatively
perfected world.

Heidegger agreed that Western history was shaped by a *telos* that culminated in industrial technology, but argued that this *telos* was a consequence of
the self-concealment of being, not of the coming-to-presence of Absolute *Geist*
or of the actualization of humanity's capacity for producing and creating. While
Heidegger's account of the various modes of beingness governing Western history thus resembles in many ways Hegel's account of the transformation of *Geist*
into various historical *Gestalten*, it also manifests significant differences. Unlike
Geist, being has no prefigured possibilities to be actualized in history. At the
beginning of Western history, being both "gave" itself and "concealed" itself at
the same time. This ontological revelation opened up to the Greeks a possibility
for understanding what things are; yet the self-concealment of being simultaneously initiated a tendency to harden, freeze, or congeal the fluidity or dynamism of the original revelation into something fixed, constricted, and rigid.
Hence, the original revelation of being as "presencing" soon devolved into the
constricted metaphysical understanding of being as something permanently
present—an ontological ground or foundation.

For Heidegger, the *completion* of the stages of crystallizing or hardening
which constitute the history of productionist metaphysics would at the same
time make possible the emergence of a *new* beginning for history. Such a new
beginning would not be bound by the nihilistic movement which culminated
in the epoch of modern technology. Heidegger distinguished between his conception of the internal dynamic of Western history and the conception of
historical *telos* shared by Hegel and Marx. Hegel argued that the various *Gestalten* of Western history were integral aspects of the *telos* of Absolute Spirit,
which strives to pro-duce itself, to bring itself to absolute presence and, thus,
to freedom. Marx argued that the various stages of the development of economic and social life were the elaboration of the *telos* of humanity, which
strives to re-produce itself, to transform itself and the natural world into a free
expression of human productivity. According to Heidegger, Hegel's metaphysics was the source for Marx's productionist attitude. For Hegel, "every entity
appears as the material of labor. [Hegel anticipates] the modern metaphysical
essence of labor . . . as the self-arranging process of unconditioned producing, which is the objectification of everything actual through man experienced
as subjectivity." [WGM: 171/220]

Heidegger portrayed Hegel and Marx as productionist metaphysicians who explained modern technology in terms of a historical foundation, substance, essence, or *telos*. For Hegel, this foundation was Absolute Spirit; for Marx, it was human productive possibility. Heidegger contended, however, that the conception of an ontological "foundation" and "goal" for history was itself the consequence of the self-concealment of the groundless origin. The "appropriating event," *Ereignis*, dispatched the originary experience of being as presencing to the ancient Greeks. Since mortals cannot long endure such a revelation, presencing soon began to become crystallized or hardened into something present: a metaphysical ground for entities. Likewise, any subsequent originary revelation—including the one Heidegger once claimed to see in Hitler's "revolution"—would also begin to crystallize. The world-determining *Gestalten* arising from this crystallizing process constituted the "destiny" and "*telos*" of the West, just as such *Gestalten* would begin to shape the course of any post-metaphysical era as it fell away from its originary revelation. For Heidegger, these *Gestalten* were in some sense the fulfillment of the possibility opened up at the beginning, but they were primarily a negative fulfillment, a decline from the originally vital origin. Hence, the one-dimensional technological understanding of being was all that remained of radical ontological presencing. For Hegel and Marx, by way of contrast, the late stages of Western history were stages of maturity, in which either Absolute *Geist* or the potentiality of the creative-productive human "species-essence" came to fruition, actualizing its potential in a way analogous to how an animal actualizes its potential, realizes its *telos*.

B. CONCEPTUAL AND POLITICAL CRITICISM OF HEIDEGGER'S DECONSTRUCTION OF FOUNDATIONALIST-PRODUCTIONIST METAPHYSICS

While influenced by Hegel's historical determinism, then, Heidegger renounced the progressive dimension of that determinism, a progressivism which was transformed by Marx into his own vision of human progress. Influenced by Nietzsche and by such reactionary Nietzscheans as Jünger, Heidegger had no confidence in the Enlightenment vision of "progress." Hence, he made both Hegel and Marx walk on their heads, in the sense of claiming that their progressive view of history would be better read as a history of decline and degeneration. For Heidegger, the liberatory promise of the Enlightenment in fact paved the way for an epoch of total human enslavement. Viewing the Enlightenment as a crucial phase in the rise of the technological Will to Power is what enabled Heidegger to equate extermination camps with mechanized agriculture and hydroelectric dams. For Heidegger, genocide was a predictable outcome of the reckless power impulse at work in the Enlightenment, an impulse given free rein in modern technology. Ernesto Laclau has objected to this view of Western history: "When the theorists of the eighteenth century are presented as the initiators of a project of 'mastery' that would eventually lead to Auschwitz, it is forgotten that Auschwitz was repudiated by a set of values that, in large part, also stem from the eighteenth century."[6] Like so many Germans of his generation, however, and like a number of recent French thinkers, Heidegger saw only the "dark side" of the Enlightenment project and also discounted the possibility of human freedom.

In one way or another, Heidegger's determinism influenced the work of a number of French structuralists. For structuralists, language—the symbolic order—is the most pervasive and powerful structure governing human behavior. At birth, we are "thrown" into a language and into a cultural discourse which shape our identity, attitudes, beliefs, and values. As Heidegger would say: We do not use language; rather, language uses us. John Sturrock has argued that

> it is the Symbolic order in which structuralism is interested. because that order is the system in which we ourselves can never be more than "events." In consequence structuralism has come to stand for a way of thinking opposed to individualism, or even to humanism, for intentional human agency is given a reduced role in its interpretations of culture. Much has been written of "the disappearance of the subject" under the structuralist dispensation, meaning that structuralism has carried its strong bias against essentialism so far as somehow to deny the existence of the human being altogether, and to see the individual as nothing better than an unstable, replaceable form within a soulless system.[7]

Heidegger's relation to structuralist determinism was ambiguous. On the one hand, he conceded that human behavior in the epochs of Western history has been shaped by the various stages of the metaphysics of presence, i.e., by the successive rise of metaphysical foundations from Plato's forms to Nietzsche's Will to Power. On the other hand, Heidegger also argued that the structuralist quest for a stable foundation, center, or cornerstone to explain the activity taking place within a given system (whether it be the human psyche or the economic realm) was itself an expression of the totalizing impulse of foundationalist-productionist metaphysics. In other words, Heidegger conceded that a *Gestalt* organizes behavior in each historical epoch, but added that this *Gestalt* is itself ultimately groundless, foundationless, without purpose, an-archic, despite all appearances to the contrary. For any given history-organizing *Gestalt* is but a crystallized version of the originary and abysmal revelation, a revelation which has no "purpose," no "reason," no "goal," no transcendent "meaning."

Heidegger, then, along with Nietzsche and other critics of foundationalist metaphysics, helped to point the way to the post-structuralism promoted by deconstructionists such as Jacques Derrida. If in fact there is no metaphysical foundation for self and world, then there is no grounding upon which structuralists can erect totalizing schemes in anthropology, psychology, philosophy, or any other domain. Some deconstructionists argue that it is precisely the disclosure of the lack of metaphysical foundations that opens up the possibility for freedom in a world currently threatened by the totalizing impulse of modern technology. There are no "master-names" or "transcendental signifiers" to which one may appeal to justify or to "ground" authoritarian regimes;

rather, there is the indefinite play of signifiers which always elude any "final" or "ultimate" reading. The self-certainty which grounds the Cartesian ego is undermined in such a way that the self is de-centered and becomes a play of signifiers. But in what sense does such deconstruction make freedom possible? How can we define "freedom" in terms other than those associated with the "selfhood" traditionally linked with the capacity for freely chosen action? Derrida seems to leave us with what Peter Dews has called the choice between "a view of the subject as an immobile center, a core of self-certainty, or the acceptance that there is no subject at all, except as an 'effect' of the play of the text."[8]

Heidegger never claimed that the deconstructed, post-Cartesian "self" would be "free" in any traditional sense; for him, "freedom" meant submitting to the necessity of being. Apart from the meditative activity of deconstructing the history of Western metaphysics, Heidegger concluded that there was nothing that could be "done" to move beyond totalitarianism in an epoch stamped by the *Gestalt* of the worker. In *One-Dimensional Man*, Herbert Marcuse expanded on Heidegger's suggestion that the technological system not only undermines all opposition but actually transforms critique into a source of strength for that system.[9] Just as Heidegger called for the "step back" from engagement within the technological system, so Marcuse spoke of the importance of the "great refusal" in the face of the totalizing impulse of the system. Both thinkers maintained some hope that great art might be able to disclose an alternative to technological nihilism, but they left us with the question: What could possibly motivate the production of such liberating works of art? If humanity is so dominated by *Gestell*, then it would appear that only a superhuman event could break its grip and initiate a new historical possibility.

Heidegger attempted to deconstruct the totalizing, foundational impulse of Western metaphysics in order to struggle against those totalitarian technological ideologies—such as Bolshevism and Americanism—which proclaimed themselves to be progressive. We have followed both this struggle against technological totalitarianism and its paradoxical denouement in Heidegger's support for one of the greatest totalitarian movements of this century. Given this outcome of Heidegger's political decision, we may perhaps understand his increasingly despairing attitude regarding hopes of changing a system which can so easily divert revolutionary critique into support for the system itself. Heidegger's political engagement in 1933-34 led him to conclude that all merely human "revolutions" and "decisions" would simply reinforce the system already in play. The question for us is: Is that conclusion tenable?

In his political decision of 1933, Heidegger acted on his belief that being free meant aligning himself with the new historical destiny dawning at that time. His political decision was motivated in part by his deconstruction of the metaphysical foundationalism which made possible the modern technology against which he fought. The political consequences of Heidegger's deconstruction of metaphysics have been dismissed by some as irrelevant to the validity of that deconstruction, but others have maintained there is an important connection between Heidegger's thought and his reactionary politics.

Some of Heidegger's defenders, for example, maintain that deconstruction of foundationalist metaphysics is the best possible defense *against* the totalitarianism which justifies itself on the basis of such metaphysics, while many of his critics assert that his deconstruction of the foundations of Enlightenment humanism in fact allowed him to support a regime which also rejected those foundations. To examine the *political* implications of Heidegger's deconstruction of Western metaphysics and to evaluate his claims about human freedom, we cannot avoid a critical examination of that deconstruction itself.

Much of Heidegger's critical account of metaphysics—its totalizing impulse, its foundationalism, its quest for complete control—is now widely shared. Many pragmatists, for example, despite important reservations, have argued that Heidegger's critique of foundationalism contributes to an appreciation of the limits of human knowledge and political power. For such pragmatists, inquiry, knowledge, and the institutions dependent upon that knowledge are always conditional in nature, open in principle to question and revision. As post-modernists and post-structuralists, they have defended Heidegger's deconstructive critique of the arrogant and totalizing dimension of Enlightenment faith in absolute rational foundations for science, philosophy, economics, psychology, and politics. Yet many pragmatists have also been wary of Heidegger's deconstructionism precisely because of its connection to his problematic interpretation of Western history and to his reactionary political vision.[10]

Consider, Joseph Margolis argues, that for Heidegger our only hope for salvation from the technological era lies in turning

> to the very source of our prodigal but dangerous ontologies. This is the key to Heidegger's extravagance. . . . The result is a peculiarly preposterous theory that entails that, however impressive it may be, very nearly the entire movement of Western philosophy from Plato to Nietzsche has gone ontologically astray *and* that Heidegger himself controls its permanent exposure. It is also what may be termed a "sly" theory, because its emphasis on the historicity of human existence and inquiry somehow manages to ignore the concrete history of actual existence and actual inquiry.[11]

Despite Heidegger's recognition of the historical character of human existence, his talk of the transcendent ontological dimension (*Ereignis*) which "gives" the epoch-shaping modes of beingness would, in Margolis's view, seem to be a transhistorical "source" for the historical—a groundless ground, a foundationless foundation, an abysmal origin not arising from the practical activity of humanity itself but "arriving" instead without warning, from "nowhere."

Much of the contemporary pragmatist appreciation and critique of Heidegger is indebted to the work of Derrida. Derrida has argued that Heidegger did not carry the deconstruction of metaphysics to its completion, because he failed to deconstruct the idea of a "primordial" epoch blessed by a direct encounter with being or of a primordial "sender" dispatching messages to the privileged recipient, human *Dasein*. To this extent, Heidegger's thought reveals aspects of metaphysical foundationalism and structuralism. What Derrida said in his critique of the structuralism of Claude Lévi-Strauss can also be applied—*mutatis mutandis*—to Heidegger:

> Turned towards the lost or impossible presence of the absent origin, this structuralist thematic of broken immediacy is therefore the saddened, *negative*, nostalgic, guilty, Rousseauistic side of the thinking of play whose other side would be the Nietzschean *affirmation*, that is the joyous affirmation of the play of the world and the innocence of becoming, the affirmation of a world of signs without fault, without truth, and without origin which is offered to an active interpretation. *This affirmation then determines the noncenter otherwise than as the loss of center.* And it plays without security. For there is a *sure* play: that which is limited to the substitution of given and existing, present, pieces. In absolute chance, affirmation also surrenders itself to genetic indetermination, to the *seminal* adventure of the trace.[12]

In *Radical Hermeneutics*, John D. Caputo agrees with Derrida that Heidegger should have extended his deconstructive method "all the way to 'Being,' 'meaning,' 'man,' and 'truth,' and the concomitant privileging of nearness, unity, ownness, rootedness, origin, and primordiality which so punctuates [his] text."[13] Late in life, Heidegger was at times able to forgo his nostalgia for "originary presencing" and to concede that there was never a time in which humans dwelt "nearer" to being. At such moments, he could affirm, like the mystic Eckhart, that things have no "origin"; rather, all things take part in a play that is "without why." Despite such resources within Heidegger's own thought for "overcoming" the metaphysical residues present therein, he never really abandoned what Derrida has described as the Platonic yearning to overcome fallenness and to attain authentic nearness to being.[14] At least one critic has argued, however, that "in his evocations of '*différance*,' 'arche-writing,' the 'trace,' Derrida [himself] still has not escaped the 'idea of the first,' even though this first cannot take the form of 'presence.' "[15]

Pursuing the critique of Heidegger's attachment to metaphysics, Richard J. Bernstein has asked: How radically different is Heidegger's conception of the privileged relation between being and humanity from the onto-theo-logical idea of the special relation between God and man, a relation which endows man alone with reason, *logos*, consciousness?[16] Was Karl Löwith right in depicting Heidegger's conception of being's relation to human *Dasein* as simply a Heideggerean version of the modern metaphysical subjectivism so deeply influenced by Christian theology?[17] Such questions demand answers before we unhesitatingly adopt either Heidegger's critique of metaphysics or the deconstructionist method used to make that critique.

In the light of these problems, Margolis has also questioned Heidegger's peculiar self-assurance that he alone was endowed with prophetic insight into the meaning of history.[18] According to Margolis, by virtue of what we know about the *social* character of all genuine knowledge, we clearly have no reason to privilege one man's contentious, if at times brilliant, assertions about the history of Western philosophy, the nature of human existence, the origins of the present age, and the direction of our future. Issues of this magnitude must be addressed—if they can be "addressed" at all—not by prophetic individuals, but rather by individuals, groups, and communities joined in a self-critical conversation which recognizes the difficulties inherent in coming to any agreement.

Bernstein, too, has inquired further into the validity of Heidegger's deconstruction of Western metaphysics, commenting that "Heidegger's narrative of metaphysics as the history of Being is powerful, seductive, and despite its provocativeness, it is—I would argue—extraordinarily perverse."[19] This perversity stems from the narrative's unjustifiable depiction of *all* great Western thinkers as driven by a totalizing quest for final knowledge, and from the narrative's related failure to acknowledge the extent to which those thinkers were also—like Heidegger himself—attempting to break free from the conventional wisdom of their predecessors. All great Western thinkers, not only Heidegger, have been motivated in part by the Socratic spirit of continued questioning and criticism. Yet under the influence of Heidegger and his (critical) follower Jacques Derrida, contemporary deconstructionists have spoken broadly of "rupture," "break," "unmasking," "transgression," "difference," and "refusal" in contemporary critical discourse, as if these were recently discovered interpretive practices and as if previous thinkers were not aware of the "slippage" or "gap" involved in the movement from "name" to "thing," or from "signifier" to "signified." Such a reading of Heidegger ignores how he himself implicitly acknowledged that "metaphysicians" such as Aristotle, Leibniz, Schelling, and Nietzsche could be "read" in a way which undermines the very "foundationalism" supposedly intrinsic to their metaphysics.

Bernstein also echoes Derrida's complaint that despite Heidegger's criticism of the totalizing, foundationalist tendency of Western metaphysics, Heidegger himself fell victim to this tendency in his *own* narrative of the history of being, in which every great metaphysician is said to be compelled by the same drive toward totality, unity, harmony, completeness, universality, foundational truth, and essence. Indeed, as Bernstein argues, the deconstructive critique of the essentialism and foundationalism of Western metaphysics "lapses into, and gains its polemical force from, an implicit essentialism when it refers to '*the* Western Mind' or '*the* cogito of Western philosophy.' "[20] Furthermore, as Derrida has argued, Heidegger's quest to "overcome" the history of productionist metaphysics was not successful because such overcoming is impossible to begin with: "There *is not* a transgression, if one understands by that a pure and simple landing into a beyond of metaphysics. . . . Now even in aggressions or transgressions, we are consorting with a code to which metaphysics is tied irreducibly, such that every transgressive gesture reencloses us . . . within this closure."[21]

At times, Heidegger spoke as if his longed-for "new beginning" would involve an entirely new era, one very different from the two thousand five hundred-year history of productionist metaphysics. At other times, however, he acknowledged that the "overcoming" of metaphysics did not simply mean leaving it behind, for to some extent both the language and the structural manifestations of metaphysics—including industrial technology—will remain with us. Yet in spite of such concessions to the lasting influence of metaphysics, Heidegger retained hopes for a radical rebirth, a rebirth which Derrida as well as many pragmatists regard as curiously utopian, naive, idealistic, and metaphysical.

Despite his criticism of Heidegger, however, Derrida has argued that deconstruction can play an important, though perhaps not a revolutionary, role within society. This claim stems in part from Derrida's participation in his own historically decisive political moment. If Heidegger's thought was shaped by the political crisis of 1933, Derrida's thought was influenced by the political discord leading up to the "days of May" in Paris in 1968. His deconstructionism, then, is motivated in part by the impulse to question the foundations of the authoritarian structures and practices of French political economy which marginalized women, non-whites, the working class, homosexuals, and the young. Working to expose the "fissure" or "gap" between signifier and signified, seeking to disclose that all totalizing "foundations" presuppose a difference which has been concealed or marginalized, Derrida revels in the play of signifiers, extolling their endless interconnectedness. He opposes the metaphysical drive to acquire "master-names" which found and justify totalizing social institutions and practices. To this onto-theo-logo-centric metaphysical quest for the ultimate signifier, Derrida opposed the practice of "dissemination," the revelation of the uncontrollable overflow of "SPERM, the burning lava, milk, spume, froth, or dribble of the seminal liquor."[22]

It is worth pointing out that there is an important difference between Derrida's celebration of the superabundance of signification and Heidegger's talk in 1935 about the necessity of "channeling" the overflowing presencing in order to "found" *the* new world envisioned by National Socialism. One can discern the metaphysical residues in such talk not only in the notion that there was only *one* way for Germany to go in 1933, but also in the privileging of the poet's voice, the "expressive substance" of things, or the founding of the new world.[23] Heidegger's totalitarian temptation, from Derrida's point of view, was heightened by the fact that he focused on only one "difference": the difference between being and entities. For deconstruction to avoid the seductiveness of metaphysical totalizing, however, it must celebrate more than one difference, and it must question all yearning for a "dwelling in the nearness of being." Caputo has effectively described the political import of "dissemination":

> Dissemination is directed at constellations of power, centers of control and manipulation, which systematically dominate, regulate, exclude. Its model is the Socratic work of showing up the contingency of every scheme. It delimits the authority of all programmers, planners, managers, and controllers of all sorts. It compromises the prestige of the expert, releases all the loose ends in every system, exposes the systematic violence of any tightly organized structure (university, church, hospital, government, etc.). And it does all this not by any show of strength of its own but by letting the system itself unravel, letting the play in the system loose.[24]

Deconstructionists endowed with pragmatic sensibility, such as Caputo, have recognized that there is an unspoken condition necessary for such "dissemination": freedom of speech. Without such freedom, no dissemination, no "loosening up" of existing structures is possible. Mahatma Gandhi is famous for having pointed out in the 1930s that his tactic of "civil disobedience" could work only within a society that was "civil," i.e., which respected the rights of people to protest existing conditions. Gandhi would not have survived long in an un-civil society, such as Nazi Germany. The father of deconstructionism, Heidegger, however, assailed the Enlightenment doctrine of "universal rights" that were the basis for such a "civil" society. He argued that the Enlightenment's confidence in progress, in "universal rights," and in the universal applicability of scientific rationality was merely a symptom of a degenerate stage of human existence, modern subjectivism. For Heidegger, then, the Enlightenment's proclamation of universal rights and truths was itself the political manifestation of the metaphysical subjectivism at work in the power-hungry French and English.

Heidegger's failure to understand the political conditions necessary for the very activity of deconstruction is one reason many pragmatists have voiced skepticism and concern about the ahistorical sensibility of some post-modernists and deconstructionists. Critical deconstruction of the Enlightenment and of modern political and scientific foundationalism has become something of a culture industry. This critique has arisen partly in response to disappointment regarding the alleged outcomes of the Enlightenment. Proponents of this critique generally agree with Heidegger that there is a direct relation between the Enlightenment's quest for ultimate foundations in the realm of science and in the domain of politics and human affairs. Political totalitarianism, then, is said to be the inevitable consequence of the quest for final truth in science because that quest was not at all disinterested, but rather was motivated by hidden power motives. The Enlightenment worshipped scientific rationality, then, because it was the kind of knowledge most consistent with acquiring power.

While reservations about meta-narratives and totalizing claims in either politics or science are justifiable at the end of a century so wracked by movements claiming to represent the "final truth," also justifiable are defenses of the political rights which make possible the agonistic discussion of political and social practices in a pluralistic, post-modern world. The dilemma is how to reconcile the deconstruction of all political foundations with the insight that "undistorted communication" (Habermas) is a "basic principle" necessary for all critique. According to many post-modernists, there is another problem as well. They allege that the technological society is so all-embracing that communication is systematically distorted; hence, the ideal of "free speech" is unattainable. Indeed, appeals to the principle of "free speech" are ideological smokescreens concealing those totalizing practices which not only distort speech but see to it that it is denied in particular to marginalized social groups. Here, we may recall Marcuse's warning: critique is readily co-opted by a totalizing system.

Let us examine briefly the extent to which the historical outcome of

Enlightenment modernity confirms the claim that even rationality, which originally purported to be critical and liberating, is inevitably transformed into a means for enhancing the power of the rising technological system. In the Enlightenment, the principles of reason—including skepticism regarding the claims of authority, demand for evidence, logical argumentation—were believed to be both the expression of and the condition necessary for human freedom. Rationalism, both political and scientific, made possible the deconstruction of the theological dogmas, metaphysical doctrines, political institutions, and social arrangements and practices which produced and justified undeniable misery and constraints on liberty.

While some Enlightenment rationalists may have viewed science as an instrument for gaining power for its own sake, many more conceived of science and technology more modestly: as necessary conditions for achieving the goal of liberating people from authoritarianism, dogmatism, ignorance, violence, hunger, disease, and other impediments to the attainment of human freedom. These Enlightenment rationalists would obviously have been surprised to see how the subsequent dynamics of capitalism would alter what they had hoped would be the proper relationship between technological-scientific rationality, on the one hand, and human liberation, on the other.[25] Increasingly (some would argue: from the very beginning), however, the critical, negative force of "rationality" was subverted in favor of the instrumental, positive application of science and technology to increasing the power and wealth of various elites. If this centralization of power and wealth distorted the political realm, this fact was justified by appeals to the "scientific" findings of Darwin, who "proved" that competition was the natural and therefore most adaptive way of organizing society.

According to Marcuse, the extraordinary material achievements of modern technology have been so successful that they have undermined interest in a critique of the social contradictions involved in those achievements. So enamored are people with the outpouring of new commodities that they not only ignore the social problems that are present for all who care to see, but they also have little interest in or access to those works of art which indicate what is *absent* in technological society.[26] Since "liberal" technocratic societies follow rational (i.e., technical-scientific) principles in producing wealth and also proclaim themselves to be based on the self-evident freedoms articulated by the Enlightenment, critique of those democracies is easily labeled either irrational or utopian. The apparently irrational practices of the society—including its engaging in a nuclear arms race, its undermining of its social fabric by a permanent war economy, and its endless quest for a higher material living standard—are justified as the only "rational" ways of protecting "freedom." As long as the "totally administered society" delivers the goods, these contradictions are tolerated. For Marcuse, the question is: "how can the administered individuals—who have made their mutilation into their own liberties and satisfactions, and thus reproduce it on an enlarged scale—liberate themselves from themselves as well as from their masters? How is it even thinkable that the vicious circle can be broken?"[27]

If Heidegger and Marcuse are correct in saying that we live in totally administered societies, then we may easily understand their despairing attitudes toward alternatives to the given. Yet in response to their grim vision, we must ask: *Is* the technological world totalitarian? Is there freedom for critical discussion, for envisioning and realizing alternatives, for effecting social and cultural change? Or, in fact, does the "technological subject" simply lack any resistance to being stamped by the technological Will to Power? Michel Foucault, who in his "final interview" acknowledged that for him "Heidegger has always been the essential philosopher," answered the last question affirmatively in most of his writings.[28] Toward the end of his life, however, he seemed to change his mind. In "The Subject and Power" (1982), he remarked: "When one defines the exercise of power as a mode of action upon the actions of others, when one characterizes these actions by the government of men by other men—in the broadest sense of the term—one includes an important element: freedom. Power is exercised only over free subjects, and only insofar as they are free."[29]

Furthermore, despite his earlier agreement with the contention that the "dialectic of the Enlightenment" (Horkheimer and Adorno) promoted subtle as well as overt forms of domination, Foucault also came to depict himself as sharing in the Enlightenment tradition of promoting liberation through critical analysis. On this astonishing "turn" in Foucault's thought, Jürgen Habermas has commented: "Up to now, Foucault traced this will-to-knowledge in modern power formations only to denounce it. Now, however, he presents it in a completely different light, as the critical impulse worthy of preservation and in need of renewal. This connects his own thinking to the beginnings of modernity."[30] Habermas argues that Foucault had based his earlier portrayal of the "totalizing" tendency of modernity on a highly selective reading of democratic societies, a reading which, for example, ignored the role of the legal system in constitutional democracies.[31] Dieter Freundlieb has, in my view, correctly concluded that early Foucault lacked "a differentiated set of categories that could account for the complexities of modern societies."[32]

The same criticism of early Foucault's "all-encompassing theory of power" may be leveled at Heidegger's all-encompassing theory of the technological Will to Power. While insightful in many ways, Heidegger's account appears to leave no room for what many people regard as the legal, political, cultural, and social expressions of resistance to the "disciplinary matrix" of modern technology. Such resistance is a manifestation of at least a measure of human freedom, even in the face of the undeniable power of the multifarious forms of repression and distraction at work in modern technological societies. Heidegger discounted the potential of constitutionally based democracies for resisting technological totalitarianism for several reasons. To begin with, because of his reading of Western history, he argued that liberal democracy and Soviet communism were "metaphysically" the same in that both were manifestations of subjectivism and humanism.

This apparent inability to see that there are important, not merely superficial, differences between Stalinism and Americanism was characteristic not only of Heidegger but also of many other Germans of his generation, including Horkheimer, Adorno, and to a lesser extent Marcuse. We have already seen that, for various historical and cultural reasons, these German-born thinkers believed that the Enlightenment was motivated not by an interest in human freedom, but rather by hidden power motives which later became clear in communist, fascist, and capitalist versions of totalitarianism. Moreover, influenced by the belief of Hegel and other German thinkers that human affairs are shaped by hidden events that transcend human ken and control, Heidegger concluded that those who spoke about human "agency" or "autonomy" or "freedom" were victims of what Marx would have called "false consciousness." The only real freedom for Heidegger involved surrendering to and affirming necessity. Hence his tendency to discount questions of "moral responsibility" regarding his own behavior during the 1930s.

Despite Heidegger's resistance to the modernist ideals of human agency and freedom, some commentators claim to have discovered in his early thinking the conceptual resources needed to defend a pragmatic version of a social and political ethics which affirms human freedom and responsibility. One pragmatist, Charles M. Sherover, has argued that—Heidegger's misgivings about the Enlightenment and his doubts about human agency notwithstanding—one can derive from *Being and Time* a workable (though not absolute) foundation for a social and political ethics that would be suitable for contemporary society. According to Sherover, moreover, commitment to promoting such an ethics is an appropriate activity for philosophers.

> A founded existential ethic would recognize historicity as a continuing condition of beings who are continually engaged in concrete situations, concrete individual and social concerns, decisions and allegiances, with specific conflicting obligations to sort out and conflicts of loyalties to resolve. It would not abrogate individual responsibility by dictating specific acts but it would show us how authentically to decide in the multifarious situations in which we find ourselves. It would delineate the considerations which are germane to our continuing task of resolving conflicts of value loyalties and evaluating the consequences of our own decisions. In its demand for authenticity, its own prime commitment must be a fundamental loyalty to the possibilities of freedom which provides its ground.[33]

Sherover believes that early Heidegger, at least, can be read in a way that defines freedom not as the affirmation of necessity, but rather as the capacity for making decisions about how one is to live, decisions which in ideal conditions allow one to express one's abilities and to realize one's vision, both for oneself and for other members of one's community. Defined negatively, freedom is liberation from those constraints which interfere with self-realization, self-expression, and the achievement of well-being for oneself and one's community. While no absolute foundation may be available to ground this vision of human freedom, lack of such a foundation will not deter those who have an intuition of the validity of such freedom from promoting and defending it.

Not all pragmatists share Sherover's conviction that critical deconstruction of oppressive political structures and engaged reconstruction of the social

community are necessary or even appropriate activities for those engaged in
the critique of metaphysical foundationalism. Richard Rorty, for example, another pragmatist influenced by Heidegger's deconstructive method, suggests
that deconstruction has and ought to have no real political effect. It is one
thing, he suggests, to speak as intellectuals about deconstructing Western
metaphysics; it is quite another to take that deconstruction into the political
domain. In other words, while we may abandon our search for the chimera of
"absolute foundations" or "final truths" in epistemology, science, and political theory, we should not confuse such abandonment with relinquishing our
social solidarity as expressed in the liberal humanism which defines the Western world.[34] Hence, while Rorty likes Heidegger's method, he accuses Heidegger's critique of modernity and industrial technology of being in some respects
naive. There is really no alternative, so Rorty insists, to increasing our commitment to the industrial technology which has come out of the Enlightenment—unless we are willing to see millions of people starve to death around
the world.[35] In a world bereft of foundations for making monumental decisions, however, we may well ask: *On what basis* are we to say that it is better to
feed starving millions, for example, than to worry about the fate of the entire
human species in the face of a population explosion which threatens the
stability of the biosphere? Moreover, are there not *empirical* questions to be
asked regarding the relationship between those starving millions, on the one
hand, and the influence of colonial-imperial economic practices—including
those sponsored by the rich industrial democracies praised by Rorty—which
helped to create the conditions for "overpopulation"?

Rorty justifies his attempt to separate the activity of deconstructing the
Western fascination with "objectivity" and "foundations" from the commitment to democratic-liberal social solidarity by saying that theory has not
played a significant role in the praxis involved in establishing and furthering
American democratic principles.[36] Unfortunately, however, we may not so
readily separate cultural criticism from its possible political consequences.
Political theory, including the Enlightenment universalism of America's
"founding fathers," played an important role in the history of American democracy. Rorty tends to downplay this particular case of a beneficial consequence of political-cultural theory, however, in order to highlight the dangers
involved in attempting to force social reality to live up to the "objective"
demands of a particular theory, whether it be as sophisticated as Marxism or
as primitive as National Socialism. Rorty, then, believes that his resistance to
foundational theoretical schemes is the best way of defending against totalizing schemes which purport to be grounded in an objective, universal
understanding of human nature and the purpose of historical existence. Unfortunately, National Socialism swept into power in part because of its attack
upon the Enlightenment principles of universalism, rationality, and objectivity. Deconstructing the theoretical foundations of Weimar parliamentarianism
helped to create a power vacuum that was quickly filled by Hitler's violent
reaction against everything decadent and "Western."

Just as the reactionary genius Heidegger purported to unmask the atomistic individualism and commercial egoism of the Enlightenment, so many of today's critics expose the class oppression, racial bias, and sexism involved in the Enlightenment's proclamation of "the rights of man." But we must learn to distinguish—as Foucault did—between the authentic emancipatory impulse of Enlightenment rationalism and the prejudices and hidden power interests which diverted at least some of that impulse into a quest for power and control. Surely many Enlightenment writers, with the benefit of hindsight, would now concede that their original visions of "man," "reason," "freedom," and "nature" were too constricted and biased. Certainly they would be shocked to hear themselves depicted as power-mongering racists, as apologists for capitalist exploitation of the masses, as sexists who excluded all but "masculinist" forms of rationality, or even as "speciesists" who reduced all of nature to the status of mere raw material for human exploitation. Enlightenment thinkers such as Voltaire believed that they were performing a profound kind of unmasking—by revealing the entrenched power motives in the guise of patriarchal domination of the young, ecclesiastical control of conscience and scientific inquiry, state control of speech, writing, assembly, and commerce. Further, despite the fact that most Enlightenment writers believed that only white, male, property-owning, educated males of European descent were deserving of fully "equal rights," subsequent political liberation movements— including the emancipation of the slaves and the granting of suffrage and other basic rights to women—succeeded primarily by claiming to be *fulfilling* the Enlightenment ideal: the achievement of *universal* human rights.

In view of the fact that Western countries (socialist and capitalist), in the past and today as well, frequently appeal to the "universal" political principles of the Enlightenment to justify their imperialistic intervention in Third World societies, we may understand why some critics conclude that those principles are really ideological in nature, i.e., intellectual apologies for oppression and domination secretly grounded on the assumption that Western values (and Western interests) are best.[37] Earlier in this century, many Germans, including Heidegger, in fact believed that their country was being invaded by the insidious influence of French and English power interests which masqueraded in the disguise of "universal" rights to political and economic liberty. National Socialism came into power by promising to rid Germany of such alien ideas as "constitutionally guaranteed freedoms." The disastrous consequences of that episode in German history should serve to warn of the dangers involved in expressions of contempt for the very political rights which make it possible for individuals to engage in a radical critique of a society supposedly "grounded" on those rights. There is a difference, of course, between questioning the metaphysical "foundation" of basic political rights and arguing that a given society hypocritically proclaims those rights while simultaneously denying them to various segments of its population.

As I mentioned above, much of deconstruction is in fact motivated by a (frequently hidden) liberatory interest. If the technological society eliminates all differences and reduces everything to the same monochromatic raw material, however, whence can arise the "fissure" which causes the authoritarian system of signifiers to tremble, to quake, to loosen up? While significant changes in a prevailing cultural paradigm cannot be explained merely in terms of arbitrary human decisions, neither can free human decisions be discounted in such an explanation. Humans are thrown at birth into a cultural discourse and, hence, into a destiny which they themselves did not choose. People cannot return to the "origins" of that discourse in order to start a new one. To a large extent, then, individuals are for the most part players in a game of institutional, social, political, economic, literary, artistic, and religious discourse, only part of which they comprehend.

Despite being shaped by such discourse, however, individuals are not merely automatons at the mercy of an inexorable destiny. Rather, they may also bring to their historical discourses unexpected insights, novel variations, new possibilities which reveal it is by no means fixed but instead is open to disrupture. It goes without saying, of course, that the possibility for such variation and novelty is greatly enhanced by political systems which both guarantee and encourage self-expression and which also promote the economic means necessary for individuals to develop the capacity for self-expression. Unfortunately, however, Heidegger regarded "self-expression" and "democratic principles" as bourgeois ideals symptomatic of the one-dimensional atomism and egoism of a modern subjectivism which was blind to the fact that the "actors" on the stage of human history were players in a drama that they did not themselves compose. If we may benefit from Heidegger's insight that modern technology is characterized by a one-dimensional way of disclosing entities, we must also be willing to criticize his presuppositions about the extent to which humans are incapable of resisting and developing alternatives to that disclosure.

C. SEXUAL DIFFERENCE VS. ONTOLOGICAL DIFFERENCE

Heidegger believed that he alone possessed the key needed to unlock the secrets of modern technology. In portraying *all* other accounts as superficial or derivative, he depicted his own account of Western history as the standard or basis against which all other accounts were to be measured. Despite his critique of the one-dimensional thinking at work in modern technology, then, Heidegger read not only Western history but also Western thinkers in a way that forced them to fit into his powerful but nevertheless limited conceptual grid. Surely, Heidegger's account of the "history of being" is an outstanding example of a "meta-narrative" which attempts to organize all phenomena in terms of an all-embracing conceptual scheme.

Heidegger was usually contemptuous and dismissive of the competing contemporary accounts of Western history: theological, liberal economic, Marxist, Hegelian, and naturalistic. In addition to these accounts, all of which have been continually updated by their proponents in the light of new discoveries in the sciences or in anthropology or in comparative religion, other accounts have arisen in recent decades. Examining one of these alternative accounts helps to de-center and to question Heidegger's supposedly privileged

and self-enclosing account of Western history and technology, thereby bringing to the fore the "difference" hidden by his own account. I have in mind the
feminist account of the origin and character of modern technology.

Feminists argue that in focusing on the ontological difference, Heidegger
concealed, marginalized, displaced, and suppressed another difference: the *sexual* difference. Heidegger's view of Western history, then, was blind to a different source for the domineering tendency of modern technology: the patriarchal
domination of woman, body, and nature. Feminists maintain that the origins of
modern technology lie not merely in developments within Greek philosophy,
and certainly not in the history of some esoteric "self-concealment of being,"
but rather in specific historical practices and institutions which constitute patriarchy. The history of patriarchy is a totalizing movement which represses and
denies the feminine "difference" which constantly threatens to undermine and
to displace the masculine "foundation" of all things.

Patriarchy is grounded upon a radical binary opposition between the
masculine (spirit, culture, knowledge, power, activity, soul, freedom, detachment, independence, transcendence, eternity) and the feminine (material,
nature, instinct, weakness, passivity, body, submission, immanence, relationship, dependence, finitude, and mortality). The patriarchal ego seeks to guarantee its own potency and immortality by dissociating itself from the female
pole of the binary opposition by virtue of which the masculinist ego manufac-
tures its own identity. Finitude, dependency, mortality, corporeality—all of
these are projected by the patriarchal ego onto the female, who is identified
with "mother nature" and the body. Intolerant of all reminders of these
modes of "otherness" or "difference" which undermine the delusional attempt to make itself God, the patriarchal ego attempts to subjugate this threatening "otherness" to the control of reason, discipline, technology.[38] Modern
technology, then, according to feminist critics, is the late expression of the
ancient patriarchal quest to transform the earth into a totally controlled environment which reflects the quest for mortal ego's paradoxical quest for immortality. As Rosemary Radford Ruether maintains,

> The nineteenth-century concept of "progress" materialized the Judeo-Chris-
> tian God concept. Males, identifying their egos with transcendent "spirit," made
> technology the progressive incarnation of transcendent "spirit" into "nature."
> The eschatological gods became a historical project. Now one attempted to realize
> infinite demand through an infinite expansion of productive power. Infinite de-
> mand incarnate in finite nature, in the form of infinite exploitation of the earth's
> resources for production, results in ecological disaster: the rapid eating up of the
> organic foundations of life under our feet in an effort to satisfy ever-growing appe-
> tites for goods. The matrix of being, which is no less the foundation of human
> being, is rapidly depleted. . . .
> . . . The patriarchal self-deception about the origins of consciousness ends
> logically in the destruction of the earth.[39]

While agreeing that the West is governed by an aggressive, hubristic,
death-denying drive for total control, Heidegger maintained that this drive

arose because of *ontological*, not gender-related, factors. Feminists would reply that Heidegger's talk of the "history of being" is but another instance of the tendency to denigrate concrete, material social interaction and the organic dimension of human existence by explaining human history as the expression or servant of some transcendent source, whether it be the father God or the self-concealing Origin. Heidegger's denial that humans arise from what Ruether calls the organic "matrix of being" is, in the view of some feminists, evidence of his contempt for the body and of his preference for defining humanity in terms of "transcendent" awareness.

Especially in view of the masculinist vocabulary he used during the 1930s, Heidegger seems to have had little sympathy with a feminist account of Western history. And yet in his notion that human existence is essentially "care," in his doctrine of authentic producing as "letting things be," and in his belief that people are not social atoms but instead are internally related with one another, Heidegger often thought in terms of non-masculinist categories. Moreover, as Derrida has suggested, one can discern an important difference between early and later Heidegger, a difference that may be described as a shift from a "masculinist" to a "feminist" conception of truth, as a gesture toward Nietzsche's notion that "truth is a woman."[40]

According to Derrida, Nietzsche regards metaphysics as a clumsy masculinist enterprise which naively believed that there was an eternal or primordial truth to be obtained by a frontal assault. For Nietzsche, however, there is no such final truth; rather, "truth" is de-centered, beguiling, abysmal, self-concealing, non-essentializing, perspectival, non-appropriative, and non-identical. In short, truth is the seductive Dionysian, the feminine which ultimately unravels and undermines all masculinist attempts to appropriate her, to make her fixed, concrete, eternal, unchanging. Early Heidegger may be said to have defined truth in a way analogous to masculinist metaphysics, namely, as the violent hermeneutical activity of disclosing or of making manifest that which is hidden: being as such. According to early Heidegger, to be "authentic" or "owned" (*eigentlich*) meant to appropriate (*sich aneignen*) what is most worthy of thinking: the being of entities.

Later Heidegger, however, recognizing the extent to which his previous quest to disclose being as such betrayed his links with the foundationalist metaphysics of which he became so critical, focused not on the thinker's appropriative activity of uncovering what is hidden, but rather on the self-concealing, dissimulating play of *Ereignis*, the event which "gives" both being and time, presencing and temporality. Heidegger, in other words, began to see that what is truly worthy of thinking is not presencing but rather the absencing which makes such presencing possible. Such absencing is "truth," understood as the self-concealing and abysmal event (*-lethe*) which "gives" both presencing and temporality, but which can never itself be made present, i.e., "true." Derrida aptly remarks that for later Heidegger

> truth, unveiling, illumination are no longer decided in the appropriation of the
> truth of Being, but are cast into its bottomless abyss as non-truth, veiling and

dissimulation. The history of Being becomes a history in which no being, nothing, happens except *Ereignis'* unfathomable process. The proper-ty of the abyss (*das Eigentum des Abgrundes*) is necessarily the abyss of proper-ty, the violence of an event which befalls without Being.[41]

This ever-receding, dissumulating, self-withdrawing, and self-veiling event of truth, that which "gives" the various modes of presencing which have stamped the history of the West, may be understood in part as *Ent-eignis*, as the event of dis-appropriation, as the "feminine" which refuses to be dis-closed or unveiled by an obtrusive metaphysical-masculinist thrust. The thinker, then, is not to appropriate being but instead to pay homage to the self-concealing abyss which "gives" being in the first place, to submit to or to be appropriated by the play which can never be controlled or understood. Perhaps the major difference between Heidegger and Derrida is that while the former insisted that in this abyss there remains a mystery worthy of contem-plation, a mystery which "gives" the ontological and temporal perimeters of the West, the latter contends that there is no such mystery, but rather only the unending, unpredictable, and ungroundable play of signifiers.[42]

That even the allegedly patriarchal Heidegger can be read as having called into question certain of the masculinist dimensions of his own ontologi-cal project reminds us of the dangers involved in applying any interpretive framework which tries to explain too much. Essentialist feminism, represented by authors such as Susan Griffin, Mary Daly, and Adrienne Rich, risks elevat-ing in an ahistorical and uncritical way the "feminine" to a privileged position in respect to the now de-centered and dis-counted "masculine."[43] Derrida, however, despite his playful depiction of the struggle between the masculinist unveilers and the dissimulating feminine, warns that "there is no such thing as a Being or an essence of *the* woman or the sexual difference. . . . "[44] Likewise, Julia Kristeva has urged us not to make the difference between male and female, masculinist and feminist, into a binary opposition in which the female pole becomes depicted as essentially superior, for "the very dichotomy man/ woman as an opposition between two rival entities may be understood as belonging to *metaphysics*."[45] While the feminist interpretation of the patriar-chal origins of modern technology may de-center important aspects of Hei-degger's account of technology and may thereby reveal the sexual "difference" that he largely ignored, at least some of those interpretations seem to privilege the "essential" female in a way analogous to how Heidegger privileged the history of being in his own account of modern technology.

Western history is overdetermined by numerous factors; no single expla-nation can hope to account for the extraordinary complexity involved in the origins, development, and future of modern technology. Heidegger contrib-uted to the contemporary discourse which depicts modern technology as a manifestation of hidden power interests. These interests, including those of state socialism and monopoly capitalism, have justified themselves on the basis of a foundationalist narrative, one of the key doctrines of which has been "universal human progress."

Those who engage in the deconstruction of this narrative do not necessarily share Heidegger's concern about the dangers of a nihilistic world. Nor
do they always discern the political perils involved in the radical critiques of
the Enlightenment's "progressive" ideals regarding human liberty. While we
may justifiably call for the dismantling of all authoritarian and hierarchical
structures, and while we may laud attempts to de-center the primacy of the
Enlightenment's atomistic, masculinist subject, we may also question the tendency to be relentlessly suspicious regarding the motives behind the aim of
protecting the "rights" of individual subjects. Events of the twentieth century
show that those movements which denigrate individual freedoms either as
bourgeois ideals based on outmoded metaphysical foundations (Marxism), or
as egocentric practices associated with the degeneration of those foundations
(fascism), usually lead to repressive, collectivist, totalitarian political regimes.
If the Enlightenment conception of the "subject" deserves redefinition, the
Enlightenment's goal of protecting individuals from oppression and abuse also
deserves being expanded and preserved.

Despite its emancipatory aims, however, the Enlightenment also promoted attitudes which, especially in conjunction with subsequent developments in the economic and industrial realms, have in some ways clearly
blunted or compromised those aims. The dark side of the Enlightenment
includes anthropocentrism, subject-object dualism, and a totalizing tendency
reflected in the overestimation of the importance of scientific and instrumental rationality (metaphysical foundationalism) at the expense of other modes
of reasoning. Instrumental rationality, when placed in the service of the power
drive present in monopoly capitalism and state socialism, tends to produce
social formations that not only repress (whether overtly or indirectly) political
freedoms but also distort the development of the kind of subjects necessary to
engage in the communicative praxis required to reveal and to criticize such
repression. The liberal commitment to protecting individuals from oppression,
then, requires a critique of how market values, technological innovations, and
concentrations of wealth in monopoly capitalism tend not only to undermine
the structure of the subject who is supposedly "free," but also to displace or to
marginalize those traditions and attitudes which promote community and solidarity, and which resist being absorbed into the domain of liberal political
economy. While we must keep in mind the dangers of Heidegger's illiberal
worldview, then, we must also take seriously his critique of the limitations of
the technocratic character and the atomistic individualism of most modern
liberal democracies. As one critic recently noted, Heidegger "dared to state
that human fulfillment is not likely to be attained through an ever-expanding
technology or in a managerial society, and that democratic individualism has
resulted in the loss of cultural specificity and in delegitimating long-estab-
lished community."[46]

In my view, if humankind is both to achieve liberation from various kinds
of political, social, cultural, and economic oppression, and to avoid destroying
the ecosphere by nuclear war or by industrial pollution, new narratives are
needed which delineate and celebrate the differences inherent in a multi-

voiced humanity, which attempt both to define and to protect the "rights" and "interests" that are arguably common to the great majority of humans at this point in history, which encourage the development of communities that do not involve regression to collectivist practices or attitudes, which develop an alternative to the dissociative and anthropocentric attitude toward nature and the human body, and which emphasize the importance of modes of reasoning other than instrumental-scientific without at the same time denigrating the latter. While these new narratives would celebrate diversity, they would possibly discover elements of a common narrative in the one now being developed by post-modern scientists, who have replaced the mechanist model of nature with one that emphasizes nature's capacity to develop novelty, complexity, uniqueness, and freedom.[47]

To some extent, the optimism of the totalizing narratives of modernity was grounded in the belief that nature is a machine which can be completely understood and controlled. The non-mechanist narratives of post-modern science suggest that the natural world is far too complex and diverse to be understood in terms of a single principle, despite the ongoing search for a "grand unified theory" in physics. Post-modern science, then, not only helps to undercut the basis for totalizing narratives which promote domination, but also sets the stage for overcoming the humanity-nature dualism which has helped to justify the "exploitation" of nature without and nature within. Such overcoming will take a long time and will require the development of different narratives in different cultural contexts. While affirming that human awareness arises from the play of natural forces, and while emphasizing the need to show respect for all forms of life, the requisite new narratives would not call for a primitive "return" to nature in the form of the recollectivization that characterized fascist movements. Rather, as I shall argue in a subsequent book, these new narratives would ideally encourage the development of authentic individuation that is consistent not only with a richly variegated human culture but also with a diverse and stable biosphere.

In many different domains, including natural science and philosophy, there is a growing consensus that the ancient search for final truths is flawed not only because humanity itself is finite, and because the very activity of interpretation changes what is being interpreted, but also because the universe itself is changing in ways that transcend the activity of interpretation. Although Heidegger may not have agreed with this particular way of describing the finite character of human understanding, he would have agreed with the conclusion that human existence is historical and thus lacks absolute foundations for its projects and attitudes. Sensitive to the dangers of nihilism posed by the dissolution of previous foundations, Heidegger attempted to find a non-absolute, historical "ground" to guide his own people. Unfortunately, this attempt ended in disaster.

Half a century ago, Heidegger faced the complex crisis of how to endure the end of an old world and how to assist in the advent of a new one. Despite the fact that Western people have now moved into the technological world which Heidegger envisioned and at first resisted, we share an aspect of Hei-

degger's own crisis, namely, how to define and to promote human well-being
in the embrace of a technocratic economic system which seems to promote a
new version of recollectivization by standardizing "experience" (patterns of
consumption), homogenizing culture, colonizing leisure time, and excluding
difference. We can learn from Heidegger's meditation on modern technology,
even if what we learn is simply that there are political perils associated with
attempts to found a post-technological world. In important respects, his question remains our own: can we develop the non-absolutist, non-foundational
categories necessary to assess, to confront, and to transform the technological
and economic mobilization of humanity and the earth at the beginning of the
twenty-first century?

Notes

INTRODUCTION

1. In *Being and Technology: A Study in the Philosophy of Martin Heidegger* (The Hague: Martinus Nijhoff, 1981), John Loscerbo has provided an insightful but completely ahistorical examination of Heidegger's concept of technology. Loscerbo makes only one footnoted reference to the political dimension of that concept.

2. Concerning the renewal of the question regarding Heidegger's political past, see my essay "The Thorn in Heidegger's Side: The Question of National Socialism," *Philosophical Forum*, XX (Summer, 1989), pp. 326–365. See also my essay "L'affaire Heidegger," *Times Literary Supplement*, No. 4462 (October 7–13, 1988), pp. 1115–1117.

3. Jeffrey Herf, *Reactionary Modernism: Technology, Culture, and Politics in Weimar and the Third Reich* (New York: Cambridge University Press, 1984). I am greatly indebted to Herf's excellent book.

4. As I discovered only after completing this study, Albert Borgmann and Carl Mitcham argue that two of the major lacunae in our understanding of Heidegger's concept of technology are (1) his relationship to Ernst Jünger and reactionary politics and (2) the role of art in providing an alternative to the technological mode of "producing." This study fills these gaps in our understanding of Heidegger's thought. Cf. Borgmann and Mitcham's outstanding survey "The Question of Heidegger and Technology: A Review of the Literature," *Philosophy Today*, No. 2, XXXI (Summer, 1987), pp. 98–194. Borgmann and Mitcham rightly note that Phillip R. Fandozzi's study *Nihilism and Technology: A Heideggerean Investigation* (Washington, D.C.: University Press of America, 1982) is important and helpful for our understanding of Heidegger's concept of modern technology.

For an attempt to show Heidegger's place in the emerging field known as "philosophy of technology," see Carl Mitcham, "What Is the Philosophy of Technology," *International Philosophical Quarterly*, XXV, No. 1 (March, 1985), pp. 73–88. Cf. also Jean-Yves Goffi, *La philosophie de la technique* (Paris: Presses Universitaires de France, 1988).

1. GERMANY'S CONFRONTATION WITH MODERNITY

1. Paul Monaco, *Modern European Culture and Consciousness, 1870–1970* (Albany: SUNY Press, 1983), p. 91.

2. My thanks to Charles M. Sherover for reminding me of the liberal dimension of German political history.

3. David B. King, "Culture and Society in Modern Germany: A Summary View," pp. 15–44, in *Essays on Culture and Society in Modern Germany*, ed. Gary D. Stark and Bede Karl Lackner (College Station: Texas A & M Press, 1982). King's essay is an

excellent survey of modern German history. Cf. also Golo Mann, *The History of Germany since 1789*, trans. Marian Jackson (New York: Frederick A Praeger, 1968).

4. Edmond Vermeil, *Germany in the Twentieth Century: A Political and Cultural History of the Weimar Republic and the Third Reich* (New York: Praeger, 1956), p. 15.

5. See, for example, George L. Mosse, "Fascism and the French Revolution," *Journal of Contemporary History*, Vol. 24, No. 1 (January, 1989), pp. 5–26.

6. Cited by Shlomo Avineri, *Hegel's Theory of the Modern State* (New York: Cambridge University Press, 1974), p. 93.

7. V. R. Berghahn, *Modern Germany: Society, Economy, and Politics in the Twentieth Century* (Cambridge: Cambridge University Press, 1982), p. 1; emphasis in quotation mine. This is an excellent survey of the economic and political circumstances which faced Heidegger and other Germans in the first part of this century. Cf. also Kenneth D. Barkin, *The Controversy over German Industrialization, 1890–1902* (Chicago: University of Chicago Press, 1970); W. O. Henderson, *The Rise of German Industrial Power, 1834–1914* (Berkeley: University of California Press, 1975); Karl Hardach, *The Political Economy of Germany in the Twentieth Century* (Berkeley: University of California Press, 1980); Gustav Stolper, *Deutsche Wirtschaft seit 1870* (Tübingen: J. C. B. Mohr, 1966).

8. Berghahn, *Modern Germany*, p. 1. I am in debt to Berghahn for several of the factual observations which I make in this chapter.

9. George L. Mosse, *The Crisis of German Ideology: Intellectual Origins of the Third Reich* (New York: Grosset & Dunlap, 1964), pp. 13–14. Mosse is famous for his wide-ranging essays on right-wing movements in Germany and Europe in the twentieth century.

10. Ibid., p. 15.

11. Cited in Rodney Stackelberg, *Idealism Debased: From Völkisch Ideology to National Socialism* (Kent, Ohio: Kent State University Press, 1981), p. 12.

12. For an excellent account of Langbehn and van den Bruck, see Fritz Stern, *The Politics of Cultural Despair: A Study in the Rise of the Germanic Ideology* (Garden City, N.Y.: Doubleday Anchor, 1965).

13. Cited in Christian Graf von Krockow, *Die Entscheidung: Eine Untersuchung über Ernst Jünger, Carl Schmitt, und Martin Heidegger* (Stuttgart: Ferdinand Enke Verlag, 1958), pp. 37–38. Von Krockow's book provides a very helpful survey and analysis of the intellectual, political, and cultural matrix in which Heidegger's early thought arose.

14. Max Scheler, *Problems of a Sociology of Knowledge*, trans. Manfred S. Frings (London: Routledge & Kegan Paul, 1980), p. 133.

15. Ibid., p. 118.

16. Ibid., pp. 56–57.

17. According to Louis Dupeux, Thomas Mann used the term "conservative revolution" in 1926, as did Hugo von Hofmannsthal, but already in 1906 Moeller van den Bruck (later, author of *The Third Reich* [1922]) was using the term. Cf. Dupeux, *Stratégie communiste et dynamique conservatrice: Essai sur les différents sense de l'expression "National-Bolschevisme" en Allemagne, sous le République de Weimar (1919–1933)* (Paris: Librairie Honore Champion, 1976), pp. 1–2.

18. Stern, *The Politics of Cultural Despair*, p. 11.

19. Cited in von Krockow, *Die Entscheidung*, p. 17.

20. For a surprisingly sympathetic treatment of Dilthey's thought, see Georg Lukács, *The Destruction of Reason*, trans. Peter Palmer (Atlantic Highlands, N.J.: Humanities Press, 1981), pp. 418–447.

21. Herf, *Reactionary Modernism*.

22. On the high-minded but misguided idealism of the *völkisch* movement, see Stackelberg, *Idealism Debased*.

23. Cited by von Krockow, *Die Entscheidung*, p. 20.

24. Ibid., p. 21.

25. Tim Keck, "Practical Reason in Wilhelmian Germany: Marburg Neo-Kantian

Thought in Popular Culture," in *Political Symbolism in Modern Europe*, ed. Seymour Drescher, David Sabean, and Allan Sharlin (New Brunswick: Transaction Books, 1982), p. 64.

26. Ibid., p. 74.

27. Vermeil, *Germany in the Twentieth Century*, p. 60.

28. Gordon A. Craig, *The Germans* (New York: G. P. Putnam's Sons, 1982), p. 143.

29. For an overview of the extraordinary times of Weimar, see Karl Dietrich Bracher, *The German Dictatorship*, trans. Jean Steinberg (New York: Praeger, 1970); Peter Gay, *Weimar Culture: The Outsider as Insider* (New York: Harper & Row, 1968).

30. Charles S. Maier, "Between Taylorism and Technocracy: European Ideologies and the Vision of Industrial Productivity in the 1920s," *Journal of Contemporary History*, 5 (1970), pp. 54–55.

31. Herf, *Reactionary Modernism*, pp. 154–155. See chapter seven, "Engineers as Ideologues," for a stimulating discussion of the role of engineers in "spiritualizing" modern technology.

32. Ibid., p. 154.

33. Ibid., pp. 170–171.

34. Ibid., p. 159.

35. Cited in ibid., p. 163.

2. POLITICAL ASPECTS OF HEIDEGGER'S EARLY CRITIQUE OF MODERN TECHNOLOGY

1. The best available account of Heidegger's life is Hugo Ott, *Martin Heidegger: Unterwegs zu seiner Biographie* (Frankfurt: Campus, 1988). For a number of observations about Heidegger's life, I am indebted to Ott's important work.

2. Michael E. Zimmerman, *Eclipse of the Self: The Development of Heidegger's Concept of Authenticity* (Athens: Ohio University Press, 1981).

3. On the ideal of the "cultured man," see Charles E. McClelland, "The Wise Man's Burden: The Role of Academicians in Imperial German Culture," in *Essays on Culture and Society in Modern Germany*, pp. 45–69.

4. For a detailed account of Heidegger's dramatic conception of philosophy, see my essay "Heidegger's 'Existentialism' Revisited," *International Philosophical Quarterly*, No. 3, XXIV (September, 1984), pp. 219–236.

5. Charles Guignon, *Heidegger and the Problem of Knowledge* (Indianapolis: Hackett Publishing Co., 1983), p. 86. This is an excellent interpretation of Heidegger's thought.

6. Ibid., pp. 41–57.

7. Stephen Eric Bronner, "Martin Heidegger: The Consequences of Political Mystification," *Salmagundi*, No. 38–39 (Summer–Fall, 1977), p. 169.

8. See Russell Nieli, *Wittgenstein: From Mysticism to Ordinary Language* (Albany: SUNY Press, 1987), pp. 15–29: "Heidegger's Nothing and the Experience of Radical Derealization-Depersonalization."

9. Lucio Coletti, *Marxism and Hegel*, trans. Lawrence Garner (London: Verso, 1979), pp. 172–173.

10. Theodor W. Adorno, *The Jargon of Authenticity*, trans. Kurt Tarnowski and Frederic Will (Evanston: Northwestern University Press, 1973), pp. 35–36.

11. Max Scheler, "Reality and Resistance: On *Being and Time*, Section 43," trans. Thomas Sheehan, in Sheehan, *Heidegger: The Man and the Thinker* (Chicago: Precedent Publishing, 1981), p. 143. For a helpful account of Scheler's views on modern technology, see Manfred S. Frings, "Zum Problem des Ursprungs der Technik bei Max Scheler," *Phänomenologische Forschungen*, Vol. 15, *Studien zum Problem der Technik* (Freiburg/München: Verlag Karl Alber).

12. Winfried Franzen, *Von der Existenzialontologie zur Seinsgeschichte: Eine Unter-suchung über die Entwicklung der Philosophie Martin Heideggers* (Meisenheim am Glan: Verlag Anton Hain, 1975), p. 71. Franzen's book offers many insightful critical analy-ses of Heidegger's thought and its relation to his political circumstances.

13. Ibid., p. 73.

14. Oswald Spengler, *The Decline of the West* (*Der Untergang des Abendlandes*), 2 vols., trans. Charles Francis Atkinson (New York: Alfred A. Knopf, 1928).

15. *Mensche und Technik: Beitrag zu einer Philosophie des Lebens* (München: C. H. Beck'sche Verlagsbuchhandlung, 1932), p. 68.

16. Ibid., p. 69.

17. Ibid., pp. 78–79; my emphasis.

18. Ibid., p. 71.

19. Ibid., p. 75.

20. Ibid., p. 81.

21. See Oswald Spengler, *Preussentum und Sozialismus* (München: Oscar Beck, 1919).

22. Ludwig Klages, *Geist als Widersacher der Seele* (Leipzig: J. A. Barth, 1929). A new edition was published by Bouvier (Bonn, 1981). For an insightful commentary on Klages, cf. Richard Hinton Thomas, "Nietzsche in Weimar Germany—and the Case of Ludwig Klages," in Anthony Phelan, ed., *The Weimar Dilemma: Intellectuals in the Weimar Republic* (Manchester: Manchester University Press, 1985).

23. Scheler, *Problems of a Sociology of Knowledge*, pp. 133–134.

24. For an examination of the topic of boredom in Heidegger's 1929–30 lectures, see Parvis Emad, "Boredom as Limit and Disposition," *Heidegger Studies,* I (1985), pp. 63–78.

25. See Friedrich Hölderlin, *Patmos*, in *Selected Verse*, ed. and trans. Michael Ham-burger (London: Anvil Press, 1986), p. 193. This edition includes the German text and translation.

26. Cited by Fritz Stern, in *Dreams and Delusions: The Drama of German History* (New York: Alfred A. Knopf, 1987), p. 154.

3. HEIDEGGER, NATIONAL SOCIALISM, AND MODERN TECHNOLOGY

1. A translation of *Der Arbeiter* is being prepared by Dirk Leach.

2. Heidegger was replying to the following essay by Jünger: "Über die Linie," in *Martin Heidegger zum 60. Gebürtstag* (Frankfurt am Main: Vittorio Klostermann, 1959).

3. Portions of the following remarks appeared in "L'affaire Heidegger" in *Times Literary Supplement* and in my essay "Philosophy and Politics: The Case of Heidegger," *Philosophy Today*, XXXIII, No. 2 (Summer, 1989). For a more in-depth treatment of this topic, see my essay "The Thorn in Heidegger's Side: The Question of National Social-ism," *Philosophical Forum* (Summer, 1989).

4. Jacques Derrida, *De l'esprit: Heidegger et la question* (Paris: Galilée, 1987).

5. See the essay on Heidegger in Jürgen Habermas, *The Philosophical Discourse of Modernity*, trans. Frederich G. Lawrence (Cambridge: MIT Press, 1987). In his intro-duction to the German edition of Farias's book, Habermas takes a few steps back from his claim that *Being and Time* anticipates Nazi political views.

6. Otto Pöggeler, "Afterword to the Second Edition," in *Martin Heidegger's Path of Thinking*, trans. Daniel Magurshak and Sigmund Barber (Atlantic Highlands, N.J.: Humanities Press International, 1987), p. 272.

7. Philippe Lacoue-Labarthe, *La fiction du politique* (Paris: Christian Bourgois, 1987), p. 38.

8. Herbert Marcuse, "Beiträge zu einer Phänomenologie des historischen Mater-ialismus," *Philosophische Hefte*, I (1928), pp. 45–68. A translation of this essay, "Con-

tributions to a Phenomenology of Historical Materialism," appears in *Telos*, No. 4 (Fall, 1969), pp. 3–34. Cf. also Marcuse, "Über konkrete Philosophie," *Archiv fur Sozialwissenschaft und Sozialpolitik*, 62 (1929), pp. 111–128; Marcuse and Olafson, "Heidegger's Politics: An Interview," *Graduate Faculty Philosophy Journal*, VI (Winter, 1977), pp. 28–40; and my essay "Heidegger and Marcuse: Technology as Ideology," *Research in Philosophy and Technology*, II (1977), pp. 245–261.

In a letter to *Der Spiegel* (March 7, 1966, pp. 10–11), Marcuse stated that *Being and Time*, which owed so much to Kierkegaard, "had no unequivocal [political] implication. . . . *Sein und Zeit*, the disgraceful acts of Freiburg, and the turbid, orphic platitudes of the Hölderlin and Trakl essays ought not to be stitched together. No one can defend his propaganda for Hitler (in the raiment of philosophy). No fanatic can destroy his epochal *Sein und Zeit*."

9. Pierre Bourdieu, *L'ontologie politique de Martin Heidegger* (Paris: Les Éditions de Minuit, 1988 [originally published 1976]).

10. Keith Bullivant, "The Conservative Revolution," in *The Weimar Dilemma*, p. 50.

11. See Hans Sluga, "The Sense of Order: Philosophy and National Socialism" (manuscript in preparation).

12. Ibid., pp. 43–44. Many of the following citations are also found in Hans Sluga, *Gottlob Frege* (London: Routledge and Kegan Paul, 1980), pp. 59–60.

13. Hans Sluga, "The Meeting at Magdeburg" (unpublished manuscript).

14. Ibid.

15. Ibid.

16. Ibid. The remarks by Rings are found in Guido Schneeberger, *Nachlese zu Heidegger: Dokumente zu seinem Leben und Denken* (Bern: Buchdruckerei AG, 1962), p. 264.

17. Cited in Simon Taylor, *The Rise of Hitler* (New York: Universe Books, 1983), p. 74.

18. Barkin, *The Controversy over German Industrialization*, p. 129. This conservative critique of abstract industrial social relations also found expression in the American South shortly before the Civil War. Apologists for slavery argued that the Northern industrial system had reduced supposedly "free workers" to the status of wage slaves to be exploited by indifferent capital interests. The solution was to transform the so-called free worker into an industrial serf, whose interests would be looked after by paternalistic capitalists much in the way that slaves were known personally and cared for by wealthy Southern planters. For a critical exposition of this defense of Southern slavery, see Eugene D. Genovese, *The World the Slaveholders Made* (New York: Pantheon Books, 1969); *Roll, Jordan, Roll: The World the Slaves Made* (New York: Pantheon Books, 1979). On the curiously conservative dimension of the Marxist Genovese's writings, cf. Paul Gottfried, "The Mind of Eugene D. Genovese," *Telos*, No. 75 (Spring, 1988), pp. 173–179.

19. Heidegger formulated his own defense in several places. There are three major sources. First, there is his de-Nazification testimony to French authorities and to the Faculty Senate at Freiburg University in 1945–46. As a result of this testimony, Heidegger was forbidden to teach until the early 1950s, but was not summarily dismissed from the faculty. Much of his testimony to French authorities is found in Karl Moehling, "Heidegger and the Nazi Party: An Examination" (Ph.D. diss., Northern Illinois University, 1971). Second, in 1966 Heidegger consented to an interview about his political past by editors from *Der Spiegel* on the condition that he have the right to revise his own remarks (which he did, extensively) and that the interview be published only posthumously (which it was, in 1976). Heidegger's decision to grant this interview was prompted by an anonymous essay about his involvement with National Socialism which appeared in *Der Spiegel*, No. 7, February 7, 1966, pp. 110–113. Heidegger wrote a letter rebutting some of the charges in *Der Spiegel*, No. 11, March 7, 1966, p. 11. Third, Heidegger instructed his son, Herrmann, to publish at a "propitious time" a defense Heidegger composed (we are told) in 1945, called "The Rectorate 1933/34:

Facts and Thoughts." This defense, published in 1983, inadvertently helped to encourage Hugo Ott's groundbreaking historical investigations which have exposed the mendacious character of many remarks made in these "defenses."

Heidegger's most consistent defender has been Francois Fédier. Cf. Fédier, Trois attaques contre Heidegger," *Critique*, XXIII, No. 234 (November, 1966), pp. 883–904; Fédier, "À propos de Heidegger: Une lecture Denoncée," *Critique* XXIII (1967), pp. 672–686; Fédier, "À propos de Heidegger," *Critique*, XXIV (1968), pp. 435–437; Fédier, "'Une grosse bêtise'," *Le Nouvel Observateur*, January 22–28, 1988, p. 50; Fédier, "L'intention de nuire," *Le débat*, No. 48 (January-February, 1988), pp. 136–141.

Prior to the present debate, there were previous waves of critical examination of Heidegger's political involvement: (1) immediately after the war; (2) during the 1960s; (3) during the 1970s. Participants in the first wave included Maurice de Gandillac, "Entretien avec Martin Heidegger," *Les temps modernes*, II (1946), pp. 713–716; Alfred de Towarnicki, "Visite à Martin Heidegger," *Les temps modernes*, II (1946), pp. 717–724; Eric Weil, "Le cas Martin Heidegger," *Les temps modernes*, III (1947), pp. 128–138; Alphonse de Waehlens, "La philosophie de Heidegger et le nazisme," *Les temps modernes*, III (1947), pp. 115–127; Karl Löwith, "Les implications politiques de la philosophie de l'existence chez Heidegger," *Les temps modernes*, II (November, 1946), pp. 343–360.

Participants in the second wave included Jean-Pierre Faye, "Heidegger et la 'revolution'," *Médiations*, No. 3 (Autumn, 1961), pp. 151–159; Faye, "Attaques Nazies contre Heidegger," *Médiations*, No. 5 (Summer, 1962), pp. 137–151; Paul Hühnerfeld, *In Sachen Heidegger: Versuch über ein deutsches Genie* (Hamburg: Ferdinand Ecke Verlag, 1958); Alexander Schwann, *Politische Philosophie im Denken Heideggers* (Köln und Opladen: Westdeutscher Verlag, 1965); Robert Minder, "Heidegger, Hebel und die Sprache von Messkirch," *Der Monat*, Vol. 18 (July 1966), pp. 13–23; Minder, "À propos de Heidegger, langage, et Nazisme," *Critique*, No. 237 (1967), pp. 289–297; Faye, "La lecture et l'énoncé," *Critique*, No. 237 (1967), pp. 288–295; Beda Allemanne, "Martin Heidegger und die Politik," *Merkur*, No. 235 (October, 1967), pp. 962–976; Jean-Michel Palmier, *Les écrits politiques de Martin Heidegger* (Paris: L'Herne, 1968); Palmier and Frederic de Towarnicki, "Entretien avec Heidegger," *L'Expresse*, No. 954 (October 20–26, 1969), pp. 78–85.

Contributions during the 1970s included Otto Pöggeler, *Philosophie und Politik bei Heidegger* (Freiburg und München: Karl Alber, 1972); Michael E. Zimmerman, "Heidegger, Ethics, and National Socialism," *Southwestern Journal of Philosophy*, V (Spring, 1976), pp. 97–106; Karsten Harries, "Heidegger as a Political Thinker," *Review of Metaphysics*, XXIX, No. 4 (June, 1976), pp. 642–669; Herbert Marcuse and Frederick Olafson, "Heidegger's Politics: An Interview," *Graduate Faculty Philosophy Journal*, 6 (Winter, 1977), pp. 28–40; Stephen Eric Bronner, "Martin Heidegger: The Consequences of Political Mystification," *Salmagundi*, No. 38–39 (Summer-Fall, 1977); Thomas Sheehan, "Philosophy and Propaganda: Response to Professor Bronner," and Bronner, "The Poverty of Scholasticism/A Pedant's Delight: A Response to Thomas Sheehan," *Salmagundi*, No. 43 (Winter, 1979), pp. 173–199; Gregory Schufreider, "Heidegger on Community," *Man and World*, 14 (1981), pp. 25–84; Howard Eiland, "Heidegger's Political Engagement," *Salmagundi*, 70–71 (Spring-Summer, 1986), pp. 267–284; Michael E. Zimmerman, *Eclipse of the Self* (Athens: Ohio University Press, 1981 [1986]), chapter six, "National Socialism, Voluntarism, and Authenticity."

20. Cf. Hugo Ott, *Martin Heidegger: Unterwegs zu seiner Biographie* (Frankfurt am Main: Campus, 1988); Ott, "Martin Heidegger als Rektor der Universität Freiburg i. Br. 1933/34," *Zeitschrift des Breisgau-Geschichtsvereins*, 102 (1983), pp. 121–136; Ott, "Martin Heidegger als Rektor der Universität Freiburg, 1933/34," *Zeitschrift für die Geschichte des Oberheins*, 132 (1984), pp. 343–358; Ott, "Martin Heidegger und die Universität Freiburg nach 1945," *Historisches Jahrbuch*, 105 (1985), pp. 95–128; "Martin Heidegger und der Nationalsozialismus," in *Heidegger und die praktische Philosophie*, ed. Annemarie Gethmann-Siefert and Otto Pöggeler (Frankfurt am Main:

Suhrkamp, 1988). Many of Ott's findings are summarized in Richard Wolin, "Recherches récentes sur la relation de Martin Heidegger au National Socialism," *Les temps modernes*, 42 (October, 1987), pp. 56–85.

21. Victor Farias, *Heidegger et le nazisme* (Paris: Verdier, 1987). Cf. my review "L'affaire Heidegger," *Times Literary Supplement*, No. 4462 (October 7–13, 1988), pp. 1115–1117, and especially Thomas J. Sheehan's outstanding critical review "Heidegger and the Nazis," *New York Review of Books*, XXXV, No. 10 (June 16, 1988), pp. 38–47. Cf. also Richard Rorty, "Taking Philosophy Seriously," *New Republic* (April 11, 1988), pp. 31–34. A summary of the controversy unleashed in France over this book is found in "L'affaire Heidegger: Peut-on encore croire les philosophes?" by Pierre Boncenne, Jean Blain, and Alain Jaubert, *Lire*, No. 153 (June, 1988), pp. 41–54, and in Richard Wolin, "The French Heidegger Debate," *New German Critique*, No. 45 (Fall, 1988), pp. 135–161. Cf. also the essays by Maurice Blanchot, Hans-Georg Gadamer, Philippe Lacoue-Labarthe, and Emmanuel Levinas in the January 22–28, 1988, issue of *Le Nouvel Observateur*, as well as the essays by Pierre Aubenque, Henri Crétella, Michel Deguy, Gérard Granel, Stéphane Moses, and Alain Renaut in *Le débat*, No. 48 (January–February, 1988). There is an excellent collection of essays in Gethmann-Siefert and Pöggeler's book *Heidegger und die praktische Philosophie*. Also see Otto Pöggeler, "Den Führer führen? Heidegger und kein Ende," *Philosophische Rundschau*, XXXII (1985), pp. 26–67.

22. Dr. Carl Ulmer, letter to *Der Spiegel*, No. 19 (May 2, 1977), p. 10.

23. Karl Löwith, *Mein Leben in Deutschland vor und nach 1933* (Stuttgart: J. B. Metzler, 1986), p. 57. For a translation of Löwith's remarks about Heidegger, see "My Last Meeting with Heidegger in Rome, 1936," *New German Critique*, No. 45 (Fall, 1988), pp. 115–116.

24. Taylor, *The Rise of Hitler*, p. 106.

25. Jürgen Habermas, "Martin Heidegger on the Publication of Lectures from the Year 1935," trans. Dale Ponikvar, *Graduate Faculty Philosophy Journal*, VI, No. 2 (Fall, 1977), pp. 155–180.

26. Christian E. Lewalter, "Wie liest man 1953 Sätze von 1935?" *Die Zeit*, No. 39 (September 24, 1953), p. 6.

27. Ibid.

28. See Heidegger's letter in support of Lewalter, *Die Zeit*, No. 39, September 24, 1953, p. 18.

29. Cf. Pöggeler, *Martin Heidegger's Path of Thinking*, p. 278. Petra Jaeger, editor of the *Gesamtausgabe* edition of *Einführung in die Metaphysik*, agrees that Heidegger interpolated the passage in question in 1953.

30. Cf. Farias, *Heidegger et le nazisme*, pp. 213–233.

31. Cited by Wolfgang Schirmacher, *Technik und Gelassenheit: Zeitkritik nach Heidegger* (Freiburg/München: Karl Alber Verlag, 1983), p. 25.

32. Derrida, *De l'esprit*, p. 119.

33. Jacques Derrida, "L'enfer des philosophes," *Le Nouvel Observateur*, November 6–12, 1987, p. 172.

34. For a review of the conservative nature of German students and faculty during the early twentieth century, see Geoffrey J. Giles, "National Socialism and the Educated Elite in the Weimar Republic," in *The Nazi Machtergreifung*, ed. Peter D. Stachura (London: George Allen & Unwin, 1983). For more extended analyses, see Fritz K. Ringer, *The Decline of the German Mandarins: The German Academic Community, 1890–1933* (Cambridge: Harvard University Press, 1969); Alice Gallin, *Midwives to Nazism: University Professors in Weimar Germany, 1925–1933* (Macon: Mercer University Press, 1986); George L. Mosse, "Fascism and the Intellectuals," in *The Nature of Fascism*, ed. S. J. Woolf (London: Weidenfeld & Nicolson, 1968).

35. On these issues, see Ott, "Martin Heidegger als Rektor der Universität Freiburg," *Zeitschrift für die Geschichte des Oberrheins*. For an explanation of the meaning of *Gleichschaltung*, see Karl Dietrich Bracher, "Stages of Totalitarian 'Integration' (*Gleichschaltung*): The Consolidation of National Socialist Rule in 1933 and 1934," in *Repub-*

lic to Reich: The Making of the Nazi Revolution, ed. Hajo Holborn, trans. Ralph Manheim (New York: Pantheon Books, 1972).

36. Cited by Ott, "Martin Heidegger als Rektor der Universität Freiburg 1933/ 34," *Zeitschrift für die Geschichte des Oberheins*, p. 356.

4. JÜNGER AND THE *GESTALT* OF THE WORKER

1. Bourdieu, *L'ontologie politique de Martin Heidegger*, p. 36.

2. Herf, *Reactionary Modernism*, p. 16. This is an outstanding essay in political, cultural, social, and intellectual history. I benefited greatly from Herf's insights and from his nearly exhaustive bibliography on a wide range of topics dealing with technology and modernity in Weimar Germany.

3. Herf provides an excellent survey of the recent critical debate about the origins, character, and meaning of National Socialism.

4. Stern, *The Politics of Cultural Despair*, p. 354.

5. Jürgen Habermas has been a key figure in criticizing his Critical School predecessors for their (misguided) critique of the Enlightenment. See, for example, "The Entwinement of Myth and Enlightenment: Re-reading *Dialectic of Enlightenment*," trans. Thomas Y. Leven, *New German Critique*, No. 26 (Spring–Summer, 1982), pp. 13–30. Cf. also Peter U. Hohendahl, "The Dialectic of Enlightenment Revisited: Habermas' Critique of the Frankfurt School," *New German Critique*, No. 35 (Spring/ Summer, 1985), pp. 3–26.

6. While Max Weber spoke of the limits of what was politically achievable, and while he is often represented as upholding "liberal" values in early twentieth-century Germany, his liberalism was not the one familiar to those raised in the American, French, or British tradition. Weber affirmed the importance of the autonomous individual but also argued that a charismatic leader might be necessary to do battle with the "iron cage" of bureaucratic rationalism which was choking Germany and other countries. In affirming the "supremacy of personality in politics," did Weber inadvertently help to promote Hitler, who held a similar view? These issues are treated in a very illuminating way by S. Turner and R. Factor in *Max Weber and the Dispute over Reason and Value* (London: Routledge & Kegan Paul, 1984). Cf. also "The Illusion of Politics: Politics and Rationalization in Max Weber and Georg Lukács," *New German Critique*, No. 26 (Spring/Summer, 1982), pp. 55–79.

7. Herf, *Reactionary Modernism*, p. 12.

8. Ernst Jünger, foreword to *Krieg und Krieger* (Berlin, 1930), cited by Ansgar Hillach, "The Aesthetics of Politics: Walter Benjamin's 'Theories of German Fascism,'" *New German Critique*, No. 17 (Spring, 1979), p. 99. This is a very insightful essay.

9. Hillach, "The Aesthetics of Politics," p. 115.

10. Paul Fussell, *The Great War and Modern Memory* (New York: Oxford University Press, 1975), p. 115. One can scarcely say enough about this extraordinarily moving document about how World War I affected the participants—and about how it shaped the rest of the century. Cf. also Modris Eksteins, *The Rites of Spring: The Great War and the Birth of the Modern Age* (New York: Houghton-Mifflin, 1989); Robert Wohl, *The Generation of 1914* (Cambridge: Harvard University Press, 1979).

11. Cited by Hillach, "The Aesthetics of Politics," pp. 108–109.

12. Lukács, *The Destruction of Reason*, p. 437.

13. See Helmut Kaiser, *Mythos, Rausch und Reaktion: Der Weg Gottfried Benns und Ernst Jüngers* (Berlin: Aufbau-Verlag, 1962), pp. 105ff.

14. For an extraordinary account of the violent, masculinist outlook of the *Freikorps*, see Klaus Theweleit, *Male Fantasies*, Vol. I, trans. Stephen Conway (Minneapolis: University of Minnesota Press, 1987). John D. Caputo has pointed out to me that Nietzsche, in section 763 of *The Will to Power*, speaks of the masculinist martial virtues of the worker-soldier. "*From the future of the worker.*—Workers should learn to feel like soldiers. An honorarium, an income, but no pay!"

15. Hans-Peter Schwarz, *Der konservative Anarchist: Politik und Zeitkritik Ernst Jüngers* (Freiburg im Breisgau: Rombach, 1962), p. 69.

16. Karl Heinz Bohrer, *Die Aesthetik des Schreckens: Die pessimistische Romantik und Ernst Jüngers Frühwerk* (München: Carl Hanser Verlag, 1978). A highly informative and insightful book.

17. See, for example, Paul Tillich, "Kairos II: Ideen zur Geisteslage der Gegenwart," *Gesammelte Werke*, VI, *Der Widerstreit von Raum und Zeit* (Stuttgart: Evangelisches Verlagswerk, 1963); "Kairos und Logos: Eine Untersuchung zur Metaphysik der Erkenntnis," *Gesammelte Werke*, IV, *Philosophie und Schicksal*.

18. On the important idea of "decisionism," see von Krockow, *Die Entscheidung*.

19. Cited by Lukás, *The Destruction of Reason*, p. 491.

20. Friedrich Nietzsche, *The Birth of Tragedy*, trans. Walter Kaufmann, in *Basic Writings of Nietzsche*, ed. Kaufmann (New York: Modern Library, 1968), p. 60.

21. Ibid.

22. Cited in René J. Muller, *The Marginal Self* (Atlantic Highlands, N.J.: Humanities Press International, 1987), p. 142.

23. F. T. Marinetti, *Selected Writings*, ed. R. W. Flint, trans. R. W. Flint and Arthur A. Coppotelli (New York: Farrar, Straus & Giroux, 1972), p. 91.

24. Ibid., p. 42.

25. John Orr, "German Social Theory and the Hidden Face of Technology," *European Journal of Sociology*, XV (1974), p. 317.

26. On the topic of "National Bolshevism," see Dupeux, *Stratégie communiste et dynamique conservatrice*.

27. Orr, "German Social Theory and the Hidden Face of Technology."

28. Schwarz, *Der konservative Anarchist*, pp. 76–81.

29. Ernst Jünger, *On the Marble Cliffs*, trans. Stuart Hood (New York: Penguin, 1970).

30. Schwarz, *Der konservative Anarchist*, p. 193.

31. Cited in ibid., p. 193.

32. Cited in ibid., p. 194.

33. Ernst Jünger, "Le Travailleur: Entretien avec Ernst Jünger," recorded by Frederick de Towarnicki, in *Martin Heidegger*, ed. Michel Haar (Paris: L'Herne, 1983), pp. 147, 149.

34. Schwarz, *Der konservative Anarchist*, pp. 84–85.

35. Bohrer, *Die Aesthetik des Schreckens*, p. 476.

36. Ibid., p. 481.

37. Ibid., pp. 482–483.

38. Ibid., pp. 486–487.

39. Ibid., p. 488.

40. Cited in Herf, *Reactionary Modernism*, p. 196.

5. HEIDEGGER'S APPROPRIATION OF JÜNGER'S THOUGHT

1. On the importance of World War I for shaping a whole generation, see Robert Wohl, "Germany: The Mission of the Young Generation," in *The Generation of 1914* (Cambridge: Harvard University Press, 1979).

2. In "The Politics of Heidegger's Rectoral Address," *Man and World*, 20 (1987), pp. 171–187, Graeme Nicholson offers some helpful insights into this troubling address.

3. Ott, *Martin Heidegger, Unterwegs zu seiner Biographie*, pp. 146–166. These pages make up a section called " 'The Peculiar Yearning for Severity and Arduousness': The Soldierly Context of the Rector's Address." The title of Ott's chapter is taken from the essay by Winfried Franzen cited in the next footnote.

4. Winfried Franzen, "Die Sehnsucht nach Härte und Schwere," in *Heidegger und die praktische Philosophie*, ed. Annemarie Gethmann-Siefert and Otto Pöggeler (Frankfurt am Main: Suhrkamp, 1988), pp. 78–92.

5. Ott, *Martin Heidegger*, p. 147. Martin H. Sommerfeldt's biography of Göring.

6. Cited in ibid., p. 231.

7. Martin Heidegger, "Wege zur Aussprache," in Schneeberger, *Nachlese zu Heidegger: Dokumente zu seinem Leben und Denken* (Bern, 1961), pp. 258–262.

8. Schneeberger, *Nachlese zu Heidegger*, p. 181. In *Prophets of Extremity: Nietzsche, Heidegger, Foucault, Derrida* (Berkeley: University of California Press, 1985), Allan Megill provides a good commentary on Heidegger's Nazi speeches. In *New German Critique*, No. 45 (Fall, 1988), pp. 96–114, William S. Lewis has translated several of Heidegger's Nazi speeches.

9. Cited by Anson G. Rabinbach, "The Aesthetics of Production in the Third Reich," *Journal of Contemporary History*, II (1976), pp. 43–74. The quotation is found on p. 51. This essay is highly recommended.

10. Cited in Schneeberger, *Nachlese zu Heidegger*, p. 181. For a brief overview of National Socialist attitudes toward work, see T. W. Mason, "Labour in the Third Reich, 1933–1939," *Past and Present*, No. 35 (December, 1966), pp. 112–141.

11. Schneeberger, *Nachlese zu Heidegger,* p. 181.

12. Ibid., p. 200.

13. Ibid.

14. Cited in J. P. Stern, *Hitler: The Führer and the People* (Berkeley: University of California Press, 1975), p. 54. This book is an excellent analysis of the "spirit" of National Socialism. Ott, in *Martin Heidegger*, p. 302, reveals that Heidegger himself confessed to having read at least part of *Mein Kampf.*

15. Franzen, *Von der Existenzialontologie zur Seinsgeschichte*, p. 87.

16. Rabinbach, "The Aesthetics of Production in the Third Reich," p. 62.

17. Cited in Schneeberger, p. 200.

18. See for example Robert Minder, "Martin Heidegger ou le conservatisme agraire," *Allemagnes d'Aujourd'hui*, No. 6 (January–February, 1967), pp. 34–49. For a sympathetic account of the radical peasant movement in pre-Nazi and Nazi Germany, see Anna Bromwell, *Blood and Soil: Richard Walther Darre and Hitler's Green Party* (Buckinghamshire: The Kensal Press, 1985); cf. also Bromwell, *Ecology in the Twentieth Century: A History* (New Haven: Yale University Press, 1989).

19. Schneeberger, pp. 216–217. Trans. Thomas J. Sheehan, in *Heidegger: The Man and the Thinker*, p. 123.

20. Schneeberger, p. 218; Sheehan, p. 124.

21. Cited in Schneeberger, p. 157. The quotation is taken from a journalist's transcription of Heidegger's speech. There is every reason to believe that the transcription is accurate.

22. Alfred Rosenberg, *The Folkish Idea of the State*, in Barbara Miller Lane and Leila J. Rupp, ed., *Nazi Ideology before 1933: A Documentation* (Austin and London: University of Texas Press, 1978).

23. Cited in Schneeberger, p. 136.

24. Ibid., p. 181.

25. Peter D. Stachura, "German Youth, the Youth Movement, and National Socialism in the Weimar Republic," in Stachura, ed., *The Nazi Machtergreifung*; Michael H. Kater, "Generationskonflikt als Entwicklungsfaktor in der NS-Bewegung vor 1933," *Geschichte und Gesellschaft*, 11 (1985), pp. 217–243.

26. Cited in Schneeberger, p. 156.

27. Ibid., p. 48.

28. Ibid., p. 150.

29. Ibid., p. 75.

6. JÜNGER'S THOUGHT IN HEIDEGGER'S MATURE CONCEPT OF TECHNOLOGY

1. Martin Heidegger, "Wege zur Aussprache," cited in Schneeberger, *Nachlese zu Heidegger*, p. 259.

2. Theweleit, *Male Fantasies*, p. 229.

3. Ibid., p. 230.

4. Cited in Schneeberger, p. 261.

5. Ibid., p. 150.

6. In the 1920s, Max Scheler wrote about the rise of the industrial army in a way that anticipated what Heidegger was to say years later: "Everyone knows and feels it day by day, that the threatening step of the worker battalions—which daily oil and attend the powerful mechanism—in whose grasping arms our life is played out, also draws a new tone into our *spiritual* life, in politics, literature, art, science, and religion. . . . " *Die Wissensformen und die Gesellschaft, Gesammelte Werke*, Vol. 8 (Bern/München: Francke Verlag, 1960), p. 447.

7. Walter Benjamin, "The Work of Art in an Age of Mechanical Reproduction," in *Illuminations*, trans. Harry Zohn (London: Fontana/Collins, 1973).

8. Cf. Guy Debord, *La société du spectacle* (Paris: Buchet-Chastel, 1967).

9. Parvis Emad, "Heidegger on Pain: Focusing on a Recurring Theme in His Thought," *Zeitschrift für Philosophische Forschung*, 36 (July–September, 1982), pp. 354–355.

10. An excellent example of one of Solzhenitsyn's essays which reveals his proximity to Heidegger's critique of the Enlightenment and humanism is *A World Split Apart*, Commencement Address at Harvard University, June 8, 1978, trans. Irina Ilovayskaya Albertil (New York: Harper & Row, 1978).

11. For an elaboration of the religious themes in Heidegger's thought, see my essay "The Religious Dimension of the 'Destiny of Being,' " in *Phenomenology and the Understanding of Human Destiny*, ed. Steven Skousgaard (Washington, D.C.: University Press of America, 1981). See also John D. Caputo, "Heidegger's God and the Lord of History," *New Scholasticism*, LVII, 4 (Autumn, 1983), pp. 439–464.

7. NATIONAL SOCIALISM, NIETZSCHE, AND THE WORK OF ART

1. Peter L. Rudnysky, *Freud and Oedipus* (New York: Columbia University Press, 1987), p. 233. A highly recommended text.

2. My thanks to Rudnysky for illuminating the connection between Heidegger's reflections on the tragic vision in *Oedipus Rex*.

3. Cited by Hillach in "The Aesthetics of Politics."

4. Nietzsche, *The Birth of Tragedy*, in *Basic Writings of Nietzsche*, p. 52.

5. For an insightful discussion of how the discovery of "perspective" transformed the medieval into the modern world, see David Michael Levin, *The Opening of Vision: Nihilism and the Postmodern Situation* (New York: Routledge, 1988), pp. 110–116. That book is a gold mine of ideas that overlap with many of the ones discussed in this work.

6. Cited by Henry Grosshans in *Hitler and the Artists* (New York and London: Holmes & Meier, 1983), p. 27.

7. I owe my knowledge of this reference to Lacoue-Labarthe, *La fiction du politique*. Goebbels's letter appeared in the *Lokal-Anzeiger*, April 11, 1933; reprinted in Hildegard Brenner, *Die Kunstpolitik des Nationalsozialismus* (Reinbeck bei Hamburg: Rowohlt, 1963), pp. 178–180.

8. Adolf Hitler, *The Speeches of Adolf Hitler*, ed. Norman H. Baynes, Vol. I (New York: Oxford University Press, 1942), p. 569. Nuremberg Parteitag address on September 11, 1935.

9. On the Nazi attitude toward "degenerate" art, cf. Franz Roh, *"Entartete" Kunst: Kunstbarbarei im Dritten Reich* (Hannover: Fackeltrager-Verlag, n.d.).

10. Hitler, *The Speeches of Adolf Hitler*, p. 571.

11. Ibid., p. 573.

12. Ibid., p. 576.

13. Albert Speer, *Inside the Third Reich*, trans. Richard and Clara Winston (New York: Macmillan Co., 1970).

14. Anson G. Rabinbach, "The Aesthetics of Production in the Third Reich," *Journal of Contemporary History*, 11 (1976), pp. 43–74. The material referred to is found on p. 43. The literature on the "aesthetics" of National Socialism is extensive. See for example Noel O'Sullivan, "Fascism as Theatrical Politics," in *Fascism* (London: J. M. Dent & Sons, 1983); Vernon L. Lidtke, "Songs and Nazis: Political Music and Social Change in Twentieth-Century Germany," in *Essays on Cultural and Society in Modern Germany*; Russell A. Berman, "Modernism, Fascism, and the Institution of Literature," in *Modernism: Challenges and Perspectives*, ed. M. Chefdor, R. Quinones, and A. Wachtel (Urbana: University of Illinois Press, 1986). For non-fascist and leftist developments in German art, cf. John Willett, *Art and Politics in the Weimar Period: The New Sobriety [Die Neue Sachlichkeit], 1917–1933* (New York: Pantheon Books, 1978).

15. Rainer Stollmann, "Fascist Politics as a Total Work of Art: Tendencies of the Aesthetization of Political Life in National Socialism," trans. Ronald L. Smith, *New German Critique*, No. 14 (Spring, 1978), pp. 41–60. The material referred to is on p. 44.

16. Hans-Jürgen Syberberg, *Die freudlose Gesellschaft* (München: Carl Hanser Verlag, 1981), pp. 74–76. I owe this reference to Lacoue-Labarthe. On the importance of film for National Socialism, cf. David Weinberg, "Approaches to the Study of Film in the Third Reich," *Journal of Contemporary History*, XIX, No. 1 (January, 1984), pp. 106–126. On the aesthetic dimension of Hitler's highway program, cf. James D. Shand, "The *Reichsautobahn*: Symbol for the Third Reich," *Journal of Contemporary History*, XIX, No. 2 (April, 1984), pp. 189–200.

17. Gottfried Benn, "The Confession of an Expressionist," in Victor H. Miesel, ed., *Voices of German Expressionism* (Englewood Cliffs, N.J.: Prentice-Hall, 1970), p. 192. Else Buddeberg, *Gottfried Benn* (Stuttgart: J. B. Metzlersche Verlagsbuchhandlung, 1961), provides an excellent treatment of the relation between Benn's artistic views and his affiliation with National Socialism. The parallels with Heidegger's case, though not explored by Buddeberg, are remarkable. Although he does not mention Heidegger, Miesel's collection of essays provides the groundwork for understanding why Heidegger's thought is so often linked with expressionism.

18. Lacoue-Labarthe, *La fiction du politique*. I am indebted to this book for its insights about Heidegger, National Socialism, and art.

19. It is interesting to note that Jakob Burckhardt, one of Heidegger's cultural heroes, entitled the first chapter of his famous book *The Civilization of the Renaissance in Italy*, trans. S. G. C. Middlemore (London: Swan Sonnenschein & Co., 1898), "The State as Work of Art" ("Der Staat als Kunstwerk").

20. Burckhardt, *The Civilization of the Renaissance in Italy*, p. 62; *Die Kultur der Renaissance in Italien* (Berlin: Verlag von Th. Knaur Nachf., 1928). Curiously enough, what Burckhardt seems to have meant by the idea of the state as a "work of art" was not so much the cosmic origin of Venice but instead the attempts by political thinkers such as Machiavelli to conceive of the state as something to be planned and constructed by human beings. Machiavelli, like other Renaissance humanists, sought to move beyond the Catholic church's doctrine of natural law, according to which the state and social forms of life were God-given and thus inviolate. National Socialists, and Heidegger in his own way, seem to have conceived of the Third Reich both as a transhuman event and as something accomplished by human will.

21. Stern, *Hitler*, p. 32; emphasis in original.

22. Stollman, "Fascist Politics as a Total Work of Art," p. 47.

23. Benjamin, "The Work of Art in an Age of Mechanical Reproductions," in *Illuminations*, pp. 243–244. On this topic, cf. Marjorie Perloff, *The Futurist Movement: Avant-Garde, Avant Guerre, and the Language of Rupture* (Chicago: University of Chicago Press, 1986), pp. 28–30.

24. Stollman, "Fascist Politics as a Total Work of Art," p. 52.

25. The debate about the futurist exhibit is discussed by Brenner, *Die Kunstpolitik des Nationalsozialismus*, pp. 75ff.

26. Benjamin, "Art in an Age of Mechanical Reproduction," p. 244.

27. Rabinbach, "The Aesthetics of Production in the Third Reich," p. 63.

28. Ibid., p. 67.

29. Klee, "Creative Credo," in Miesel, *Voices of German Expressionism*, p. 83.

30. Cited in Grosshans, *Hitler and the Artists*, p. 121.

31. On the twin themes of decay in European culture and artistic resistance to such decay, see *Degeneration: The Dark Side of Progress*, ed. J. Edward Chamberlin and Sander L. Gilman (New York: Columbia University Press, 1985).

8. HÖLDERLIN AND THE SAVING POWER OF ART

1. My thanks to Hubert L. Dreyfus, Michel Haar, and Kathleen Wright for the insights they shared with me during our conference on "Heidegger and the Work of Art" at the University of California, Berkeley, in September, 1988. Particular thanks to Kathleen Wright for revealing to me the importance of Heidegger's 1934–35 lectures on Hölderlin for understanding his concept of technology.

2. For an important critique of Heidegger's "interpretation" of Hölderlin, see Annemarie Gethmann-Siefert, "Heidegger und Hölderlin. Die Überforderung des 'Dichters in dürftiger Zeit'," in *Heidegger und die praktische Philosophie*, ed. Gethmann-Siefert and Pöggeler. For a more sympathetic approach to Heidegger's interpretation of Hölderlin's poetry, see Fred R. Dallymayr, "Heidegger, Hölderlin, and Politics," *Heidegger Studies*, 2 (1986), pp. 81–95, and David Haliburton, *Poetic Thinking: An Approach to Heidegger* (Chicago: University of Chicago Press, 1981).

3. For an excellent treatment of Heidegger's conception of moods and their relation to the technological age, see Michel Haar, "*Stimmung* et pensée," in F. Volpi et al., eds., *Heidegger et l'idée de la phenomenologie* (Kluwer, 1988), pp. 265–283. See also Michel Haar, *Le chant de la terre* (Paris: L'Herne, 1987), for many helpful analyses of Heidegger on nature, art, and technology.

4. On this topic, cf. Haar, "*Stimmung* et pensée."

5. In an unpublished essay, "The Mo(u)rning of History," Kathleen Wright discusses the theme of endowment and anointed task.

6. It should be clear that Heidegger wanted to contrast *his* conception of a new encounter with the Dionysian with the *Nazi* conception of such an encounter. While Heidegger conceived of the Dionysian in ontological terms, the Nazis conceived of it in instinctual terms. The former wanted a renewal of German *Geist*; the latter, a renewal of German *Blut*.

7. Wright, p. 19.

8. Ibid., p. 20.

9. Ibid.

10. The first version of "The Origin of the Work of Art" was recently published by Emmanuel Martineau in a bilingual (German-French) edition: *De l'origine de l'oeuvre d'art* (Paris: Authentica, 1987).

11. I owe this insight to Michel Haar, personal conversation. I do not want to overemphasize the distinction between *schaffen* and *schöpfen*, since the violent "creating" (*schaffen*) Heidegger described in *An Introduction to Metaphysics* is to be construed not as a "subjective" activity of a human agent, but instead as a disclosive event compelled by being itself. Put otherwise, the creator does what is required for entities to show themselves and thus "to be."

12. To some extent, Heidegger was faced with the same issue posed today in the conflict between those who say that language discloses what is already there and those who say that language somehow constructs what we encounter. For a helpful survey of this topic, see Edward Pols, "After the Linguistic Consensus: The Real Foundation Question," *Review of Metaphysics*, 40 (September, 1986), pp. 17–40.

13. Michel Haar drew these distinctions in his presentation in Berkeley, September, 1988. Also see his book *Le chant de la terre*, to which I am indebted for a number of ideas in this chapter.

14. For a very insightful treatment of the idea of *Riss* in Heidegger's concept of art, see Christopher Fynsk, *Heidegger: Thought and Historicity* (Ithaca: Cornell University Press, 1986), pp. 124–157. Also see the account found in Joseph Kockelmans, *Martin Heidegger on Art and Art Works* (Dordrecht and Boston: M. Nijhoff, 1985).

15. Max Scheler, among others, also noted the fact that Christianity—especially in the Protestant Reformation—attributed all spiritual power to God, thereby stripping the earth of its animating forces and delivering it over to the technological Will to Power. Cf. Scheler, *Die Wissensformen und die Gesellschaft*, p. 444.

16. Cited by Lacoue-Labarthe, *La fiction du politique*, p. 129.

17. For an excellent study of Heidegger's interpretation of Hölderlin's concept of the Holy, see Michel Haar, "Heidegger et le dieu de Hölderlin."

18. Lacoue-Labarthe, *La fiction du politique*, pp. 62–63. Cf. also Philippe Lacoue-Labarthe, *L'imitation des Modernes* (Paris: Galilee, 1986).

19. Ibid., pp. 64ff.

20. Ibid., p. 63.

21. Jean-François Lyotard, *Heidegger et "les juifs"* (Paris: Galilee, 1988).

22. Ibid., pp. 45ff.

23. Ibid., p. 55.

24. I am indebted to Hubert L. Dreyfus for this insight.

25. Edith Wyschogrod, *Spirit in Ashes: Hegel, Heidegger, and Man-made Death* (New Haven: Yale University Press, 1985), p. 188.

26. Charles M. Sherover has written several excellent essays on "existential ethics." For example, cf. "Founding an Existential Ethic," *Human Studies*, 4 (1981), pp. 223–236. Cf. also Frank Schalow, *Imagination and Existence: Heidegger's Retrieval of the Kantian Ethics* (Washington, D.C.: University Press of America, 1986).

27. Cited in Ott, *Martin Heidegger*, p. 20. Heidegger originally made this statement in his 1942 lectures *Hölderlins Hymne "Der Ister,"* but it was not included in the edited edition, Vol. 53, *Gesamtausgabe*.

28. Ott, *Martin Heidegger*, p. 157.

29. Ibid.

30. Ibid., p. 159.

31. Concerning the postwar relationship of Bultmann to Heidegger, cf. ibid., pp. 162–163.

9. EQUIPMENT, WORK, WORLD, AND BEING

1. Gerold Prauss, *Erkennen und Handeln in Heideggers "Sein und Zeit"* (Freiburg/München: Verlag Karl Alber, 1977). This book is the best critical study of early Heidegger's concept of the distinction between acting and knowing, readiness-to-hand and presence-at-hand.

2. Ibid., pp. 71ff.

3. For an insightful account of how early Heidegger's investigation of meaning and truth shaped his interpretation of human understanding, see Steven Crowell, "Meaning and the Ontological Difference," in *The Thought of Martin Heidegger*, ed. Michael E. Zimmerman, Tulane Studies in Philosophy, XXXII (1984), pp. 37–44.

4. For examples of Karl Löwith's important criticisms of Heidegger's anthropocentrism, see *Der Weltbegriff der neuzeitlichen Philosophie* (Heidelberg: Carl Winter, Universitätsverlag, 1960); "Mensch und Geschichte" and "Nature und Humanität des Menschen," in *Gesammelte Abhandlungen: Zur Kritik der geschichtlichen Existenz* (Stuttgart: W. Kohlhammer Verlag, 1960); "Diltheys und Heideggers Stellung zur Metaphysik," in *Zum Kritik der christlichen Überlieferung* (Stuttgart: W. Kohlhammer, 1966); "Zu Heideggers Seinsfrage: Die Natur des Menschen und die Welt der Natur," in *Aufsätze und Vorträge, 1930–1970* (Stuttgart: W. Kohlhammer Verlag, 1971); *Heidegger: Denker in dürftiger Zeit* (Frankfurt am Main: S. Fischer Verlag, 1953). The last book remains one of the best interpretations of Heidegger's thought.

5. Thomas J. Sheehan, "Heidegger's Philosophy of Mind," in *Contemporary Philosophy: A New Survey*, ed. Guttorm Folistad (The Hague: Martinus Nijhoff, 1983), p. 307.

6. Ibid., p. 309. For more of Sheehan's excellent analyses of Heidegger's debt to Aristotle, see "On Movement and the Destruction of Ontology," *Monist*, Vol. 64, No. 4 (October, 1981), pp. 533–542.

7. For a penetrating analysis of Heidegger on "authenticity," see Hubert L. Dreyfus and Jane Rubin, "You Can't Get Something for Nothing: Kierkegaard and Heidegger on How Not to Overcome Nihilism," *Inquiry*, 30 (1987), pp. 33–75.

8. Concerning the change in Heidegger's concept of authenticity, see my book *Eclipse of the Self*.

9. Charles Guignon, *Heidegger and the Problem of Knowledge*; Guignon, "The Twofold Task: Heidegger's Foundational Historicism in *Being and Time*," in *The Thought of Martin Heidegger*, pp. 53–59.

10. Ibid., p. 58. For further analysis of this topic, cf. Jeffrey Andrew Barash, *Martin Heidegger and the Problem of Historical Meaning* (Dordrecht: Martinus Nijhoff, 1988).

10. *BEING AND TIME*: PENULTIMATE STAGE OF PRODUCTIONIST METAPHYSICS?

1. Hubert L. Dreyfus, "Between *Techne* and Technology: The Ambiguous Place of Equipment in *Being and Time*," in *The Thought of Martin Heidegger*, pp. 23–35. This essay is one of the most instructive in all the literature on Heidegger's concept of technology.

2. Ibid., p. 31.

3. Ibid., p. 32.

4. Manfred S. Frings, "Is There Room for Evil in Heidegger's Thought or Not?" *Philosophy Today*, 32 (Spring, 1988), pp. 79–92. The quotation is on p. 88.

5. On this topic, see Paul Farwell, "Can Heidegger's Craftsman Be Authentic?" *International Philosophical Quarterly*, XXIX, No. 1 (March, 1989), pp. 77–90.

6. Prauss, *Erkennen und Handeln in Heideggers "Sein und Zeit"*, p. 22.

7. Karel Kosik, *Dialectics of the Concrete*, trans. Karel Kovanda and James Schmidt, Vol. LII of Boston Studies in the Philosophy of Science, ed. Robert S. Cohen and Marx W. Wartofsky (Dordrecht/Boston: D. Reidel Publishing Co., 1976), p. 86.

8. Ibid., p. 39.

9. Ibid. Cf. my essay "Karel Kosik's Heideggerian Marxism," *Philosophical Forum*, XV, No. 3 (Spring, 1984), pp. 209–233.

10. Scheler, *Problems of a Sociology of Knowledge*, p. 92.

11. Ibid., p. 103.

11. THE HISTORY OF PRODUCTIONIST METAPHYSICS

1. The best available account of Heidegger's concept of the history of being is found in Reiner Schürmann, *Heidegger on Being and Acting: From Principles to Anarchy*, trans. Christine-Marie Gros, in collaboration with the author (Bloomington: Indiana University Press, 1987). A fine brief account is found in Michael Allen Gillespie, *Hegel, Heidegger, and the Ground of History* (Chicago: University of Chicago Press, 1984), pp. 134–164. Cf. also Werner Marx, *Heidegger and the Tradition*, trans. Theodore Kisiel and Murray Greene (Evanston: Northwestern University Press, 1971). For a sympathetic but critical reading of Heidegger's idea of the history of being, see John D. Caputo, "Demythologizing Heidegger: *Aletheia* and the History of Being, *Review of Metaphysics*, 41 (March, 1988), pp. 519–546.

2. On this topic, see my essay "Heidegger on Nihilism and Technique," *Man and World*, VIII (November, 1975), pp. 399–414.

3. Erich Fromm, in *Escape from Freedom* (New York: Avon, 1964), provides a helpful review not only of Weber's thesis concerning the relation between Protestantism and capitalism, but also of contributions made to that thesis by R. H. Tawney and other contemporaries of Weber. For a trenchant critique of Weber's thesis, see Luciano Pelliconi, "Weber and the Myth of Calvinism," *Telos*, No. 75 (Spring, 1988), pp. 57–85.

4. Max Weber, *The Protestant Ethic and the Spirit of Capitalism*, trans. Talcott Parsons (New York: Charles Scribner's Sons, 1930), pp. 181–182.

5. Ibid., p. 182.

6. For a critique of the idea that Christianity is responsible for the development of the modern, instrumentalist worldview, cf. Carl Mitcham, "The Religious and Political Origins of Modern Technology," in *Philosophy and Technology*, ed. Paul T. Durbin and Friedrich Rapp (Boston: D. Reidel, 1983).

7. For an excellent account of the objectifying tendencies of modern technology, see Joseph P. Fell, "The Crisis of Reason: A Reading of Heidegger's *Zur Seinsfrage*," *Heidegger Studies*, Vol. 2 (1986), pp. 41–65. See also Parvis Emad, "Technology as Presence: Heidegger's View," *Listening*, Vol. 16, No. 2 (1981), pp. 133–144: Jean-François Jobi, "Heidegger et la Technique," *Studia Philosophica* (Switzerland), 38 (1975), pp. 81–127; William Lovitt, "A *Gespräch* with Heidegger on Technology," *Man and World*, VI, No. 1 (February, 1973), pp. 44–62; Harold Alderman, "Heidegger: Technology as Phenomenon," *Personalist*, 51 (Fall, 1970), pp. 535–545; Simon Moser, "Toward a Metaphysics of Technology," *Philosophy Today*, 15 (Summer, 1971), pp. 129–156.

8. Once again, Max Scheler was already asking in 1925: Which came first— science or technology? "For example, is it permitted to say whether the modern technology of production goods and communication is a merely subsequent practical application of purely speculative natural knowledge—or whether the discovery of the 'laws of nature' arises from a newly awakened will to work, a zeal—unknown to the Middle Ages—to make nature into a fit human dwelling-house, beforehand, such that the new nature-knowledge since Galileo and Leonardo—as, e.g., E. Mach supposes in his history of mechanics—from the start was born historically from the spirit of technology and from technical problems?" Scheler, *Die Wissensformen und die Gesellschaft*, p. 450. Cf. also Don Ihde, "The Historical-Ontological Priority of Technology over Science," in *Philosophy and Technology*, ed. Paul T. Durbin (Dordrecht: Reidel, 1983), pp. 235–252.

For insight into Heidegger's account of science, see John D. Caputo, "Heidegger's Philosophy of Science: The Two Essences of Science," in *Rationality, Relativism, and the Human Sciences*, ed. J. Margolis, M. Krausz, and R. M. Burian (Dordrecht: Martinus Nijhoff, 1986), pp. 43–60; Drew Leder, "Modes of Totalization: Heidegger on Modern Technology and Science," *Philosophy Today*, XXIX (Fall, 1985), pp. 245–256; Joseph Kockelmans, *Heidegger and Science* (Washington, D.C.: University Press of America, 1985); David A. Kolb, "Heidegger on the Limits of Science," *Journal of the British Society for Phenomenology*, 14 (January, 1983), pp. 50–64; Theodore Kisiel, "Heidegger and the New Images of Science," *Research in Phenomenology*, 7 (1977), pp. 162–181.

9. Cited by Ott, *Martin Heidegger*, p. 216.

10. For an example of such a reinterpretation·of Christianity, cf. Matthew Fox, *The Coming of the Cosmic Christ* (New York: Harper & Row, 1988).

11. Cf. Paul Ricoeur's one-page introduction to *Heidegger et la question de Dieu*, ed. Richard Kearney and Joseph Stephen O'Leary (Paris: Éditions Grasset & Fasquelle, 1980). Ricoeur says, in part: "What has often astonished me about Heidegger is that he seems to have systematically eluded the confrontation with the block of Hebraic thinking. It has sometimes occurred to him to think beginning with the evangelicals or with Christian theology, but always in avoiding the Hebraic *massif*, which is the absolute stranger to Greek discourse. . . . "

12. See, for example, Alan D. Schrift, "Violence or Violation? Heidegger's Thinking 'about' Nietzsche," in *The Thought of Martin Heidegger*, pp. 79–86.

13. I have explored Heidegger's debt to Nietzsche in "A Comparison of Nietzsche's Overman and Heidegger's Authentic Self," *Southern Journal of Philosophy*, XIV (Spring, 1976), pp. 213–231, and in "Heidegger and Nietzsche on Authentic Time," *Cultural Hermeneutics*, IV (1977), pp. 239–264. Cf. also *Eclipse of the Self*, pp. 89–99.

14. Friedrich Nietzsche, section 866, *The Will to Power*, trans. Walter Kaufmann and R. J. Hollingdale (New York: Vintage Books, 1968), pp. 463–464.

12. PRODUCTION CYCLES OF THE "LABORING ANIMAL": A MANIFESTATION OF THE WILL TO WILL

1. Sigmund Freud, *Civilization and Its Discontents*, trans. James Strachey (New York: W. W. Norton & Co., 1961), p. 92.

2. Cf. Max Scheler's last book, *Die Stellung des Menschen im Kosmos* (Bern: A. Francke A. G., 1928); *Man's Place in Nature*, trans. Hans Meyerhoff (New York: Noonday Press, 1961).

3. Karl Marx, *The German Ideology*, in *The Marx-Engels Reader*, ed. Robert C. Tucker (New York: W. W. Norton & Co., 1978), p. 150.

4. Karl Marx, *Capital*, trans. Ben Fowkes (New York: Vintage Books, 1977), p. 284.

5. Cited by Stuart C. Gilman, "Political Theory and Degeneration: From Left to Right, from Up to Down," in *Degeneration: The Dark Side of Progress*, ed. J. Edward Chamberlin and Sander L. Gilman (New York: Columbia University Press, 1985), p. 178.

6. For a review of philosophical themes in ethology, cf. Jonathan Bennett, "Thoughtful Brutes," Presidential Address to Eastern Division Meeting of the American Philosophical Association, December 29, 1987, in *Proceedings of the APA*, 67, No. 1 (September, 1988), pp. 197–210.

7. Don Ihde explores some of these themes in *Techniques and Praxis*.

8. For a helpful critical study of Heidegger's comments on the human hand, see Jacques Derrida, "Geschlecht II," trans. John P. Leavey, Jr., in *Deconstruction in Philosophy*, ed. John Sallis (Chicago: University of Chicago Press, 1987), pp. 161–196.

9. Hannah Arendt, *The Human Condition* (Garden City, N.Y.: Doubleday, 1959). Max Scheler argued that the pragmatic technological Will to Power had so transformed the human species that humanity should no longer be called *homo sapiens* but instead *homo faber*, the "working being" (*Arbeitswesen*). Cf. Scheler, *Die Wissensformen und die Gesellschaft*, p. 447.

10. Arendt, *The Human Condition*, p. 76.

11. Ibid., p. 293.

12. Ibid., p. 294.

13. Ibid., p. 295.

14. Ibid.

15. Important contributions to futurology and world-order studies include Ervin Laszlo, *A Strategy for the Future: The Systems Approach to World Order* (New York: George Braziller, 1974); and Richard Falk and Saul Mendlovitz, *The Strategy of World Order* (New York: World Law Fund, 1966, 1967).

16. Friedrich Georg Jünger, *The Failure of Technique: No Perfection without Purpose*, trans. F. D. Wieck (Hinsdale, Ill.: H. Regnery, 1949), originally written in 1939.

17. Jacques Ellul, *The Technological Society*, trans. John Wilkinson (New York: Vintage Books, 1964).

18. Langdon Winner, *The Whale and the Reactor: A Search for Limits in an Age of High Technology* (Chicago: University of Chicago Press, 1986), pp. 47–48.

19. For a critique of the claim that humanity has somehow "mastered" technology, cf. David Ehrenfeld, *The Arrogance of Humanism* (New York: Oxford University Press, 1978).

20. C. S. Lewis, "The Abolition of Man," in *Philosophy and Technology*, ed. Carl Mitcham and Robert Mackey (New York: Free Press, 1972), p. 147.

21. Hubert L. Dreyfus and Paul Rabinow, *Michel Foucault: Beyond Structuralism and Hermeneutics*, 2d ed. (Chicago: University of Chicago Press, 1983).

22. Ibid., p. 159.

23. Cited in ibid., p. 187.

24. Ibid., p. 187.

25. David Michael Levin, "Introduction," in *Pathologies of the Modern Self*, ed. Levin (New York: New York University Press, 1987), p. 8.

26. See my essays "Humanism, Ontology, and the Nuclear Arms Race," *Research in Philosophy and Technology*, VI (1983), pp. 157–172; "Anthropocentric Humanism and the Arms Race," in *Nuclear War: Philosophical Perspectives*, ed. Michael Fox and Leo Groarke (New York: Peter Lang Publishers, 1985); "The Incomplete Myth: Reflections on the 'Star Wars' Dimension of the Arms Race," in *Consciousness Evolution*, ed. Stanislav Grof (Albany: SUNY Press, 1988).

13. HOW MODERN TECHNOLOGY TRANSFORMS THE EVERYDAY WORLD—AND POINTS TO A NEW ONE

1. Alvin Toffler, *Future Shock* (New York: Bantam Books, 1972).

2. For an interesting philosophical reflection on computing and word processing, cf. Michael Heim, *Electric Language: A Philosophical Study of Word Processing* (New Haven: Yale University Press, 1987).

3. Albert Borgmann, *Technology and the Character of Everyday Life* (Chicago: University of Chicago Press, 1984). This is an important book, written from a Heideggerean perspective, about modern technology.

4. Thanks to Hubert L. Dreyfus for this distinction between the phonograph and the compact disc player.

5. Herbert Marcuse, *One-Dimensional Man* (Boston: Beacon Press, 1964), p. 59. On the relation between Heidegger and Marcuse, cf. my essay "Heidegger and Marcuse: Technology as Ideology," in *Research in Philosophy and Technology*, ed. Paul T. Durbin, Vol. II (Greenwich, Conn.: JAI Press, 1979), pp. 245–261. In *The Domination of Nature* (Boston: Beacon Press, 1974), William Leiss has provided a classic account of the origins and character of modern technology as seen from the perspective developed by Marcuse, Horkheimer, and Adorno.

6. I owe the wood stove vs. gas furnace example to Borgmann, *Technology and the Character of Everyday Life*.

7. Thanks to Kathleen Wright for pointing out this reference. Heidegger's seminars for Boss's psychiatry students and teachers make for very interesting reading for those concerned with the relation between philosophy and psychotherapy.

8. For this and other insights into Heidegger's life and attitudes, cf. Heinrich Wiegand Petzet, *Auf einen Stern zugehen* (Frankfurt am Main: Societät, 1983).

9. Simone Weil, *The Need for Roots*, trans. Arthur Wills (New York: Putnam, 1952); Wendell Berry, *The Unsettling of America: Culture and Agriculture* (New York: Avon, 1977).

10. Karl Marx, *Grundrisse: Foundations of the Critique of Political Economy*, trans. Martin Nicolaus (New York: Vintage Books, 1973), p. 410.

11. Marcuse, *One-Dimensional Man*, p. 59.

12. Harry Braverman, *Labor and Monopoly Capital: The Degradation of Work in the Twentieth Century* (New York: Monthly Review Press, 1974), p. 125. A powerful and disturbing book.

13. On the worker's changing attitude toward his/her equipment. See Dreyfus, "Between *Techne* and Technology," in *The Thought of Martin Heidegger*.

14. These citations are taken from an excellent essay by Bernard Gendron and Nancy Holmstrom, "Marx, Machinery, and Alienation," in *Research in Philosophy and Technology*, ed. Paul T. Durbin (Greenwich, Conn.: JAI Press, 1979), pp. 119–136. For an excellent neo-Marxist account of how changes in workplace and society lead to alienation, cf. Bruce Brown, *Marx, Freud, and the Critique of Everyday Life: Toward a Permanent Cultural Revolution* (New York: Monthly Review Press, 1973).

15. For a vigorous defense of the view (unpopular with many Marxists) that Marx himself was a technological determinist, cf. William H. Shaw, " 'The Handmill Gives You the Feudal Lord': Marx's Technological Determinism," *History and Theory*, 18, No. 2 (1979), pp. 155–176.

16. Jürgen Habermas, *Toward a Rational Society*, trans. Jeremy J. Shapiro (Boston: Beacon Press, 1970), p. 89.

17. Ibid., p. 105.

18. See, for example, Jürgen Habermas, "The Entwinement of Myth and Enlightenment: Re-reading *Dialectic of Enlightenment*," trans. Thomas Y. Levin, *New German Critique*, No. 26 (Spring–Summer, 1982), pp. 13–30; Peter U. Hohendahl, "The Dialectic of Enlightenment Revisited: Habermas' Critique of the Frankfurt School," *New German Critique*, No. 35 (Spring–Summer, 1985), pp. 3–26.

19. *Martin Heidegger im Gespräch*, ed. Richard Wisser (Freiburg/München: Karl Alber Verlag, 1970), p. 73.

20. On this topic, cf. Zimmerman, *Eclipse of the Self*, chapter eight.

21. Wolfgang Schirmacher, "From the Phenomenon to the Event of Technology (A Dialectical Approach to Heidegger's Phenomenology)," in *Philosophy and Technology*, ed. Paul T. Durbin and Friedrich Rapp (Boston: D. Reidel, 1983), p. 286. Cf. also Schirmacher, *Technik und Gelassenheit: Zeitkritik nach Heidegger* (Freiburg/München: Karl Alber Verlag, 1983).

14. AUTHENTIC PRODUCTION: *TECHNE* AS THE ART OF ONTOLOGICAL DISCLOSURE

1. See my essay "On Vallicella's Critique of Heidegger," *International Philosophical Quarterly*, forthcoming. On this topic, cf. also Haar, *Le chant de la terre*, and Cahoone, *The Dilemma of Modernity*.

2. William F. Vallicella, "The Problem of Being in the Early Heidegger," *Thomist*, 45, 3 (July, 1981), pp. 388–406; "Heidegger and the Problem of the Thing in Itself," *International Philosophical Quarterly*, XXIII, 1 (March, 1983), pp. 35–43; "Heidegger's Reduction of Being to Truth," *New Scholasticism*, LIX, 2 (Spring, 1985), pp. 156–176.

3. On the topic of the need for an appropriate "measure" in the technological world, see Werner Marx, *Is There a Measure on Earth? Foundations for a Nonmetaphysical Ethics*, trans. Thomas J. Nenon and Reginald Lilly (Chicago: University of Chicago Press, 1987), and Hans Jonas, *The Imperative of Responsibility: In Search of an Ethics for the Technological Age* (Chicago: University of Chicago Press, 1984).

4. On the idea that scientific method involved a receptive attitude toward the things being studied, see Evelyn Fox Keller, *A Feeling for the Organism: The Life and Work of Barbara McClintock* (San Francisco: W. H. Freeman & Co., 1983).

5. Plato, *Symposium*, 205c. My thanks to Ronna Burger for her reference to this passage and for advice about translating it.

6. See John D. Caputo's remarkable book *The Mystical Element in Heidegger's Thought* (Athens: Ohio University Press, 1978).

7. On this topic, see Edward G. Laury, "The Work-Being of the Work of Art in

Heidegger," *Man and World*, 11 (1978), pp. 186–198. This theme is also discussed in a thoughtful way by Haar, *Le chant de la terre*.

8. Kathleen Wright, "The Place of the Work of Art in the Age of Technology," *Southern Journal of Philosophy*, 22 (Winter, 1984), pp. 565–583.

9. See Gerhard Glaser, *Das Tun ohne Bild: Zur Technikdeutung Heideggers und Rilke* (Münich: Maander Verlag, 1983). Cf. Christoph Jamme's helpful review of this book, "Die Rettung der Dinge—Heidegger und Rilke," *Philosophische Jahrbuch*, 92, No. 2 (1985), pp. 408–415.

10. Rainer Maria Rilke, Letter to Clara Rilke, October 8, 1907, in *Briefe aus den Jahren 1906 bis 1907*, ed. Ruth Sieber-Rilke and Carl Sieber (Leipzig: Insel-Verlag, 1930), p. 364. For a translation of this and other letters, see Rilke, *Letters on Cézanne*, ed. Clara Rilke, trans. Joel Agee (New York: Fromm International Publishing Corp., 1985), p. 34. See Walter A. Strauss, "Rilke and Ponge: L'objet c'est la poétique," in *Rilke: The Alchemy of Alienation*, ed. Frank Baron, Ernst S. Dick, and Warren R. Maurer (Lawrence: Regents Press of Kansas, 1980), pp. 63–93.

11. On this topic, see Herbert Lehnert, "Alienation and Transformation: Rilke's Poem 'Der Schwan,' " in *Rilke: The Alchemy of Alienation*, pp. 95–111.

12. Rilke, Letter to Clara Rilke, October 8, 1907; *Briefe*, p. 361; *Letters on Cézanne*, pp. 32–33.

13. Rilke, *Briefe*, p. 393; Rilke, *Letters on Cézanne*, p. 67.

14. Rilke, *Briefe*, p. 405; Rilke, *Letters on Cézanne*, p. 80, my emphasis.

15. Lehnert, "Alienation and Transformation," in *Rilke: The Alchemy of Alienation*, p. 104.

16. For a contemporary account of work in a German Mercedes-Benz factory by an author inspired by Heidegger and Jünger, see Dirk Leach, *Technik*, trans. Brice Matthieussent (Gris Banal Editeur, 1986).

17. "Deep ecology" was first defined by the Norwegian philosopher and environmental activist Arne Naess, in "The Shallow and the Deep, Long-Range Ecology Movement: A Summary," *Inquiry*, 16 (1973), pp. 95–100. Subsequently, Naess, George Sessions, Bill Devall, Alan Drengson, Warwick Fox, I, and others explored and expanded upon the initial definition of deep ecology. A selection of pertinent essays includes Arne Naess, "Identification as a Source of Deep Ecological Attitudes," in *Deep Ecology*, ed. Michael Tobias (San Diego: Avant Books, 1985); George Sessions, "Shallow and Deep Ecology: A Review of the Literature," in *Ecological Consciousness: Essays from the Earthday X Colloquium*, ed. Robert C. Schultz and J. Donald Hughes (Washington, D.C.: University Press of America, 1981); Sessions, "The Deep Ecology Movement: A Review," *Environmental Review*, 11 (1987), pp. 105–125; Bill Devall, "The Deep Ecology Movement," *Natural Resources Journal*, 20 (1980), pp. 299–322; Devall, *Simple in Means, Rich in Ends: Practicing Deep Ecology* (Layton, Utah: Gibbs M. Smith, 1988); Devall and Sessions, *Deep Ecology: Living as if Nature Mattered* (Layton, Utah: Gibbs M. Smith, 1985); Alan R. Drengson, *Beyond Environmental Crisis: From Technocrat to Planetary Person* (New York: Peter Lang, 1989); Warwick Fox, "Deep Ecology: A New Philosophy of our Time?" *Ecologist*, 14 (1984), pp. 194–200. The best available account of deep ecology is Warwick Fox's dissertation: "Toward a Transpersonal Ecology: The Context, Influence, Meanings, and Distinctiveness of the Deep Ecology Approach to Ecophilosophy" (Murdoch University [Australia], 1988).

18. Lynn White, Jr., "The Historical Roots of Our Ecologic Crisis," originally printed in *Science*, CLV (March 10, 1967), reprinted in *Philosophy and Technology*, ed. Carl Mitcham and Robert Mackey (New York: Free Press, 1972), pp. 259–265. The cited passage is on p. 263. A huge literature has developed in response to White's critique of Christianity. For bibliographical references, see Carl Mitcham and Jim Grote, ed., *Theology and Technology: Essays in Christian Analysis and Exegesis* (Lanham, Md.: University Press of America, 1984).

19. For an insightful account of the role played by Stoicism in shaping Western

anthropocentrism, see John Passmore, *Man's Responsibility for Nature: Ecological Problems and Western Traditions* (London: Duckworth, 1980), especially chapter two.

20. On the role played by Bacon in the "domination of nature," see William Leiss, *The Domination of Nature* (Boston: Beacon Press, 1972), and Evelyn Fox Keller, *Reflections on Gender and Science* (New Haven: Yale University Press, 1985).

21. Réné Scherer, "Le dernier des philosophes," in *Heidegger ou l'experience de la pensée*, by Réné Scherer and Arion Lothar Kelkel (Paris: Éditions Seghers, 1973), p. 5.

22. See my essays "Toward a Heideggerean *Ethos* for Radical Environmentalism," *Environmental Ethics*, V (Summer, 1983), pp. 99-131; and "Implications of Heidegger's Thought for Deep Ecology," *Modern Schoolman*, LXIV (November, 1986), pp. 19-43.

23. Murray Bookchin, a severe critic of deep ecology, has warned that Heidegger's thought is tainted with fascism and is thus unsuitable for providing guidance about the ecological crisis. See, for example, Bookchin, "A Reply to My Critics," *Green Synthesis*, No. 29 (December, 1988), pp. 5-7. Bookchin, founder of the "social ecology" movement in the United States, is the author of several books on the environmental crisis, including *The Ecology of Freedom: The Emergence and Dissolution of Hierarchy* (Palo Alto: Cheshire Books, 1982). On the relation between reactionary German philosophy and environmentalism in the first part of the twentieth century, cf. Anna Bromwell, *Ecology in the Twentieth Century* (New Haven: Yale University Press, 1989).

24. On this topic, cf. Dolores LaChapelle, *Sacred Land, Sacred Sex—Rapture of the Deep: Concerning Deep Ecology—and Celebrating Life* (Silverton, Colo.: Finn Hill Arts, 1988). This is a remarkable and far-seeing critique of Western humanity's dissociation from nature, and an account of the revival of the rituals and attitudes necessary to transform the human relationship to nature. LaChapelle discussed Heidegger's importance for deep ecology in another important book, *Earth Wisdom* (Los Angeles: Guild of Tutors Press, 1978).

25. Arne Naess, "The Arrogance of Anti-humanism," *Ecophilosophy*, VI (1984), p. 8.

26. David Michael Levin, *The Body's Recollection of Being: Phenomenological Psychology and the Destruction of Nihilism* (London: Routledge & Kegan Paul, 1985); Levin, *The Opening of Vision: Nihilism and the Postmodern Situation* (New York: Routledge, 1988).

27. Marx, *Is There a Measure on Earth?*

CONCLUSION

1. Reinhold Niebuhr, *Faith and History* (New York: Charles Scribner's Sons, 1951), p. 121.

2. Robert Tucker, *Philosophy and Myth in Karl Marx* (Cambridge: Cambridge University Press, 1972), p. 146.

3. Karl Marx, *Capital*, trans. Ben Fowkes (New York: Vintage Books, 1976). Cf. my essay "Marx and Heidegger on the Technological Domination of Nature," *Philosophy Today*, XXIII, No. 2 (Summer, 1979), pp. 99-112. For a Heideggerean interpretation of Marx's views on technology, see Kostas Axelos, *Alienation, Praxis, and Technē in the Thought of Karl Marx,* trans. Ronald Bruzina (Austin: University of Texas Press, 1976).

4. On Heidegger's relation to Hegel, see Michel Haar, "Structures hegeliennes dans la pensée heideggerienne de l'histoire," *Revue de metaphysique et de morale*, No. 1 (1980), pp, 48–59; David Kolb, *The Critique of Pure Modernity* (Chicago: University of Chicago Press, 1986).

5. Robert Tucker, in *Philosophy and Myth in Karl Marx*, p. 66, notes that Hegel's doctrine of "the cunning of Reason" seems to have been influenced by Adam Smith's

concept of the "invisible hand" governing the workings of the marketplace. Liberal economists generally do not regard the "invisible hand" as a manifestation of a metaphysical agency such as *Geist* in the world, however, but instead as a market mechanism which ensures that the economic decisions of countless individuals will lead, quite independently of the conscious intentions of those individuals, to the optimal relation among production, price, and demand.

6. Ernesto Laclau, "Politics and the Limits of Modernity," in *Universal Abandon? The Politics of Postmodernism*, ed. Andrew Ross (Minneapolis: University of Minnesota Press, 1988), p. 66.

7. John Sturrock, ed., *Structuralism and Since: From Lévi-Strauss to Derrida* (New York: Oxford University Press, 1979), p. 13.

8. Peter Dews, *Logics of Disintegration: Post-structuralist Thought and the Claims of Critical Theory* (London: Verso, 1987), p. 32. A highly recommended text.

9. Herbert Marcuse, *One-Dimensional Man: Studies in the Ideology of Advanced Industrial Society* (Boston: Beacon Press, 1964).

10. For a helpful account of the issues involved in the debate between modernism and post-modernism, see Garry M. Brodsky, "Postmodernity and Politics," *Philosophy Today*, XXXI (Winter, 1987), pp. 291–305.

11. Joseph Margolis, "Pragmatism, Transcendental Arguments, and the Technological," in *Philosophy and Technology*, ed. Paul T. Durbin and Friedrich Rapp (Boston: D. Reidel Publishing Co., 1983), p. 294.

12. Jacques Derrida, "Structure, Sign, and Play in the Discourse of the Human Sciences," in *Writing and Difference*, trans. Alan Bass (Chicago: University of Chicago Press, 1978), p. 292. My reading of Derrida has been assisted by John D. Caputo's excellent book *Radical Hermeneutics: Repetition, Deconstruction, and the Hermeneutic Project* (Bloomington: Indiana University Press, 1987).

13. Caputo, *Radical Hermeneutics*, p. 154.

14. Jacques Derrida, "*Ousia* and *Gramme*: Note on a Note from *Being and Time*," in *Margins of Philosophy*, trans. Alan Bass (Chicago: University of Chicago Press, 1982).

15. Dews, *Logics of Disintegration*, p. 40.

16. Richard J. Bernstein, "Metaphysics, Critique, and Utopia," *Review of Metaphysics*, 42 (December, 1988), pp. 255–273.

17. For references to Löwith's critique of Heidegger, see note 4, chapter nine.

18. Margolis, "Pragmatism, Transcendental Arguments, and the Technological."

19. Bernstein, "Metaphysics, Critique, and Utopia."

20. Ibid.

21. Jacques Derrida, *Positions*, trans. Alan Bass (Chicago: University of Chicago Press, 1981), p. 12; cf. also Derrida, *Writing and Difference*, p. 280.

22. Jacques Derrida, *Dissemination*, trans. Barbara Johnson (Chicago: University of Chicago Press, 1981), p. 266. I owe this reference to John D. Caputo.

23. Derrida, *Positions*, p. 11.

24. Caputo, *Radical Hermeneutics*, p. 260.

25. Leo Marx made this point in a lecture presented at Tulane University, January 26, 1989.

26. Marcuse, *One-Dimensional Man*.

27. Ibid., pp. 250–251.

28. Michel Foucault, "Final Interview," trans. Thomas Levin and Isabelle Lorenz, *Raritan*, V, No. 1 (Summer, 1985), pp. 1–13. Citation is from p. 8.

29. Michel Foucault, "The Subject and Power," in *Michel Foucault: Beyond Structuralism and Hermeneutics*, 2d ed., ed. Hubert L. Dreyfus and Paul Rabinow (Chicago: University of Chicago Press, 1983), p. 221.

30. Jürgen Habermas, "Taking Aim at the Heart of the Present," in *Foucault: A Critical Reader*, ed. David C. Loy (Oxford: Basil Blackwell, 1986), p. 107. Cited in Dieter Freundlieb, "Rationalism v. Irrationalism? Habermas's Response to Foucault," *Inquiry*, 31 (1988), pp. 171–192. Cf. footnote 8 on p. 190.

31. Cf. Freundlieb, "Rationalism v. Irrationalism?" Citation is from pp. 182–183. This is an excellent essay, to which I am indebted.

32. Ibid.

33. Charles M. Sherover, "Founding an Existential Ethic," *Human Studies*, 4 (1981), pp. 223–236. The quotation is from p. 235.

34. Rorty makes this point in a number of places, including his recent essay regarding Heidegger's affiliation with National Socialism, "Taking Philosophy Seriously," *New Republic* (April 11, 1988), pp. 31–34.

35. Richard Rorty, "Overcoming the Tradition: Heidegger and Dewey," *Review of Metaphysics*, XXX, No. 2 (December, 1976), pp. 280–305. For a critique of Rorty, see John D. Caputo, "The Thought of Being and the Conversation of Mankind: The Case of Heidegger and Rorty," *Review of Metaphysics*, XXXVI, No. 3 (March, 1983), pp. 661–685.

36. Richard Rorty, "Objectivism and Solidarity," in *Post-analytic Philosophy*, ed. John Rajchman and Cornel West (New York: Columbia University Press, 1985).

37. Those who say that, because there are no universal or cross-cultural foundations for evaluating moral and political behavior, we must thus adopt a *laissez faire* attitude toward such behavior (apparently no matter how foreign to our standards) must be more willing than I am to suspend judgment about customs such as the one still practiced in parts of India which dictates that widows be thrown on the funeral pyres of their dead husbands. (My thanks for this example to a stimulating conversation with Martha Nussbaum, Ronna Burger, and Eric Mack.) Moreover, the *laissez faire* attitude would not provide one with much leverage to criticize the political practices of Stalin or Hitler.

38. On the topic of the God project, see Ken Wilber, *Up from Eden: A Transpersonal View of Human Evolution* (Boulder: Shambhala, 1983).

39. Rosemary Radford Ruether, *New Woman, New Earth: Sexist Ideologies and Human Liberation* (New York: Seabury Press, 1975), pp. 194–195. See also Marilyn French, *On Power: Women, Men, and Morality* (New York: Summit Books, 1985). See also my essays "Heidegger, Feminism, and the Technological Domination of Nature," to appear in *Research in Philosophy and Technology*, and "Feminism, Deep Ecology, and Environmental Ethics," *Environmental Ethics*, 9 (Spring, 1987), pp. 21–44.

40. Jacques Derrida, *Spurs/Éperons*, trans. Barbara Harlow (Chicago: University of Chicago Press, 1979). On this topic, see John D. Caputo, " 'Supposing Truth to be a Woman . . .': Heidegger, Nietzsche, Derrida," *Tulane Studies in Philosophy*, XXXII (1984), pp. 15–21.

41. Derrida, *Spurs/Éperons*, p. 119.

42. See Caputo, *Radical Hermeneutics*, chapter seven, "Cold Hermeneutics: Heidegger/Derrida."

43. See Susan Griffin, *Woman and Nature: The Roaring inside Her* (New York: Harper & Row, 1978); Mary Daly, *Gyn/Ecology: The Metaethics of Radical Feminism* (Boston: Beacon Press, 1978); Adrienne Rich, *Of Woman Born* (New York: W. W. Norton & Co., 1986). For an excellent critical yet sympathetic study of contemporary feminism, see Jean Grimshaw, *Philosophy and Feminist Thinking* (Minneapolis: University of Minnesota Press, 1986).

44. Derrida, *Spurs/Éperons*, p. 121.

45. Julia Kristeva, "Women's Time," trans. Alice Jardine and Harry Blake, *Signs*, VII, No. 1 (Autumn, 1981), pp. 13–35. Citation is from p. 33. While Kristeva's attempt to dissolve the binary opposition between male and female is motivated by her desire to eliminate the metaphysical dualism that is the basis for sexual oppression and marginalization, critics charge that without appropriate gender categories one cannot effectively criticize the existing social and political arrangements which oppress and exploit women, and which thereby distort all human relationships. While one may understandably wish to overcome absolutizing and essentializing gender categories, then, one may also wish to retain a nuanced, historically limited, and flexible set of categories which make possible effective political action.

46. Paul Gottfried, "Heidegger on Trial," *Telos*, No. 74 (Winter, 1987–1988), pp. 147–151. Citation is from p. 151. In reply to his own question as to why there is such an uproar about Heidegger's politics at this particular time, Gottfried replies that liberals are concerned about his critique of their project, which has turned into the technocratic-managerial "democratic" state. Given Heidegger's critique of the entire Enlightenment tradition, critique of his position from Marxists and liberals alike is not surprising.

47. See *The Reechantment of Science: Postmodern Proposals*, ed. David Ray Griffin (Albany: SUNY Press, 1988).

Index

Actualitas: translation of *energeia*, 170
Aeschylus: fate and humanity, 119
Aesthetics: National Socialism, 99–106; Nietzsche and metaphysics, 107. *See also* Art
Agrarianism: Heidegger and peasantry, 70–71; humanity's relationship to earth, 196
Aitia: Roman interpretation, 170
Aition: Heidegger's use of term, 232
Aletheia: Heidegger's translation, 145, 225; truth as unconcealment, 175
Americanism: Heidegger on dangers of, 41, 42; Heidegger's equation with Bolshevism, 90, 264–65; and Enlightenment, 266
Anglo-American: modern technology and language, 215
Angst: primal moods, 115
Animals: humanity and rationality, 192–95
Anorexia nervosa: the body and nature, 245–46
Anthropocentrism: earth and deep ecology, 242; Heidegger and deep ecology, 243
Antigone: Heidegger's tragic view of life, 105; Heidegger on Germany's tragic destiny, 118–19, 120
Anti-Semitism: *völkisch* movement, 9; assassination of Rathenau, 14–15; Heidegger and National Socialism, 37, 43. *See also* Auschwitz; Holocaust; Jews and Judaism
Anxiety: and depersonalization, 24; as historical, 25
Apollo: art and the Dionysian, 108, 109; anointed tasks of Greeks and Germans, 116
Architecture: National Socialism, 100
Arendt, Hannah: humanity as laboring animal, 197–99
Aristotle: Heidegger's debt to, 143–46; productionist metaphysics, 157, 161, 196; social structure, 158; definition of man and *kosmos*, 159; essence-existence distinction, 168–69; definition of being, 170–71; concept of truth, 174

Art: productionist metaphysics, xx, 159–60; history and revolution, xxiv; technology, 46, 78–80, 110; Nietzsche, 53–54, 76, 77; as *techne*, 94, 223, 229; National Socialism as national aestheticism, 99–106; Nietzsche, 106–12, 113; earth, world, and authentic creating, 121–26; purpose, 234–35; the "thing" and authentic producing, 236–37; Rilke, Cézanne, and disclosure of the "thing," 237–41
Artisan: art and technology, 230–31; existence in technological world, 241
Assertions: grounding in everyday practices, 141–42
Atomic age: *Gestalt* and humanity, 198–99
Auschwitz: as radical break in history, 127–28; artistic laws of tragedy, 128–29; Heidegger and Enlightenment, 255–56
Authoritarianism: Germany and industrialization, 8; Jünger and *Gestalt* of the worker, 61; deep ecology, 243

Being: defined, xix; role of temporality in understanding, 146–49; history of and human existence, 167–68; transformation from *energeia* to *actualitas* to the Will to Power, 168–73; as self-emergence and as appearing, 223–29; *Geist*, 254; Heidegger and deconstructionism, 259
Being and Time: critique of modernity and mass culture, 21–26; analysis of the workshop, 137; instrumentalist orientation, 150; utilitarianism, 152–53, 155; ethics, 265
Benn, Gottfried: National Socialism and art, 101
Besitzburgertum: *völkisch* movement, 10
Blindness: theme of Jünger and Heidegger, 95, 96
Body: Jünger and objectification, 57; Heidegger on "letting things be," 244–47
Bolshevism: Heidegger on dangers of, 42; Germany's fear of, 78; equated with Americanism, 90

Boredom: theme of homesickness, 31–32; basic mood of Germans, 116

Burckhardt, Jakob: admired by Heidegger, 107; idea of the state, 286n

Calvinism: work ethic and industrialization, 178–80

Capitalism: National Socialism, 39–40; reactionary modernism, 48; Protestant work ethic and industrialization, 178–80; Marxism on good and evil, 252–53

Catholicism: Germany and modernity, 6; Heidegger's biography, 18; Roman concept of truth, 177; Heidegger's opposition, 182–83

Certainty: concept of truth, 187

Cézanne, Paul: art and the disclosure of the "thing," 237–41

Chaos: Nietzsche and art, 106–12

Christianity: nihilism, 78; Heidegger and Overman, 92; Hölderlin's poetry and Greek tragedy, 126; Greco-Roman metaphysics, 171; Roman concepts of truth and falsity, 177–78; salvation and modernity, 180–81; Heidegger's opposition, 182–83; anthropocentrism, 242; Heidegger's determinism, 252; Marxism, 252–53; Scheler on God and the earth, 288n

Class: Germany and industrialization, 8; *völkisch* movement, 13; worker *Gestalt*, 58; Heidegger and National Socialism, 69; Greek social structure, 158

Cold War: Heidegger's view of world affairs, 91

Commercialism: *völkisch* movement, 10

Communism: National Socialism, 39; Heidegger and democracy, 264. *See also* Bolshevism; Stalinism

Concentration camps: Heidegger's awareness of, 75–76

Conservatives: appeal of *Being and Time*, 25–26

Craftworkers: authentic production, 240–41

Creativity: art and National Socialism, 109–10; poets and language, 114

Culture: Heidegger and the academic ideal, 19. *See also* Mass culture

Cybernetics: character of modern technology, 199–202

Dasein: translation, xxvi; definition, 140; structure of being, 144–45; ontological structure, 225; world as fourfold, 240

Deconstructionism: Heidegger and post-structuralism, 256–57; politics, 257–58, 261–62, 267–68

Deep ecology: Heidegger, 241–44

Democracy: degeneration of Europe, 107; Heidegger and communism, 264

Depersonalization: descriptions and explanations, 24–25

Derrida, Jacques: Heidegger and deconstructionism, 258–59; politics and deconstructionism, 261

Descartes, René: idea of *Erlebnis*, 106; metaphysics and mathematics, 152; definition of man, 159; humanity and subjectivism, 171–72; concept of space, 211; anthropocentrism and nature, 242

Determinism: historical, 251–55; Heidegger and progressivism, 255; Heidegger and French structuralists, 256

Deutsche Philosophische Gesellschaft (DPG): National Socialism, 38–39

Difference: sexual and ontological, 268–74

Dilthey, Wilhelm: influence on Heidegger, 22, 146

Dionysius: art and the Apollinian, 108, 109; anointed task of Greeks, 115–16; Heidegger and National Socialism, 287n

Discipline: Jünger's new man, 56

Dissemination: deconstructionism and politics, 261–62

Drama: Heidegger's view of history, 97–98

Dynamis: Heidegger's use of term, 145

Earth: the world and authentic creating, 121–26; as distinct from world, 226–27; self-concealment, 228; deep ecology, 241–42; technological destruction and human finitude, 246

Ecology: body and concern with nature, 245. *See also* Deep ecology

Economics: Marxist interpretation of Western history, 10, 253–54; depersonalization, 24

Eidos: definition of *Gestalt*, 109; Plato, 168

Einstein, Albert: concept of space, 211

Elitism: Jünger, 54–55, 88; Heidegger, 88

Energeia: Aristotle's use of term, 169

Enlightenment: reactionary politics, 4; Germany and modernity, 5, 6; Heidegger, 19–20, 38, 92, 255–56, 262; Spengler, 28; reactionary modernism, 47; Germany's history, 48–49; rationality, 218; nature and anthropocentrism, 242; Heidegger and deconstructionist critics, 250–51; politics and society, 262–67, 272

Entelecheia: definition, 169

Ent-fernen: Heidegger's use of term, 151

Environment: Spengler on modern society, 28

Environmentalism: Heidegger and deep ecology, 241–44

Episteme: *techne*, 231–32

Equipment: defined in terms of usefulness, 160–61; technology and functionalization, 213

Ereignis: translation of *dynamis*, 145

Erlebnis: Germany and Cartesianism, 106

Ethics: as secondary, 130–31; *Being and Time*, 265

Ethology: humanity and animals, 194

Ethos: Heidegger and ethics, 131
Europe: future and technological *Gestalt*, 92–93
Everydayness: industrial urban society, 22–24
Expressionism: National Socialism and art, 101; character of technological age, 237

Facism: Heidegger's philosophy, 37; Jünger, 63
Falling: *Dasein* and temporality, 146–47
False: Roman concept of truth, 175–76, 177
Feminism: Heidegger and modern technology, 251; Heidegger's view of history and ontological difference, 269–71
Fichte, Johann Gottlieb: French Revolution and German self-image, 6
Fiction: Jünger and *Gestalt* of the worker, 64–65
Film: as technological instrument, 86; Hitler and art, 101
Foucault, Michel: quest for bio-power, 202–4
Foundationalism: Heidegger's critique of, 258
Frankfurt Critical School: technology and society, 217
Freedom: Heidegger and National Socialism, 45, 265; *Dasein* and entities, 164; deconstructionism, 257; Enlightenment and politics, 264, 272; Heidegger's early definition, 265
Freikorps: Jünger, 52
French Revolution: Germany and modernity, 6–7; National Socialism, 102
Functionalism: technology and work, worker, and equipment, 212–16
Future: Heidegger and Jünger's vision of, 83–90
Futurism: National Socialism, 103–4

Gandhi, Mahatma: deconstructionism and politics, 262
Geist: Hegel's interpretation of history, 254
Germany: Heidegger and National Socialism, xxi; industrialism and modernity, xxi–xxii; reactionary politics, 4–5; emergence of modern, industrial, 5–8; impact of World War I, 13–14; prevailing mood, 26; decline of West, 29; boredom and regeneration, 32; conservative revolution, 47; fear of communism, 78; Greece and aestheticism, 101–2; poet's role in transforming *Geist*, 113–15; anointed task and Hölderlin's poetry, 115–18; Heidegger's changing vision of tragic destiny, 118–21; Heidegger's interpretation of history, 126–27; industrial production, 156–57; language and technology, 215; Enlightenment, 265, 267
Gestalt: Jünger's concept of the worker, 57–65, 80; Heidegger and Jünger's concept, 80–83; defined in terms of *eidos*, 109; art and the artist, 111; *Gestell*, 124;

Hegel's interpretation of Western history, 254, 256
Gestell: use of term and *Gestalt*, 124; modern technology, 216–19
Gods: Heidegger and new, 92; Nietzsche, 107–8, 186; departure of old, 117; death of and World War II, 127; Christian theology and Greco-Roman metaphysics, 171; Roman and Greek concepts, 176; Christian tradition and subjectivism, 177–78; Heidegger on Christian concept, 182–83
Goebbels, Joseph: Jünger's vision, 65
Göring, Hermann: Heidegger and propaganda, 68
Great Depression: Weimar Republic, 16
Greece: history of West, xix; reactionary politics, 4; technological heritage and new beginning, 93; Heidegger on art, 99–100; national aestheticism in Germany, 101–2; Heidegger on National Socialism, 105; anointed task, 115–16; Hölderlin and *physis*, 124; productionist orientation and modern technology, 148–49, 166; handicraft and productionist metaphysics, 156–63, 223; entities and humanity, 188; being as appearing, 224–25; production and metaphysics, 230; metaphysics and anthropocentrism, 242

Hand: animals and humanity, 195–96, 244
Handicraft: Greeks and origins of productionist metaphysics, 156–63; authentic production, 222–23, 240–41
Hegel, Georg Wilhelm Friedrich: view of history, xviii; history of productionist metaphysics, 173; influence on Heidegger, 250, 254; as productionist metaphysician, 255; influence of Adam Smith, 295–96n
Heraclitus: influence on Hölderlin, 125; humanity and entities, 188; *physis* and concealment, 229
Hero: Greek tragic and *Volk*, 119–20
Heroism: praised by Heidegger, 74–75
Historicism: transcendental aspects of foundational, 147–48; concept of human existence, 166
History: productionist metaphysics and modern technology, xix; Heidegger's historical circumstances, xx–xxi; Jünger's view of world, xxiv; Marxist economic interpretation, 10, 253–54; *völkisch* movement and transcendental meaning, 11–12; Spengler's concept, 27; Heidegger and Enlightenment, 38; Jünger and Spengler, 51; Jünger and *Gestalt* of the worker, 61; Heidegger's view of as tragedy, 95–98; primal moods, 115; Heidegger's historical judgment, 131–32; as ontological play, 166; Heidegger and deep ecology, 242–43, 244; Heidegger, Hegel, and

Marxism, 250; Heidegger's historical
determinism, 251–55; Heidegger and
competing accounts, 268–69
Hitler, Adolf: Germans and reactionary
politics, 5; World War I, 14; failure of
Weimar Republic, 16; Heidegger's call for a
leader, 33; Heidegger's support, 37, 73;
world opinion, 38; support of German
philosophers, 38–39; Heidegger and
National Socialism, 43–44; *völkisch*
rhetoric, 48; Jünger, 63; national
aestheticism, 99, 100–101; futurism, 104
Hölderlin, Friedrich: compared to Jünger, 77;
art and nature, 111; role in transforming
German *Geist*, 113–15; Germany's
anointed task, 115–18; Heidegger and
Germany's tragic destiny, 118–21; Greeks
and insight, 124; influenced by Heraclitus,
125; Heidegger's non-instrumentalism,
154; space and time, 211; purposelessness
of technology, 235; Heidegger and human
salvation, 246–47
Holocaust: Heidegger and anti-Semitism, 43;
Heidegger and condemnation of, 131. *See
also* Auschwitz
Homesickness: Heidegger and philosophy,
23; metaphysical boredom, 31
Horror: basic mood of Germans, 116
Hubris: Germany and Auschwitz, 127–28;
Heidegger's vision of history, 132;
arrogance of modern man, 188
Humanism: technology, 167; Plato's doctrine
of truth, 174; rationality, 192; Heidegger
and image of humanity, 203;
anthropocentrism and deep ecology, 242,
244
Humanity: clever animal, 165, 221; rational
animal, 174, 191–97; Arendt's laboring
animal, 197–99; deep ecology, 244
Husserl, Edmund: definition of philosophy,
154–55
Hypokeimenon: definition, 182

Idea: Roman and Greek translations, 170
Imperium: Roman concept of command, 176
Individuation: Heidegger and National
Socialism, 44–45
Industrialism: meaning of technology, xvii;
Germany and modernity, xxi–xxii; *völkisch*
movement, 8–9; Weimar era, 16
Industrialization: National Socialism and
aestheticism, 104
Industrial Revolution: in Germany, 7–8
Institutions: power and technological era,
202–3
Instrumentalism: Heidegger's early, 150–53,
164; Heidegger and non-instrumentalist
ontology,153–56; Greeks and productionist
metaphysics, 158, 166; science, 159;
authentic production as *techne*, 230

Jews and Judaism: Heidegger and tragedy,
128; Western image, 128–29; Heidegger's
insensitivity, 129–30; concept of human
existence, 183. *See also* Anti-Semitism
Jünger, Ernst: influence on Heidegger, xxii,
66, 248, 255; view of history, xxiv, 34;
Heidegger on importance of, 35–36; war
experience, 49–54; concept of worker
Gestalt, 57–65, 80–83; Nietzsche, 76, 189,
190; purposelessness of technology, 235
Justice: values and Will to Power, 187

Kant, Immanuel: task of philosopher, xviii;
Enlightenment, 20; Heidegger's debt to,
143–46; history of productionist
metaphysics, 172–73
Kinesis: being of living things, 144
Klages, Ludwig: Heidegger on, 29–30
Koan: Zen Buddhism and modern technology,
219–20
Kultur: use of term, 19

Language: industrial urban society, 23;
Jünger, 59–60, 78–79; work and
technology, 86; poets and creativity, 114;
use of tools, 142; humanity and animals,
194; typewriter and word processor, 206;
technology and production process, 215;
structuralism, 256. *See also* Rhetoric
Latin: translation of Greek terms and
productionist metaphysics, 170. *See also*
Romans
Leibniz, Baron Gottfried Wilhelm:
metaphysics of subjectivity, 172
Lethe: Greek concept of truth, 175
Logic: cybernetic-computer revolution,
199–200
Logos: poet and *physis*, 125; rationality,
191–92; translation, 229; production, 233
Luther, Martin: Germany and modernity, 5–6;
subjective consciousness and modernity,
179–80

Machines: functionalization of work and
workers, 213–14; earth as machine and
deep ecology, 241–42
Marxism: economic interpretation of Western
history, 10, 253–54; depersonalization, 24;
Heidegger and facism, 37; proletariat and
human universality, 90; Heidegger's
analysis of workshop, 155; Heidegger on
critiques of industrial production, 156; god
and man, 182; humanity and animals,
193–94; fetishism of commodities, 207;
rural life, 210; machines, 214; Heidegger's
concept of history, 250; Heidegger's
historical determinism, 251–52; conflict
between good and evil, 252–53; Hegel's
interpretation of history, 254; productionist
metaphysics, 255
Mass culture: critiqued in *Being and Time*,

21–26; evaluations of Jünger and
Heidegger, 87–88; as threat to "high
culture," 206
Meaning: transcendental conception of in
Being and Time, 144
Metaphysics: history of productionist, xix,
224; Plato and temporality, 146; Greek and
origins of productionist, 156–63; history of
being, 167–68; transformation of being
from *energeia* to *actualitas* to the Will to
Power, 168–73; change of truth from
aletheia to certainty, 173–82; change of
substance from *hypokeimenon* to subject,
182–90; authentic mode of production,
222–23; Heidegger and deconstructionism,
250–51, 255–68; Hegel and Marx as
productionist metaphysicians, 255
Modernity: meaning of technology, xvii;
Germany and industrialism, xxi–xxii;
Heidegger's concept of technology,
xxiii–xxiv; Germany and religion, 5–6;
völkisch movement, 8–13; critiqued in *Being
and Time*, 21–26; one-dimensional
ontology, 151; Nietzsche and arrogance,
186; bodily experience, 245
Modern technology. *See* Technology
Moods: Germany and prevailing, 26; poetry
and primal, 114–15; Greek and German,
116–17; ontological function, 141
Morality: Heidegger and ethics, 130;
Heidegger and National Socialism, 265
Music: technological innovation, 206
Mystery: Heidegger and Hölderlin's poetry,
118
Mythology: Jünger and World War I, 50–51;
Germany and National Socialism, 102;
Heidegger's self-mythification, 132

Nationalism: German and progressivism, 11;
Heidegger and Jünger, 71–72
National Socialism: Heidegger's support, xxi,
36–40; Jünger, xxii, 63, 65; Heidegger's
philosophy, 34–35; Heidegger and Jünger,
36; Heidegger's "official story," 40–45,
279–80n; reactionary modernism, 48, 84;
Heidegger's political speeches, 67–69,
69–71; as generational phenomenon, 74;
as national aestheticism, 99–106;
Dionysian and Apollinian views of art,
108–9; view of poetry, 114; death of God
at Auschwitz, 126–33; Heidegger and
nature of handicraft, 137; Heidegger's new
man, 190; racism, 191; art and production,
231; Heidegger and deep ecology, 243;
Germany and the Enlightenment, 267
Naturalism: Heidegger on Nietzsche and
Rilke, 193; Heidegger's criticism, 243–44
Nature: earth and world, 122; Hölderlin's
poetry and *physis*, 124–26; Heidegger's
instrumentalist depiction of, 162; science
and modernity, 184–85; humanity and

animals, 196; productionist view, 224;
body and ecology, 245
Nazism. *See* National Socialism
Nietzsche, Friedrich Wilhelm: Spengler and
Scheler, 30–31; Heidegger on racism, 44;
art, 53–54, 76, 77, 106–12; Jünger, 76,
81; history of productionist metaphysics,
173, 186; Overman and concept of justice,
187–88; Heidegger's interpretation,
189–90; purposelessness of technology,
235; human finitude, 246; Heidegger and
progress, 255; metaphysics and sexual
difference, 270
Nihilism: Jünger, 35, 36; Heidegger and
National Socialism, 45; Nietzsche, 186,
189–90; personal and social pathology,
204
Noein: Heidegger's interpretation, 229
Nuclear power: technological innovation, 208

Occasioning: Heidegger's use of term, 232
Oedipus: Heidegger's depiction of self and
history, 95–96; German *Volk*, 119
Ontology: definition of phenomenology, 138;
Heidegger's early instrumentalist, 150–53;
non-instrumentalist in Heidegger's early
thought, 153–56; sexual difference, 268–74
Ouisa: Aristotle and Plato, 169
Overman: Heidegger's attitude toward
Jünger's, 88; Heidegger's description of,
91–92; limits and chaos, 111; Nietzsche
and arrogance of modernity, 186;
Nietzsche's concept of justice, 187–88;
Heidegger's new man, 190

Pain: Jünger on bourgeois age, 89
Parmenides: being and thinking, 223–24;
apprehending and being, 229
Patriarchalism: contempt for the body and
the female, 245–46; sexual and ontological
difference, 269; Heidegger and sexual
difference, 271
Peasantry: Heidegger and nature of work,
70–71
Phenomenology: defined as ontology, 138;
science as authentic disclosing and
producing, 163–64
Philosophy: history of German, xviii; German
philosophy and National Socialism, 38–39;
work and working, 67–68; Heidegger on
importance of, 84; as science, 163;
usefulness and science, 185; contribution
of Heidegger, 248. *See also* Metaphysics
Photography: Jünger and temporality, 57; as
technological instrument, 86
Physics: concepts of space and time, 211
Physis: Greek definition of nature, 125, 224;
Heidegger's use of term, 145; being as
appearing, 225; authentic producing as
poiesis, 233–34; purpose of art, 234–35
Picture: German translation, 86–87

Plato: productionist metaphysics, xix, 157, 196, 222; temporality and metaphysics, 146; social structure, 158; definition of man, 159; *eidos*, 168; *ouisa*, 169; subjectivism and truth, 173–74; ambiguity of *poiesis*, 231; denial of the body, 245

Poe, Edgar Allan: aesthetic tradition and Jünger, 53

Poetry: Heidegger's view of and National Socialism, 114; "earth" and "world" dimensions, 124; mass culture, 206–7

Poets: role in transforming German *Geist*, 113–15. *See also* Hölderlin, Friedrich

Poiesis: authentic production and *techne*, 229–36

Politics: assessment of Heidegger's philosophy, xxiii; Heidegger's ambiguous relationship to reactionary, 3–5; Germany and industrialization, 8; Weimar Republic, 14–15; art, 102–3; Heidegger's Hölderlin lectures, 117; Heidegger's concept of technology, 132–33; society and technological innovation, 202; Heidegger's critique of metaphysical decline of West, 249–50; Heidegger and deconstructionism, 257–58, 261–67. *See also* Facism; National Socialism

Practice: priority over theory, 140–41

Pragmatism: deconstructionism and politics, 265–66

Primates: humanity and animals, 194

Primitive: Heidegger's use of term, 156

Production: authentic mode and metaphysics, xx, 222–23; world of the workshop, 137–43; Germany and industrial, 156–57; authentic as *techne* and *poiesis*, 229–36; authentic and art work, 236–37

Progress: technology and humanity, 196, 215; Heidegger and Enlightenment, 255–56

Prophet: Heidegger and history, 259–60

Protestant Reformation: Germany and modernity, 5–6

Protestantism: work ethic and industrialization, 178–80

Pseudos: Greek concept of truth, 175, 176, 177

Psychology: pathology and nihilism, 204

Publishing: technology and art, 87; mass culture and poetry, 206–7

Racism: *völkisch* movement, 9, 12; Heidegger on Nietzsche, 44; National Socialism, 191

Rathenau, Walther: assassination and Weimar era, 14–15

Rationalism: *völkisch* movement, 12; Enlightenment and politics, 263

Rationality: modern view and naturalistic humanism, 191–92; Enlightenment, 218; technology and purpose, 235

Reactionary modernism: conservative revolution and technology, 47–49

Reason: Roman concept of truth, 175

Religion: Heidegger's biography, 18–19; Jünger and *Gestalt* of the worker, 61

Revolution: Jünger and *Gestalt* of the worker, 61–62

Rhetoric: Heidegger and Jünger, 67, 71–76, 89–90

Rilke, Rainer Maria: art and the disclosure of the "thing," 237–41

Romans: reactionary politics, 4; concept of truth, 175, 176; conquest and deceit, 176–77; concept of nature, 224. *See also* Latin

Romanticism: German and Enlightenment, 20; art and aesthetics, 107

Sacrifice: Jünger's new man, 56–57; Heidegger and Jünger's language, 73–74

Sado-masochism: Jünger's man of steel, 56

Salvation: Christianity and modernity, 180–81

Schaffen: translation, 122

Scheler, Max: Heidegger on, 30–31; human evolution, 193; Heidegger and industrialism, 285n; Christianity and the earth, 288n; science and technology, 290n; Will to Power, 291n

Schlageter, Leo: Heidegger on heroism, 75

Schöpfen: translation, 122

Science: Scheler, 30, 290n; Heidegger's definition of, 67; Nietzsche and art, 113; instrumentalism, 153, 159; phenomenology as authentic disclosing and producing, 163–64; commanding science and technology, 176; power and technological character, 181–82; nature and modernity, 184–85; structure and truth, 227–28; Enlightenment rationalism, 263

Science fiction: Jünger and *Gestalt* of the worker, 64

Selfishness: Heidegger on National Socialism, 131

Sexism: difference and ontology, 268–74; gender categories, 297n

Silesius, Angelius: purpose of art, 234

Slavery: Greek social structure and instrumentalism, 158; American South, 279n

Society: technological innovation and politics, 202

Soldier: Jünger's war experience, 51–52

Sophocles: Jünger and Heidegger, 95, 96; Heidegger and *Antigone*, 105, 119

Soviet Union: Jünger and *Gestalt* of the worker, 62; Heidegger and *Gestalt* of the worker, 90

Space: technological transformation, 209–12; technology and art, 236–37

Speer, Albert: architecture and National Socialism, 100

Spengler, Oswald: influence on Heidegger,

17; Heidegger's critical appropriation of, 26–29, 30–31
Sphallo: Roman concept of truth, 175, 176
Stalinism: equated with Americanism, 264–65
Structuralism: Heidegger and determinism, 256
Subjecthood: Jünger and technology, 81
Substance: change from *hypokeimenon* to subject, 182–90

Taylor, Frederick: Taylorism and Weimar era, 15; functionalization of work, 213
Techne: art as, 94; translation and meaning of, 110; art and production, 222, 223; authentic production and *poiesis*, 229–36; purpose of art, 234–35
Technology: interrelated meanings, xvii; naturalistic anthropology, xviii; Heidegger's concept as criticism of modernity, xxiii–xxiv; reactionary modernism, 12; Heidegger's critical appropriation of Spengler, 26–33; Heidegger and National Socialism, 41, 43; Jünger, 46–47, 60, 61; as aesthetic phenomenon, 54–57; art, 78–80, 110; Heidegger and Jünger, 83; Heidegger's postwar attitude, 90–93; political dimensions of Heidegger's concept, 132–33; Greek productionist orientation, 148–49; Heidegger's early instrumentalist ontology, 150–53; as outcome of "history of being," 166–67; commanding technology and science, 176; post-Darwinian definition, 191; cybernetic character, 199–202; Foucault and bio-power, 202–4; innovation and everyday life, 205–9; space and time, 209–12, 236; functionalization of work, worker, and equipment, 212–16; *Gestell*, 216–19; releasement, 219–21; craftworker and authentic production, 240–41; bodily experience, 245; criticism since Industrial Revolution, 248–49; Enlightenment rationalism, 263; totalitarianism, 264; patriarchy and sexual difference, 269
Television: technology and time, 87, 209
Telos: definition, 169, 232; Western history, 254
Temporality: role in understanding of being, 146–49
Theatre: Heidegger's view of history, 97–98
Theology: Heidegger's studies, 18; Christian and Greco-Roman metaphysics, 171
Theory: priority over practice, 140–41
Time: technological transformation, 209–12; functionalization of work and worker, 212
Tools: phenomenology and interpretation, 138–40; language, 142; Heidegger on animals, 194–95; machines, 213
Totalitarianism: Heidegger on freedom, 257; Enlightenment and politics, 262–63; technology, 264

Tragedy: Heidegger's view of life, 95–98, 167; Heidegger on Germany's destiny, 118–19; Heidegger and Auschwitz, 127–28
Transcendentalism: Heidegger, Kant, and Aristotle, 143–46; foundational historicism, 147–48
Translation: Heidegger's key terms, xxvi; Roman and Greek metaphysical terms, 170
Truth: as *aletheia*, 145; change from *aletheia* to certainty, 173–82; Heidegger and sexual difference, 270

United Nations: Heidegger and Germany's withdrawal, 68
United States: decline of the West and capitalism, 178. *See also* Americanism
University: Heidegger on modern, 20–21; Heidegger and politics, 40; Heidegger and anti-Semitism, 43; Heidegger and National Socialism, 44, 66, 67–69
Using: Heidegger's definition of *chre*, 161
Utilitarianism: *Being and Time*, 152–53, 155

Values: Heidegger and ethics, 130–31; Will to Power, 187
Violence: National Socialism and art, 102; Heidegger on Germany's tragic destiny, 120–21; Greeks and instrumentalism, 162–63
Vision: theme of Jünger and Heidegger, 95, 96
Volk: Greek tragic hero, 119–20; earth, 123
Völkisch movement: attack on modernity, 8–13; Heidegger and major themes, 23; Hitler and rhetoric, 48
Vorstellen: translation, 172

Wagner, Richard: art and the Dionysian spirit, 108–9
War: Jünger's experience, 49–54; Jünger on total mobilization, 55
Weimar Republic: politics, 14–15
West: historical theory of decline, 26–27, 29; Roman misinterpretations of truth and falsity, 177
Wilde, Oscar: aesthetic tradition and Jünger, 53
Will to Power: Jünger's and Heidegger's views of history, xxiv, 34, 83; aesthetic experience, 78; transformation from *energeia* to *actualitas*, 168–73; values, 187; Scheler on humanity, 291n
Work and working: productionist metaphysics and nature of, xx; Heidegger's political speeches, 69–71; *Gestalt* and definition of, 81–82; transformation and technology, 85–86; authentic and inauthentic, 152–53; concept of rest, 153–54; hand and thinking, 195–96; Greek view of labor, 197; technological innovation and labor-

saving, 208; space and time, 211–12; technology and functionalization, 212–16

Worker: Jünger's concept of *Gestalt*, 57–65; Heidegger and Jünger, 66–67

Workshop: Heidegger and productionist metaphysics, 137–43; totality, 150–51

World: the earth and authentic creating, 121–26; understanding of the phenomenon, 140; as distinct from earth, 226–27; self-concealment, 228–29; fourfold, 239–40

World War I: impact on Germany, 13–14; Heidegger's experiences, 18, 66; Jünger's mythologizing, 50–51; modern technology, 191

World War II: Heidegger's postwar view, 41, 90–91, 105; Heidegger's interpretation of German history, 127; Auschwitz, 129; Heidegger on progress, 196

Youth Movement: Germany and romantic idealism, 13

Zen Buddhism: releasement from modern technology, 219–20

Zoon: definition, 192

Michael E. Zimmerman, Professor of Philosophy at Tulane University, is the author of *Eclipse of the Self: The Development of Heidegger's Concept of Authenticity*. He has published articles on environmental ethics, the nature of technology, and the relationship of philosophy to the fields of medicine and psychiatry.